DIFFERENTIAL EQUATIONS: GEOMETRIC THEORY

SECOND EDITION

SOLOMON LEFSCHETZ
PRINCETON UNIVERSITY
THE NATIONAL UNIVERSITY OF MEXICO
AND
RIAS, BALTIMORE

DOVER PUBLICATIONS, INC.
New York

Published in Canada by General Publishing Company, Ltd., 30 Lesmill Road, Don Mills, Toronto, Ontario.

Published in the United Kingdom by Constable and Company, Ltd., 10 Orange Street, London WC2H 7EG.

This Dover edition, first published in 1977, is an unabridged republication of the second (1963) edition of the work originally published by Interscience Publishers in 1957 as Volume VI of the Interscience Pure and Applied Mathematics Series.

International Standard Book Number: 0-486-63463-9
Library of Congress Catalog Card Number: 76-53978

Manufactured in the United States of America
Dover Publications, Inc.
180 Varick Street
New York, N.Y. 10014

Preface

The surprisingly warm reception which greeted the author's monograph *Lectures on Differential Equations* (Princeton University Press 1946) provided the incentive to enlarge the monograph to the present volume. However, the general plan has remained the same, and readers acquainted with the monograph will readily find their way here. We continue to lean more heavily than ever on vectors and matrices and to utilize topological notions.

In relation to the monograph the present scheme is the following. The first four chapters are about as before: more or less preparatory and standard. The next three deal with point stability and contain much additional information on the work of Liapunov and his Soviet successors, notably on the so-called direct method of Liapunov. With Chapter VIII — periodic solutions — the n dimensional part is concluded. This chapter is admittedly fragmentary, but then this is unavoidable, given the present state of this general question. The remaining chapters deal with two-dimensional systems. Two chapters are devoted to the results of Poincaré and Bendixson: critical points, the index, behavior at infinity, some special systems. An introduction is also given to the important notion of structural stability and the contributions of Andronov-Pontrjagin and DeBaggis. The last two chapters deal with equations of the second order: notably the recent work of Cartwright-Littlewood, of Levinson-Smith and of Levinson, the application of the perturbation method, and other related topics. After two appendices on vectors and matrices, and on topology, the volume concludes with a small list of problems which are not exercises but rather topics appropriate for further discussion and which had to be omitted owing to lack of space.

Once in possession of the general theory two roads lay open

before us. We could follow Poincaré, Levi-Civitá, Birkhoff and study "two degrees of freedom" and the extensive doctrine centering around the three body problem. The second and more modest road, which we have selected, led to nonlinear differential equations of the second order: the group of problems stirred up a generation ago by Van der Pol. No applications whatsoever are taken up in the volume, but it is hoped that mathematically inclined physicists and engineers may find much to interest them within the covers of this book.

In a work of this scope it has not been possible to range the topics according to difficulty. For the convenience of the reader we have marked with an asterisk the more arduous parts which may be reserved for a second reading. The "first reading" parts include the existence theorem and a few consequences in Chapter II; all of Chapter III and IV; the first halves of Chapter V and VI; the two dimensional part: from Chapter IX to the end.

The author wishes to express his appreciation to Messers. Courtney Coleman, Ralph Gomory, and Felix Haas who have read significant portions of the manuscript; also, to Messers. Robert Bass who read part, and Henry Antosiewicz who read all the proof and made many valuable suggestions. I also wish to express my thanks to the Office of Naval Research as preparation of the manuscript was done in part under Office of Naval Research Contract N6-ORI-105, Task Order V.

S. LEFSCHETZ

Preface to the Second Edition

The main differences between the two editions are these: correction of many misprints and defective proof pointed out by Jose Massera; improvement in exposition suggested by Jane Scanlon; considerable extension of the material on Liapunov's direct method and its converse as required by its growing importance, in the revision of which Antosiewicz was most helpful; revision of the treatment of Chapter VI, *Stability in Product Spaces*, with the constant assistance of my colleague J. K. Hale of RIAS. To all these friendly critics the author wishes to express his deep appreciation.

S. LEFSCHETZ

Contents

CHAPTER I

Preliminary Questions

In the present chapter we summarize briefly a number of questions which will play a basic role throughout the present book: elementary notions of topology, vectors and matrices, functions of several variables.

References: For topology, Lefschetz [1]; for vectors and matrices, Bellman [4]; Gantmacher [1]; Wedderburn [1]; for functions of several variables, Bochner-Martin [1].

Notations. We shall make use throughout of the standard set theoretic symbols:

$a \in A$: a is a member of the set A;

$A \cup B$: union of the sets A, B or set of elements contained in one or the other;

$\bigcup A_\alpha$: union of the sets A_α;

$A \cap B$: intersection of the sets A, B or set of all elements contained in both;

$\bigcap A_\alpha$: intersection of the sets A_α;

$A \subset B$, $B \supset A$: A is contained in B;

$A - B$: the complement of B in A or set of all elements of A which are not in B.

The symbol $\rightarrow$ stands for "implies."

§ 1. Elements of Topology

1. By a *space* one generally understands a *set* with some structural properties. Since all our spaces will be endowed with a distance, in fact, will be subsets of Euclidean spaces, it will be most appropriate to take distance as the basic concept and define the other classical concepts such as open sets, closed sets, etc., in terms of distance.

(1.1) *Metric spaces*. A set or space $\mathfrak{R}$ is *metrized* whenever there is defined among its points $x, y, \ldots$, a real valued function $d(x, y)$, the *distance* with the following properties:

(a) $d(x, y) \geqq 0$ and $d(x, y) = 0$ when and only when x and y coincide;

(b) $d(x, z) \leqq d(y, x) + d(y, z)$ (triangle axiom).

From (a) and (b) follows $d(x, y) \leqq d(y, x)$ and since the converse manifestly holds we have:

(c) $d(x, y) = d(y, x)$ (symmetry law).

A metrizable space $\mathfrak{R}$, i.e., on which a distance may be defined, is said to be *metric*. We also say: a metric is defined on $\mathfrak{R}$.

The distance on $\mathfrak{R}$ assigns of course a distance on every subset of $\mathfrak{R}$. Thus when $\mathfrak{R}$ is metric so is every subset of $\mathfrak{R}$.

In terms of distances one may define various simple notions:

Spheroid $\mathfrak{S}(x_0, \varrho)$ *of center* x_0 *and radius* ϱ: the set of all points x at a distance less than ϱ from x_0;

distance $d(x, A)$ from a point x to a set A: $\inf d(x, y)$ for all y in A;

diameter of a set A, written diam A: $\sup d(x, y)$ for all x, y in A;

limit of a sequence $\{x_n\}$, *convergence* of the sequence, *Cauchy-sequence* are defined in the standard manner of real analysis.

An example: The space is the Euclidean plane with its usual distance and the spheroids are the circular regions.

When a set is metrizable its metrization is not unique. Thus if $d(x, y)$ is a distance so is $2\,d(x, y)$. Two metrizations say by $d(x, y)$ and $d'(x, y)$ with spheroids $\mathfrak{S}(x, \varrho)$ and $\mathfrak{S}'(x, \sigma)$ are *equivalent* whenever every $\mathfrak{S}(x, \varrho)$ contains and is contained in an $\mathfrak{S}'(x, \sigma)$. This guarantees among other things that both distances specify the same sequences as convergent and assign the same limits to convergent sequences.

(1.2) *Product spaces.* Let $\mathfrak{R}_1$, $\mathfrak{R}_2$ be two metric spaces with distances $d_1(x, y)$, $d_2(u, v)$. Let the pairs of points (x, u), (y, v), $\ldots$, one from each space be considered as points of a new space $\mathfrak{R}$. It is easy to verify that

$$d((x, u), (y, v)) = d_1(x, y) + d_2(u, v)$$

is a suitable distance in $\mathfrak{R}$. The space $\mathfrak{R}$ with this or any equivalent distance is known as the *product of* $\mathfrak{R}_1$ *by* $\mathfrak{R}_2$, written $\mathfrak{R}_1 \times \mathfrak{R}_2$. This extends at once to a product $\mathfrak{R}_1 \times \ldots \times \mathfrak{R}_n$.

The following are simple examples: (a) The product of two segments is a square and of three segments a cube. (b) The product $C_1 \times C_2$ of two circumferences is a torus. This is seen at once: — If C_1 is parametrized by θ: $0 \leq \theta < 2\pi$, and C_2 by ψ: $0 \leq \psi < 2\pi$, then $C_1 \times C_2$ is parametrized by θ, ψ varying between the same limits. Thus θ corresponds to the meridian angle and ψ to the parallel angle of the torus. (c) The product $\omega \times C$ of a closed circular region ω: $x^2 + y^2 \leq 1$, by a circumference C is a solid torus.

(1.3) *Euclidean spaces.* The line l: $-\infty < x < +\infty$ is metrized by $d(x, y) = |x - y|$ which is a suitable distance. It is thus turned into Euclidean one-space $\mathfrak{E}^1$. Euclidean n-space $\mathfrak{E}^n$ is merely the product of n lines assigned the "product" metric of (1.2) or any equivalent metric. Let the n lines l_h, $h = 1, 2, \ldots, n$, be parametrized by coordinates x_h. The points of the product $l_1 \times \ldots \times l_n$ are the sets of n real coordinates $x_1, \ldots, x_n$ and the product metric assigns to $\mathfrak{E}^n$ the distance

$$d(x, y) = \Sigma\, |x_h - y_h|.$$

An equivalent metric is the more commonly used distance

$$d'(x, y) = [\Sigma\, (x_h - y_h)^2]^{1/2}.$$

Another is the very convenient metric based on the distance

$$d''(x, y) = \sup\{|x_h - y_h|\}.$$

A further metric may be obtained as follows. Apply to $\mathfrak{E}^n$ the affine transformation

$$(1.4) \qquad x_{1h} = \Sigma\, a_{hk}\, x_k + b_h$$

where the determinant $|a_{hk}| \neq 0$. Then take as distance one of the two functions

$$\Sigma\, |x_{1h} - y_{1h}|, \quad [\Sigma\, (x_{1h} - y_{1h})^2]^{1/2}$$

both of which are equivalent to one another and to those defined by $d(x, y)$, $d'(x, y)$, $d''(x, y)$.

(1.5) *Other products.* Consider first the plane S of a complex variable z. One may metrize S by assigning to it the distance $|z - z'|$.

Consider now a product of p planes $S_1, \ldots, S_p$ of complex variables $z_1, \ldots, z_p$ and q lines $l_1, \ldots, l_q$ of real variables $x_1, \ldots, x_q$. This space $\mathfrak{R} = S_1 \times \ldots \times S_p \times l_1 \times \ldots \times l_q$ is assigned through the product metric the distance

$$d((z, x), (z', x')) = \Sigma |z_h - z_h'| + \Sigma |x_k - x_k'|.$$

The space $\mathfrak{R}$ occurs naturally in the study of functions of p complex and q real variables which we will have to consider later.

2. *Topological considerations.* The more strictly topological notions are best defined in terms of *open sets*. An open set U is simply a union of spheroids. Observe now the following two properties.

(a) If $x \in \mathfrak{S}(x_0, \varrho)$ then by the triangle axiom there is an $\mathfrak{S}(x, \sigma) \subset \mathfrak{S}(x_0, \varrho)$. Hence this characterization of an open set U: if $x \in U$ then some spheroid $\mathfrak{S}(x, \varrho) \subset U$.

(b) If $d'(x, y)$ and $\mathfrak{S}'(x, \sigma)$ are a distance equivalent to $d(x, y)$ and its spheroids then some $\mathfrak{S}'(x, \sigma) \subset \mathfrak{S}(x, \varrho)$. Hence the open set U is likewise a union of spheroids $\mathfrak{S}'$. Conversely if it is a union of spheroids $\mathfrak{S}'$ then it is likewise a union of spheroids $\mathfrak{S}$. Since all the topological properties to be discussed are expressed in terms of open sets they are the same for equivalent metrics.

The open sets U containing a point or more generally a set are called *neighborhoods* of the point or set.

We now define a *closed set* F as the complement $\mathfrak{R} - U$ of an open set. It is convenient to include the whole space $\mathfrak{R}$ and the null-set among the open and hence also among the closed sets. One verifies then that the union of any number and the intersection of a finite number of open sets is open. Hence the intersection of any number and the union of a finite number of closed sets is closed.

Two more notions are now to be defined:

the *closure* $\bar{A}$ is the closed set which is the intersection of all the closed sets containing A; it is thus the "smallest" closed set containing A;

the *boundary* $\mathfrak{B}A$ of A: $\mathfrak{B}A = \bar{A} \cap \overline{\mathfrak{R} - A}$, and it is manifestly closed.

The closure $\bar{A}$ is the set of all points at zero distance from A while the boundary $\mathfrak{B}\,A$ is the set of all points at zero distance from both A and its complement.

A *condensation point* x of A is defined by the condition that every neighborhood U of x contains points of $A - x$. The closure $\bar{A}$ is the union of A and all its condensation points.

If $A \subset \mathfrak{R}$ the open or closed sets of A are the intersections with A of the corresponding sets of $\mathfrak{R}$.

Let $\mathfrak{R}$, $\mathfrak{R}'$ be metric spaces. A *function* f on $\mathfrak{R}$ to $\mathfrak{R}'$, or *transformation* of $\mathfrak{R}$ into $\mathfrak{R}'$, written $f\colon \mathfrak{R} \to \mathfrak{R}'$, is said to be *continuous* whenever if U' is an open set of $\mathfrak{R}'$ then $f^{-1}\,U'$ is an open set of $\mathfrak{R}$. This definition is readily shown to be equivalent to the well known ε, δ definition. A continuous transformation is also known as a *mapping*.

If f is one-one and bicontinuous (both f and f^{-1} continuous) then f is said to be *topological* or a *homeomorphism*. One says also that $\mathfrak{R}$ and $\mathfrak{R}'$ are *topologically equivalent* or *homeomorphic*.

Example. Consider the system (1.4) as assigning to the point x with the coordinates x_i a new point x_1, whose coordinates in the same system as the x_i, are the x_{1h} given by (1.4). There is thus obtained a one-one transformation T of $\mathfrak{E}^n$ onto itself and T is topological.

Cells and spheres. These sets are of particular importance. The *n-cell* is the topological image of the set of points of $\mathfrak{E}^n$ represented by

$$\Sigma\, x_i^2 < 1. \tag{2.1}$$

The Euclidean space $\mathfrak{E}^n$ itself is an n-cell.

The $(n - 1)$-*sphere* S^{n-1} is the topological image of the boundary of the set (2.1), i.e. of the subset of $\mathfrak{E}^n$ given by

$$\Sigma\, x_i^2 = 1. \tag{2.2}$$

The n-cell is then the topological image of a spherical region in $\mathfrak{E}^n$. It is also the topological image for example of the sets: $|x_i| < a_i$, where the a_i are positive.

The topological image of the "solid" sphere

$$\Sigma\, x_i^2 \leq 1 \tag{2.3}$$

is sometimes called a *closed n-cell*.

Exception must be made for the dimension zero: Conventionally the zero-cell and the closed zero-cell both consist of just one point.

The open one-cell is also known as an *arc*. Thus the arc is the topological image of an interval $a < t < b$. The closed one-cell is sometimes referred to as a *closed arc*. Notice that the zero-sphere consists thus of two points. The one-sphere, the homeomorph of a circumference is called a *Jordan curve*.

3. *Compact spaces*. The term "compact" conveys quite well the general idea of what these spaces are. For the more formal definition we must introduce three properties:

Property I. From every covering $\{U_\alpha\}$ *of the space* $\mathfrak{R}$ *by open sets (the union of the* U_α *is* $\mathfrak{R}$*) there may be extracted a finite subcovering.*

Property II. If a collection $\{F_\alpha\}$ *of closed sets of* $\mathfrak{R}$ *is such that every finite subcollection intersects then the whole collection intersects.*

Property III. Every infinite subset of $\mathfrak{R}$ *has a point of condensation.*

It may be shown that for a metric space all three properties are equivalent. A metric space that possesses one of the three properties and hence all three is said to be *compact*. Noteworthy properties of these spaces are:

(3.1) *Compactness is a topological property.*

(3.2) *A compact subset of a space* $\mathfrak{R}$ *is closed and bounded.*

(3.3) *A closed subset of a compact metric space is compact (and of course also metric).*

(3.4) *Let* $f\colon \mathfrak{R} \to \mathfrak{R}'$ *be a mapping of one metric space into another. If* $\mathfrak{R}$ *is compact so is its image* $f\mathfrak{R}$.

(3.5) *The compact subsets of an Euclidean space are its closed and bounded subsets.*

On the other hand a line in an Euclidean space is closed but not compact. This brings out the distinction between the two concepts.

A noteworthy consequence of (3.4) and (3.5) is:

(3.6) *A real valued function on a compact metric set is bounded and actually assumes its maximum and minimum values on the set.*

We will now consider some useful examples of compact sets.

(3.7) *The closed cell and the sphere*. Since both are topological.

images of bounded and closed Euclidean sets they are both compact. In particular a Jordan curve and a closed arc are compact.

(3.8) *The real projective plane.* It is classically defined as a set $\mathfrak{P}^2$ of points x in one-one correspondence with all triples of real numbers (x_0, x_1, x_2) not all zero, where one does not distinguish between this triple and (kx_0, kx_1, kx_2), for any real $k \neq 0$. Moreover one allows the one-one correspondence with the triple x_i' derived from the x_i by a real linear homogeneous transformation

$$x_i' = \Sigma\, a_{ij}\, x_j, \quad |a_{ij}| \neq 0.$$

The $x_i, x_i', \ldots$, are the *projective coordinates* of the same point x in the various coordinate systems.

Take in particular the set $\mathfrak{Q}$ of the points of $\mathfrak{P}^2$ for which $x_0 \neq 0$. One may divide the coordinates of a point $x \in \mathfrak{Q}$ by x_0 so that they will become $(1, x_1, x_2)$. Thus $\mathfrak{Q}$ is in one-one correspondence with the points (x_1, x_2) of an Euclidean plane $\mathfrak{E}^2$ and one may identify $\mathfrak{Q}$ with $\mathfrak{E}^2$. As a consequence $\mathfrak{Q}$ is assigned the metric of $\mathfrak{E}^2$.

Take any ray ϱ of $\mathfrak{E}^2$ through the origin. If r denotes the radius vector then the transformation $r' = r/(1+r)$ sends ϱ into the part $0 \leq r < 1$ of the ray. Any line l through the origin when completed in $\mathfrak{P}^2$ acquires just one point ξ. To ξ there correspond two points ξ_1, ξ_2 of the circumference C of radius one. Thus to the whole line l of the projective plane $\mathfrak{P}^2$ there corresponds the segment $\xi_1\, \xi_2$ where ξ_1 and ξ_2 both correspond to ξ. One restores the one-one correspondence by "identifying" ξ_1 with ξ_2. This may be carried out effectively by plunging $\mathfrak{E}^2$ say in $\mathfrak{E}^5$ and deforming the circular region Ω bounded by C so that ξ_1 and ξ_2 are made to coincide throughout. Thus the projective plane is identified with a closed subset of $\mathfrak{E}^5$: a closed circular region when endpoints of each diameter are identified. It follows (3.5) that $\mathfrak{P}^2$ is compact and metric.

(3.9) *Projective spaces.* Everything that has been said may be extended with ease to any dimension. One thus obtains the projective n-space $\mathfrak{P}^n$ and it is also compact.

4. *Connectedness.* A set or space $\mathfrak{R}$ is *connected* whenever one cannot have $\mathfrak{R} = U \cup V$ where U, V are disjoint non-void open

sets and hence also closed sets of $\mathfrak{R}$. A connected open set is also known as a *region*.

The following properties are more or less immediate consequences of the definition of connectedness.

(4.1) *A union of connected sets with a common point is connected.*

(4.2) *The union of a sequence $\{C_n\}$ of connected sets such that consecutive sets intersect, is itself connected.*

For if $C = \bigcup C_n$ then also $C = \bigcup D_n$ where $D_n = C_1 \cup \ldots \cup C_n$, and the D_n have C_1 in common.

(4.3) *If A is connected so is its closure $\bar{A}$.*

Less obvious is the following:

(4.4) *The open cell, the closed cell, the n-sphere, $n > 0$, the projective n-space are all connected.*

The case of the zero-cell is trivial so let us consider the one-cell. First let us show that the closed arc or equivalently the segment l: $0 \leqq t \leqq 1$ is connected. Let it be otherwise and let $l = U \cup V$ where U, V are disjoint and both open and closed. Only one say U contains $t = 0$. Let $a = \inf V$. Every neighbourhood of a contains points of V and since V is closed $a \in V$. Hence $a > 0$. For every $\varepsilon > 0$ there are points of U in $a - \varepsilon < t < a$, and since U is closed $a \in U$. Thus U and V are not disjoint, a contradiction which proves our assertion.

Let A be any one of the four sets considered in (4.4) and let x be a fixed point of A. Every point y of A can be joined to x by a closed arc xy. Hence by (4.1) A is connected.

(4.5) *Components.* Given a set A and a point x of A there is a maximal connected subset B of A containing x. This set is readily shown to be determined in the same manner by any one of its points and B is known as a *component* of A. Thus a diametral plane P decomposes a sphere S into a set $S - P$ which has two components namely the two hemispheres making up $S - P$.

We shall now prove a special case, required later, of a much more general proposition.

(4.6) *Let Ω be a bounded closed region of the Euclidean plane or merely a closed region of a 2-sphere or of the projective plane. Let $\{F_\lambda\}$ be a collection of connected closed subsets of Ω such that of any two F_λ, F_μ one contains the other. Then $F = \bigcap F_\lambda$ is a non-empty closed connected subset of Ω.*

We notice first that Ω is compact. If $\{F_{\lambda_1}, \ldots, F_{\lambda_r}\}$ is any finite subcollection one of its sets is contained in all the rest and hence $\bigcap F_{\lambda_h}$ is non-empty. Consequently F is closed and non-empty. Let U be any neighborhood of F. Some $F_\lambda \subset U$. For in the contrary case if $G_\lambda = F_\lambda \cap (\Omega - U)$ then $\{G_\lambda\}$ behaves like $\{F_\lambda\}$. Hence $G = \bigcap G_\lambda$ is non-empty and is contained in every F_λ but not in F, which is absurd.

Suppose now that F is not connected. Then $F = F' \cup F''$ where F', F'' are disjoint and closed, hence also compact. Let $x' \in F'$ and $x'' \in F''$. Since $d(x', x'')$ for x' fixed is a continuous function of x'' it has a positive lower bound $2\varrho'$ on F''. Hence the spheroid $\mathfrak{S}(x', \varrho')$ does not meet F''. Thus $\{\mathfrak{S}(x', \varrho')\}$ is an open covering of the compact set F' and so it has a finite sub-covering $\{S_1', \ldots S_p'\}$. Thus $S' = \bigcup S_i'$ is a neighborhood of F' and $\bar{S}'$ is a closed set disjoint from F''. By the same reasoning F'' has a neighborhood S'' disjoint from $\bar{S}'$ and hence from S'. Thus F' and F'' have disjoint neighborhoods S' and S''.

Since $S' \cup S''$ is a neighborhood of F some $F_\lambda \subset S' \cup S''$. If $F'_\lambda = S' \cap F_\lambda$, $F''_\lambda = S'' \cap F_\lambda$, we have $F_\lambda = F_\lambda' \cup F_\lambda''$ where F_λ', F_λ'' are open disjoint subsets of F_λ. Hence F_λ is not connected, contrary to assumption and (4.6) follows.

We terminate our general topological considerations with the exact statement of the all important

(4.7) THEOREM OF JORDAN-SCHOENFLIES. *Every Jordan curve J on a two-sphere S divides S into two component regions R_1, R_2 with the common boundary J, and $R_i \cup J$ is a closed 2-cell. The same property holds for a Jordan curve J in an Euclidean plane P save that only one of the two regions say R_1 is bounded and $R_1 \cup J$ is now the only closed 2-cell.*

For the proof see for instance Kerekjarto [1] pp. 65 and 69.

§ 2. Vectors and Matrices

5. The reader is assumed familiar with the usual elementary notions on vectors and matrices and we merely recall briefly those needed later. We shall only deal with real or complex finite dimensional vector spaces. Such a space $\mathfrak{V}$ is then an additive group with the group of reals (real vector space) or complexes (complex vector space) as group of operators (the scalars). Furthermore it possesses a *base* $e = \{e_1, \ldots, e_n\}$ that is to say

a set of linearly independent vectors such that every vector x of $\mathfrak{B}$ satisfies a unique relation

$$x = \Sigma\, x_h\, e_h, \tag{5.1}$$

where the x_h are real or complex scalars according as the space itself is one or the other. The x_h are the *coordinates* or the *components* of x. The number n is the *dimension* of the space $\mathfrak{B}$ which is often denoted accordingly by $\mathfrak{B}^n$. To indicate its coordinates the space is also written $\mathfrak{B}_x, \mathfrak{B}_y, \ldots$. The vectors e_h are also said to *span* $\mathfrak{B}$.

Let $\mathfrak{B}', \mathfrak{B}''$ be subspaces of $\mathfrak{B}$. We will say that $\mathfrak{B}$ is the *direct sum* of $\mathfrak{B}'$ and $\mathfrak{B}''$, written $\mathfrak{B} = \mathfrak{B}' \oplus \mathfrak{B}''$, whenever for every $x \in \mathfrak{B}$ one may write a unique relation $x = x' + x''$, $x' \in \mathfrak{B}'$, $x'' \in \mathfrak{B}''$. Similarly for a direct sum of several subspaces $\mathfrak{B} = \mathfrak{B}_1 \oplus \ldots \oplus \mathfrak{B}_r$. Evidently $\dim \mathfrak{B} = \Sigma \dim \mathfrak{B}_h$.

Returning to $\mathfrak{B}^n$ any set $e' = \{e_1', \ldots, e_n'\}$ of n linearly independent vectors of $\mathfrak{B}^n$ is a base for $\mathfrak{B}^n$. We have then relations

$$e_j' = \Sigma\, a_{hj}\, e_h \tag{5.2}$$

where the square $n \times n$ matrix $A = (a_{hj})$ is *non-singular*, i.e., its determinant $|A| \neq 0$.

Let E_n, or often merely E, denote the $n \times n$ unit-matrix: the terms in the main diagonal are unity and the others zero. The terms of E_n (any n) are written δ_{hj} and known as *Kronecker deltas*. We recall that when A is non-singular it has a unique inverse A^{-1}, i.e., a matrix such that $A^{-1}A = AA^{-1} = E_n$.

Returning to (5.2) if x_h' are the coordinates of x as to the base e' then

$$x_h = \Sigma\, a_{hj}\, x_j'. \tag{5.3}$$

Hence if $A^{-1} = (a_{hj}^*)$ then

$$x_h' = \Sigma\, a_{hj}^*\, x_j. \tag{5.4}$$

The relations (5.3) may receive another interpretation. Whether A is or is not singular (5.3) may be considered as assigning to the vector x' of components x_j' relative to the first base e the vector x with components x_h relative to the same base. Thus (5.3) defines a linear transformation T of the vector space $\mathfrak{V}$ into itself.

6. *Norms.* One may metrize the vector space $\mathfrak{V}$ by means of a base e as follows. One assigns a norm $||x||$, a real function of the vector x such that

(a) $||x|| \geq 0$ and $||x|| = 0$ when and only when $x = 0$;
(b) $||kx|| = |k| \cdot ||x||$ for scalars k;
(c) $||x + y|| \leq ||x|| + ||y||$.

Evident norms are

$$(\Sigma\ |x_j|^2)^{1/2}, \quad \Sigma\ |x_j|, \ \sup\ \{|x_j|\}.$$

Under one or the other $||e_j|| = 1$. For this reason the e_j are sometimes called *units*. — Our preferred norm, and the one always used unless otherwise stated, is

$$||x|| = \sup\ \{|x_j|\}.$$

The properties of the norm show that $||x - y||$ is a distance for $\mathfrak{V}$ and this vector space is thus metrized. It may be noticed that either one of the above norms and related distances turn $\mathfrak{V}$ into a complete metric space: every Cauchy sequence has a limit in the space.

(6.1) One may consider an $m \times n$ matrix $A = (a_{ij})$ as a vector with coordinates a_{ij}. The appropriate space has the dimension mn and the base relative to which the a_{ij} are coordinates consists of a set $\{e_{ij}\}$ where e_{ij} is the $m \times n$ matrix with zeros everywhere except at the element in the row i and column j, whose value is unity. One may then define the norm $||A||$ of A as

$$||A|| = \sup\ \{|a_{ij}|\}.$$

(6.2) Returning to vectors, by means of the norm one may define the continuity of a vector function $x(t)$ of a real variable t, hence also the derivative $dx(t)/dt$ and one finds that the latter is the vector whose components are dx_h/dt. The integral

$$\int_{t_0}^{t} x(t)\, dt$$

is best defined directly as the vector whose components are

$$\int_{t_0}^{t} x_h(t)\, dt$$

and one verifies at once that

$$x(t) = \frac{d}{dt}\int_{t_0}^{t} x(t)\, dt.$$

The application to a matrix $X = (x_{ij}(t))$ yields at once:

$$\frac{dX}{dt} = \left(\frac{dx_{ij}}{dt}\right), \int_{t_0}^{t} X dt = \left(\int_{t_0}^{t} x_{ij}(t)\, dt\right).$$

7. We shall now consider more specific properties of matrices. By way of notation A' denotes the transpose of the matrix A and we recall the relations

$$(AB)' = B'A', (AB)^{-1} = B^{-1} A^{-1};\ (A^{-1})' = (A')^{-1},$$

where in the last two relations A, B are non-singular, and square.

A *triangular* matrix is a square matrix with all terms above or below the main diagonal zero.

The matrix diag $(A_1, \ldots, A_r)$ is

$$\begin{pmatrix} A_1, & 0, & \ldots & \\ 0, & A_2, & \ldots & \\ & & \ddots & \\ & & & A_r \end{pmatrix}$$

where the A_i are square matrices and the zeros stand for zero matrices. Noteworthy special case: diag $(a_1, \ldots, a_n)$ stands for a square matrix of order n with the a_i down the main diagonal and the other terms zero.

If $A = (a_{hj})$ is a square matrix then the expression Σa_{jj} is known as the *trace* of A, written trace A.

A *nilpotent* matrix is a square matrix A such that some power $A^m = 0$.

Let $f(\lambda) = a_0 + a_1 \lambda + \ldots + a_r \lambda^r$ be a polynomial in λ with real or complex coefficients according as A is real or complex.

Then $a_0 E_n + a_1 A + \ldots + a_r A^r$ is a uniquely defined matrix written $f(A)$.

The polynomial $\varphi(\lambda) = |A - \lambda E|$ is known as the *characteristic polynomial* and its roots are the *characteristic roots* of A.

8. Reverting to the interpretation of the system (5.3) as a linear transformation let the base e be replaced by a new base f, and let y_h, y_h' be the new coordinates of x, x' as to f. If $e = fP$ then

$$y = Px, y' = Px', y = PAP^{-1} y'.$$

Thus referring to the new base f the transformation T is represented by PAP^{-1}. Conversely every such matrix represents T in terms of a certain base. We will say that two real [complex] square matrices A, B of same order n are *similar in the real [complex] domain* if there can be found a non-singular square real [complex] matrix P of order n such that $B = PAP^{-1}$. This relation is an equivalence. For if we denote it by $\sim$ then the relation is

symmetric: $A \sim B \rightarrow B \sim A$, since
$B = PAP^{-1} \rightarrow A = P^{-1} BP$;
reflexive: $A \sim A$, since $A = EAE^{-1}$;
transitive: $A \sim B, B \sim C \rightarrow A \sim C$. For if $A = PBP^{-1}$, $B = QCQ^{-1}$ then $A = PQCQ^{-1} P^{-1} = (PQ) C (PQ)^{-1}$.

(8.1) *If $A \sim B$ and $f(\lambda)$ is any polynomial then $f(A) \sim f(B)$. Hence $f(A) = 0 \rightarrow f(B) = 0$.*

For if $B = PAP^{-1}$ then $B^r = PA^rP^{-1}$, $kB = P(kA)P^{-1}$, and $P(A_1 + A_2) P^{-1} = PA_1P^{-1} + PA_2 P^{-1}$.

(8.2) *Similar matrices have the same characteristic polynomial.*

For $B = PAP^{-1} \rightarrow B - \lambda E = P(A - \lambda E) P^{-1}$, and therefore also $|B - \lambda E| = |A - \lambda E.|$

Since the characteristic polynomials are the same, their coefficients are also the same. Only two are of interest: the determinants, manifestly equal, and the traces. If $\lambda_1, \ldots, \lambda_n$ are the characteristic roots then a ready calculation yields

$$\Sigma \lambda_i = \Sigma a_{ii} = \text{trace } A.$$

Therefore

(8.3) *Similar matrices have equal traces.*

It is convenient to denote by $C(\lambda)$ a "block" matrix of type

$$C(\lambda) = \begin{pmatrix} \lambda & & & & \\ 1 & \lambda & & & \\ & 1 & \lambda & & \\ & & & \ddots & \\ & & & 1 & \lambda \end{pmatrix},$$

all unwritten terms being zero. Sometimes the order r is indicated by a subscript as $C_r(\lambda)$. The matrix $C'(\lambda)$ is simply $C(\lambda)$ with the 1, ..., 1 diagonal above the main diagonal.

The proof of the following classical theorem will be found in Appendix I.

(8.4) FUNDAMENTAL THEOREM. *Let $\mathfrak{V}$ be a complex vector space and T a complex transformation $\mathfrak{V} \to \mathfrak{V}$. Then*: (a) *There is a direct sum decomposition $\mathfrak{V} = \mathfrak{V}_1 \oplus \ldots \oplus \mathfrak{V}_p$ such that $T\mathfrak{V}_h \subset \mathfrak{V}_h$;* (b) *there is a base $e = \{e_{hk}\}$ for $\mathfrak{V}$ such that the e_{hk} (h fixed) span $\mathfrak{V}_h$;* (c) *the transformation T_h induced by T in $\mathfrak{V}_h$ is represented by a matrix $C_r(\lambda_h)$, where λ_h is a characteristic root of the initial matrix A of T. Thus $A \sim \operatorname{diag}(C(\lambda_1), \ldots, C(\lambda_p))$ (this last type is called the "Jordan normal form");* (d) *the $\mathfrak{V}_h$ and blocks C_h are unique to within their ordering.*

(8.5) *Remark.* If one reversed the order of the e_{hk} (h fixed) $C(\lambda_h)$ would be replaced by $C'(\lambda_h)$. Hence also $A \sim \operatorname{diag}(C'(\lambda_1), \ldots, C'(\lambda_p))$.

Example. When $n = 2$ and $\lambda_1 = \lambda_2 = \lambda$ there are the two distinct types

$$\begin{pmatrix} \lambda & 0 \\ 0 & \lambda \end{pmatrix}, \quad \begin{pmatrix} \lambda & 0 \\ 1 & \lambda \end{pmatrix}.$$

9. *Real vector spaces and transformations.* Let $\mathfrak{V}$ and the linear transformation T be both real. That is to say $\mathfrak{V}$ has a real base e, and relative to that base T is represented by a real matrix A. The characteristic roots are then real or complex conjugates.

Let $\bar{z}$ denote generally the conjugate of the complex number z, and if $Z = (z_{hj})$ is a complex matrix, let $\bar{Z} = (\bar{z}_{hj})$.

Since $|A - \lambda E|$ is a real polynomial its distinct roots consist say of q real roots $\lambda_1, \ldots, \lambda_q$ and of p conjugate pairs $\mu_1, \bar{\mu}_1, \ldots, \mu_p, \bar{\mu}_p$. And now we can state the following noteworthy complement to Theorem (8.4) (for the proof see Appendix I, 4):

(9.1) *If the vector space $\mathfrak{V}$ and the transformation T are both real then the block matrices C_h corresponding to real roots λ_h will be real and so will be the associated spaces $\mathfrak{V}_h$. On the other hand the matrices C_k and spaces $\mathfrak{V}_k$ corresponding to a pair $\mu_k, \bar{\mu}_k$ can be associated in complex conjugate pairs. That is to say the spaces of an associated couple $\mathfrak{V}_k, \mathfrak{V}_k$ will possess conjugate bases.*

The vectors of the base corresponding to the normal form of the matrix make up a set $e_1, \bar{e}_1, \ldots, e_p, \bar{e}_p, e_{2p+1}, \ldots, e_{2p+q}$ where the e_{2p+j} are real and the $e_h, \bar{e}_h, h \leq p$, are conjugate pairs. Any vector x will thus have coordinates

$$x_1, x_1^*, \ldots, x_p, x_p^*, x_{2p+1}, \ldots, x_{2p+q}.$$

(9.2) *If the vector x is real then its coordinates x_{2p+j} are real and $x_h^* = \bar{x}_h$ for $h \leq p$.*

For we will have

$$x = \sum_{h \leq p} (x_h e_h + x_h^* \bar{e}_h) + \sum x_{2p+j} e_j.$$

$$= \sum (\bar{x}_h \bar{e}_h + \bar{x}_h^* e_h) + \sum \bar{x}_{2p+j} e_j.$$

Since the elements of the base are linearly independent we must have $x_h^* = \bar{x}_h, \bar{x}_{2p+j} = x_{2p+j}$, and this proves our assertion.

10. *Limits and series.* Once one has norms and limits, convergence of sequences and series are defined for vectors and matrices in the usual way. They are equivalent for both to limits and convergence of the individual components of the vectors or terms of the matrices. Let us examine a little more closely the notions for matrices. If $\{A_p\}$ is a sequence of matrices and $A_p = (a_{ij}^p)$ then $\lim A_p$ exists when and only when for every i, j: $a_{ij} = \lim a_{ij}^p$ exists and $A = (a_{ij}) = \lim A_p$. Hence the infinite series ΣA_p is convergent when and only when the series

(10.1) $$a_{ij} = \Sigma a_{ij}^p$$

are all convergent and the sum of the series is by definition the matrix $A = (a_{ij})$. Absolute convergence of ΣA_p is defined by that of every series (10.1).

If the a_{ij}^p are functions of a parameter t and the n^2 series Σa_{ij}^p are uniformly convergent as to t over a certain range then ΣA_p is said to be *uniformly convergent* as to t over the same range.

Let us apply to the A_p's the simultaneous operation $B_p = PA_pQ$ where P, Q are non-singular and fixed. If we set

$$S_{ij}^{mr} = \sum_{p=m+1}^{r} |a_{ij}^p|$$

then the corresponding T_{ij}^{mr} for the B's is

$$T_{ij}^{mr} = \sum_{p=m+1}^{r} \left| \sum_{h,k} p_{ih}\, a_{hk}^p\, q_{kj} \right| .$$

Now a n.a.s.c. for the absolute convergence of (10.1) may be phrased thus: for every $\varepsilon > 0$ there is an N such that $m > N \rightarrow S_{ij}^{mr} < \varepsilon$ whatever r, i, j. If $\alpha = \sup \{|p_{ih}|, |q_{kj}|\}$ then

$$T_{ij}^{mr} \leq n^2 \alpha^2 \varepsilon.$$

Hence the absolute convergence of all the series (10.1) implies that of

$$\Sigma B_p = \Sigma PA_pQ$$

whose limit is clearly $B = PAQ$. In particular

(10.2) *If ΣA_p converges absolutely to A and if $B_p = PA_pP^{-1}$ then ΣB_p converges absolutely to $B = PAP^{-1}$.*

Consider a power series with complex coefficients

$$f(z) = a_0 + a_1 z + a_2 z^2 + \ldots$$

whose radius of convergence $\varrho > 0$. If $X = (x_{ij})$ is a square matrix of order n we may form the series

$$a_0 E + a_1 X + a_2 X^2 + \ldots \qquad (10.3)$$

and if it converges its limit will be written $f(X)$.

(10.4) Suppose that $X = \text{diag}(X_1, X_2)$. If $g(z)$ is a scalar polynomial then $g(X) = \text{diag}(g(X_1), g(X_2))$. Hence in this case

(10.3) converges when and only when the same series in the X_i converge and its limit is then $f(X) = \mathrm{diag}\,((f(X_1), f(X_2))$.

(10.5) *Sufficient conditions for the absolute convergence of* (10.3) *are that X is nilpotent or else that its characteristic roots are all less than* ϱ *in absolute value.*

Whatever the radius of convergence, when X is nilpotent the series is finite and hence evidently absolutely convergent. In the general case, X is similar to a matrix in normal form. Remembering (8.4) we only need to consider the type $C(\lambda)$. In other words, we may assume that the matrix is $C(\lambda)$ itself. Thus $\lambda_j = \lambda$ is its sole characteristic root and so we will merely have to prove:

(10.6) *If* $X = C(\lambda)$ *and* $|\lambda| < \varrho$ *then* (10.3) *is absolutely convergent.*

Let us set $Z = C(0)$. We verify by direct multiplication that Z^r is obtained by moving the diagonal of units so that it starts at the term in the $(r+1)^{st}$ row and first column (the term $z_{r+1,1}$). Hence

$$Z^n = 0.$$

Now $X = \lambda E + Z$, and since E commutes with every matrix:

$$X^p = \lambda^p E + \binom{p}{1} \lambda^{p-1} Z + \dots + \binom{p}{n-1} \lambda^{p-n+1} Z^{n-1}.$$

Hence

$$f(X) = \begin{pmatrix} f(\lambda), & 0, & \dots & & 0 \\ \dfrac{f'(\lambda)}{1!}, & f(\lambda), 0 & & \dots & 0 \\ \dfrac{f''(\lambda)}{2!}, & \dfrac{f'(\lambda)}{1!}, & f(\lambda), 0, & \dots & 0 \\ \dots & \dots & \dots & \dots & \dots \\ \dfrac{f^{(n-1)}(\lambda)}{(n-1)!}, & \dots & \dots & \dots & f(\lambda) \end{pmatrix}.$$

Since $|\lambda| < \varrho$, $f(z)$ and all its derivatives converge absolutely for $z = \lambda$. Hence so does $f(X)$. This proves (10.6) and therefore also (10.5).

(10.7) Returning now to the arbitrary matrix X we note that we can define in particular

$$e^X = E + \frac{X}{1!} + \frac{X^2}{2!} + \cdots$$

for every X, and also

$$\log (E + X) = \frac{X}{1} - \frac{X^2}{2} + \cdots$$

when the characteristic roots are less than one in absolute value.

(10.8) The usual rules for adding, multiplying, and generally combining series in X hold here also. However those for multiplying series in X by series in Y hold only when X and Y commute. Thus we may prove $e^{X+Y} = e^X \cdot e^Y$ when X and Y commute, but not so in the contrary case.

(10.9) *If* $f(X)$ *converges absolutely and* $Y = PXP^{-1}$, $|P| \neq 0$, *then* $f(Y)$ *does also and* $f(Y) = P(f(X))P^{-1}$.

(10.10) *If* $|X| \neq 0$ *there is a* Y *such that* $e^Y = X$.

Since Y need not be unique we do not insist on designating it by $\log X$.

Referring to (10.4) if $X = \operatorname{diag}(X_1, X_2)$ and if we can find Y_1, Y_2 such that $X_i = e^{Y_i}$ then $X = \operatorname{diag}(e^{Y_1}, e^{Y_2})$ and so $Y = \operatorname{diag}(Y_1, Y_2)$ answers the question. Hence as in the proof of (10.5) we need only consider X of the type $C(\lambda)$. Here $|X| \neq 0 \rightarrow \lambda \neq 0$. In the same notations as before $X = \lambda E + Z$ $= \lambda (E + \frac{1}{\lambda} Z)$. Since Z is nilpotent so is $\frac{1}{\lambda} \cdot Z$, and therefore we may define by (10.5) the function

$$Y_1 = \log (E + 1/\lambda \cdot Z).$$

By (10.8):

$$E + 1/\lambda \cdot Z = e^{Y_1},\ X = \lambda E + Z = \lambda e^{Y_1}.$$

Since $\lambda \neq 0$ we may find a scalar μ such that $\lambda = e^\mu$. Then

$$X = (Ee^\mu)\, e^{Y_1} = (e^{\mu E}) \cdot e^{Y_1} = e^{\mu E + Y_1}.$$

Therefore $Y = \mu E + Y_1$ answers the question.

(10.11) *If the series* (10.3) *for* $f(X)$ *converges absolutely then the determinant*

(10.11a) $$|f(X)| = \Pi f(\lambda_j).$$

Hence

$$|e^X| = e^{\Sigma \lambda_j} = e^{\text{trace } X}.$$

Thus $Y = e^X$ is never singular, and so it has an inverse Y^{-1}. Now e^{-X} exists likewise and $e^X \cdot e^{-X} = e^{-X} \cdot e^X = E$, so that $Y^{-1} = e^{-X}$.

Referring to (8.4) X is similar to a triangular matrix

$$Y = \begin{pmatrix} \lambda_1 & & & \\ a_{21}, \lambda_2 & & & \\ & \ddots & & \\ a_{n1}, \ldots, a_{n,n-1}, \lambda_n \end{pmatrix}.$$

The λ_j are the characteristic roots each repeated as often as its multiplicity. This follows immediately from

$$|Y - \lambda E| = \Pi (\lambda_j - \lambda).$$

By (10.9) $f(X) \sim f(Y)$ and so we may replace everywhere X by Y, i.e. we merely need to prove (10.11) for Y. Now $f(Y)$ is of the same form as Y with λ_j replaced by $f(\lambda_j)$. This implies that the $f(\lambda_j)$ are the characteristic roots of $f(Y)$. Since the determinant is the product of the characteristic roots, (10.11a) holds for Y, hence also for X, and the rest follows.

11. *Matrix functions of scalars. Let* $X = (x_{ij}(t))$ be an $m \times n$ matrix whose terms are [real or complex] scalar functions of a [real or complex] variable t differentiable over a certain range R. We have already defined dX/dt and $Y = \int_{t_0}^{t} X\, dt$, (6.2). If the x_{ij} are continuous on the path of integration then evidently $dY/dt = X$.

Suppose now that $X(t)$, $Y(t)$ are any two square matrices of the same order, differentiable over the same range R. Then XY is differentiable over the same range and an elementary argument yields

$$(11.1) \qquad d(XY)/dt = X\, dY/dt + dX/dt \cdot Y.$$

Care must be taken here to keep X, Y always in the same order. From (11.1) we deduce readily

(11.2) $d(X_1 \ldots X_r)/dt = \Sigma\, X_1 \ldots X_{q-1}\, dX_q/dt \cdot X_{q+1} \ldots X_r$

and therefore

(11.3) $$dX^r/dt = \Sigma\, X^{q-1}\, dX/dt \cdot X^{r-q}.$$

If we differentiate both sides of $XX^{-1} = E$, $|X| \neq 0$, we obtain

$$d(X^{-1})/dt = -X^{-1}\, dX/dt \cdot X^{-1}.$$

If all the $x_{ij}(t)$ are continuous or analytic at a point or on a given set, we will say that $X(t)$ is *continuous* or *analytic* at the same point or on the same set.

Let A be a constant square matrix and set

(11.4) $X(t) = e^{tA} = e^{At} = E + t/1! \cdot A + t^2/2! \cdot A^2 + \ldots .$

By differentiation we obtain

(11.5) $$dX/dt = Ae^{At}.$$

Notice that owing to the form of (11.3) we can only prove $d/dt \cdot e^{X(t)} = e^{X(t)}\, dX/dt$, when X and dX/dt commute.

Combining (11.5) with (11.1), and setting for convenience $d/dt = D$, we obtain for any matrix X the analogue of the known elementary relation:

(11.6) $$(D - A)\, X = e^{At} \cdot D \cdot e^{-At}\, X.$$

As an application consider the matrix differential equation

(11.7) $$dX/dt = AX, \; A \text{ constant}.$$

Owing to (11.6) it reduces to

$$e^{At} \cdot D \cdot e^{-At}\, X = 0.$$

Multiplying both sides by e^{-At} (see 10.11) we have

$$D \cdot e^{-At} \cdot X = 0$$

and hence $e^{-At} \cdot X = C$, an arbitrary constant matrix. Hence the complete solution of (11.7) is $X = e^{At} \cdot C$. We will return to this later.

§ 3. Analytic Functions of Several Variables

12. We shall have repeated occasion to consider analytic functions of several real or complex variables as well as mixed functions analytic in some, but not in all the variables.

Consider first $\mathfrak{V}_x$ complex. Write $x = y + iz$, *viz.* $x_j = y_j + iz_j$ for $j = 1, \ldots, n$. A function $f(x)$ is said to be *analytic* in a region Ω of $\mathfrak{V}_x$ if it has first partial derivatives relative to all y_j and z_j which are continuous in y and z at all points of Ω and if it satisfies the Cauchy-Riemann differential equations relative to each pair of y_j and z_j at all points of Ω. The function f is said to be *holomorphic* in Ω if it is analytic and single-valued in Ω; f is said to be *analytic* or *holomorphic* in a closed set F in $\mathfrak{V}_x$ if it is analytic or holomorphic in some neighborhood of F.

A n.a.s.c. for analyticity in Ω is that f may be expanded in Taylor series around each point ξ of Ω. The series will be convergent in a set $||x - \xi|| < a$, which will be contained in Ω.

Suppose now $\mathfrak{V}_x$ real and let $[\mathfrak{V}_x]$ be its complex extension, i.e. the complex vector space obtained by allowing the coordinates x_j to take complex values. A real function $f(x)$ will be said to be *analytic* or *holomorphic* in a region Ω of $\mathfrak{V}_x$ under the following conditions: there exists a region $[\Omega]$ of $[\mathfrak{V}_x]$ and a function $[f]$ on $[\mathfrak{V}_x]$ analytic or holomorphic in $[\Omega]$, and such that: (a) $[\Omega] \supset \Omega$; (b) the values of $[f]$ on Ω are those of f.

The definition just given has the following consequence.

(12.1) *If a series of complex functions analytic or holomorphic in a fixed* $[\Omega]$, *is uniformly convergent in* $[\Omega]$, *then the limit is analytic or holomorphic in* $[\Omega]$. *If a series of real functions analytic or holomorphic in a region* Ω *is uniformly convergent in a fixed complex extension* $[\Omega]$, *then the limit is analytic or holomorphic in* Ω.

For the complex case this is a standard theorem due to Weierstrass and for the real case it is a consequence of our definition.

An *analytic* or *holomorphic* vector or a vector *power series* in x is a vector $f(x)$ whose components $f_i(x_1, \ldots, x_n)$ are analytic or holomorphic or power series in the x_h.

Given two series

$$a = a_1 + a_2 + \ldots, \quad b = b_1 + b_2 + \ldots,$$

the second is said to be a *majorante* of the first, written $a \ll b$

(Poincaré's notation) whenever $|a_m| \leq |b_m|$ for every m. More generally, if the multiple series

$$a = \Sigma\, a_{m_1, \ldots, m_p},\; b = \Sigma\, b_{m_1, \ldots, m_p}$$

are such that $|a_{m_1, \ldots, m_p}| \leq |b_{m_1, \ldots, m_p}|$ for every combination $m_1, \ldots, m_p$ then b is called a majorante of a, written as before $a \ll b$.

If $m = \Sigma m_i$ then it is often convenient to denote by (m) the set $\{m_1, \ldots, m_p\}$. Thus we would write above:

$$a = \Sigma\, a_{(m)},\; b = \Sigma\, b_{(m)}.$$

Suppose that $F(x_1, \ldots, x_n)$ is holomorphic in the closed region $\Omega: |x_i| \leq A_i,\ i = 1, 2, \ldots, n$, where the A_i are positive constants. Since Ω is compact $|F|$ has an upper bound M in Ω. It is then shown in the treatises on the subject (see for instance Picard [1], II, Ch. 9) that F admits in Ω the McLaurin expansion

$$F = \Sigma\, F^{(m)} x_1^{m_1} \ldots x_n^{m_n}$$

with the following estimate for the coefficients:

$$|F^{(m)}| \leq \frac{M}{A_1^{m_1} \ldots A_n^{m_n}}. \tag{12.2}$$

If we identify F with the series we have therefore

$$F \ll \frac{M}{\Pi\left(1 - \dfrac{x_i}{A_i}\right)}. \tag{12.3}$$

If $A = \inf A_i$ then another useful relation of the same type is

$$F \ll \frac{M}{1 - 1/A\, \Sigma\, x_i}. \tag{12.4}$$

It is in fact a simple matter to show that

$$\frac{1}{\Pi\,(1 - x_i/A_i)} \ll \frac{1}{1 - 1/A\, \Sigma\, x_i}$$

from which (12.4) follows.

(12.5) *Notation.* If $\psi_1, \ldots, \psi_p$ are differentiable functions of $u_1, \ldots, u_q$ then $\partial\psi/\partial q$ will stand for the Jacobian matrix $(\partial\psi_j/\partial u_k)$ and for $p = q$, $|\partial\psi/\partial u|$ will stand for the Jacobian determinant $|\partial\psi_j/\partial u_k|$.

13. Consider a system

(13.1) $\varphi_i (x_1, \ldots, x_n ; y_1, \ldots, y_p) = 0, \ i = 1, 2, \ldots, n,$

or as we shall write it more simply

(13.2) $$\varphi (x ; y) = 0$$

in obvious vector notations. Let the system be analytic in $(x; y)$, holomorphic at $(x^0; y^0)$ and let $\varphi (x^0; y^0) = 0$. We have now the well known

(13 3) IMPLICIT FUNCTION THEOREM. *If the Jacobian*

$$J = |\partial\varphi/\partial x|$$

does not vanish at $(x^0; y^0)$ *then the system* (13.2) *has a unique solution* $x = \psi (y)$ *holomorphic at* $y = y^0$ *and such that* $x^0 = \psi (y^0)$.

The usual proof given for complex variables applies as well to the real field.

Consider a transformation φ from the real space $\mathfrak{V}_x$ to the real space $\mathfrak{V}_y$ represented by

(13.4) $$y_i = \varphi_i (x_1, \ldots, x_n), \ i = 1, 2, \ldots, n.$$

We will say that φ is *regular* at x^0 whenever the φ_i are analytic at the point and the Jacobian

$$J = |\partial\varphi/\partial x|$$

does not vanish there. The implicit function theorem for the system (13.4) yields immediately the following property:

(13.5) *Let* φ *be regular at* x^0 *and let* $y^0 = \varphi (x^0)$. *There can be chosen in* $\mathfrak{V}_x$, $\mathfrak{V}_y$ *n-cells* $E_x{}^n$, $E_y{}^n$ *containing respectively* x^0, y^0 *and such that* φ *maps topologically* $E_x{}^n$ *onto* $E_y{}^n$.

Consider a system of relations with $p \leq n$:

(13.6) $$x_i - x_i{}^0 = \varphi_i (u_1, \ldots, u_p), \ (i = 1, 2, \ldots, n)$$

where the φ_i are real, vanish at $u = 0$, are holomorphic in a certain p-cell $E_u{}^p$: $||u|| < \varrho$, and have a Jacobian matrix of rank p at $u = 0$. We consider (13.6) as defining a mapping φ of $E_u{}^p$ into $\mathfrak{B}_x$. It is a consequence of our assumptions that φ is a topological mapping of $E_u{}^p$ into a p-cell E^p through x^0 in $\mathfrak{B}_x$. The cell E^p is known as *an analytical p-cell* through x^0. When $p = 1$ we also speak of an elementary *analytical arc* through the point. An *analytical curve* λ in $\mathfrak{B}_x$ is a connected set such that every point has a neighborhood which is an elementary analytical arc. That is to say every point x^0 of λ has a neighborhood E^1 which admits a parametric representation

$$x_i - x_i{}^0 = \psi_i(v),\ \psi_i(0) = 0,\ |v| < a,$$

where the ψ_i are analytical on their range and the derivatives $\psi_i'(v)$ never vanish simultaneously on the range. When the analytical curve λ is compact it is a Jordan curve, otherwise it is an arc.

The following intersection property will be found useful later.

(13.7) *Let the analytical cell E^p and the analytical curve λ have an infinite number of intersections with a condensation point P contained in both. Then P has a neighborhood v in the intersection μ which is a subarc of λ.*

We may suppose the representations as above with P as x^0. Under the assumption we may solve (13.6) for $u_1, \ldots, u_p$ in terms of p of the differences $x_i - x_i{}^0$, say in terms of the first p differences, and thus replace the system in the vicinity of P by another of the form

$$x_{p+i} - x^0{}_{p+i} = f_i(x_1 - x_1{}^0, \ldots, x_p - x_p{}^0)$$

where the $f_i(w_1, \ldots, w_p)$ are analytic at the origin $w = 0$. As a consequence the analytic functions of v

$$F_i(v) = f_i(\psi_1(v), \ldots, \psi_p(v)) - \psi_{p+i}(v)$$

have an infinity of zeros in the vicinity of $v = 0$. It follows that F_i vanishes on some interval $|v| < a_i$. If a is the least a_i then P has for neighborhood in μ the subarc of λ corresponding to $|v| < a$.

(13.8) *Application. Two analytical curves λ, μ in an Euclidean space, which have an infinite number of intersections with a condensation point P necessarily coincide.*

Let ξ be the intersection. The point P is in a subcell E^1 of λ, and μ intersects E^1 in an infinite number of points. Hence P is contained in a subarc ν of μ which is in the intersection ξ and is a neighborhood of P in both λ and μ. Let η be the largest subarc of both λ and μ which contains ν and suppose that η has an end point Q. Since ξ is closed in both λ and μ, Q is in ξ. By the above argument Q is contained in some subarc ν' of both λ and μ. Since this contradicts the maximal property of η, Q cannot exist. Hence $\eta = \lambda = \mu$. This proves (13.8).

(13.9) The implicit function theorem holds as well when the functions $\varphi(x; y)$ are merely differentiable with non-vanishing Jacobian. The solutions $\psi(y)$ are then merely continuous. See notably Goursat [1]. The other considerations of the present section, except (13.7) and (13.8), are likewise valid with analyticity replaced by the condition that the functions are *of class* C^r, that is possess continuous partial derivatives of order $\leq r$, where $r \geq 1$. One obtains thus cells, curves, etc., ..., said to be *of class* C^r.

(13.10) *Important remark.* The considerations of (9), especially (9.2) will cause us to consider repeatedly complex transformations of a real vector space $\mathfrak{V}_x$ into a complex vector space $\mathfrak{V}_y$ referred to coordinates y_1, $y_1{}^*$, ..., y_p, $y_p{}^*$, y_{2p+1}, ..., y_n in which however the *real* points of $\mathfrak{V}_y$ will have coordinate y_1, $\bar{y}_1$, ..., y_p, y_{2p+1}, ..., y_n. Upon setting $y_h = y_h{}' + iy_h{}''$, $h \leq p$, one may consider the real points as referred to the real coordinates $y_h{}'$, $y_h{}''$, $h \leq p$, y_{2p+j} and thus "assimilate" the transformation to one from real $\mathfrak{V}_x$ to real $\mathfrak{V}_y$ (the real part of $\mathfrak{V}_y$). The statement for instance that a function $f(y)$ is analytic means analytic in the y_h, $y_h{}^*$, $h \leq p$ and the y_{2p+j}. The modifications in the various statements made in the present number are elementary and may be left to the reader.

A function $f(y)$ will be said to be real whenever $\overline{f(y)} = f(\bar{y})$. It is said to be *analytic* at the real point $\xi(\xi = \bar{\xi})$ whenever $f(y_1, y_1{}^*, ..., y_{2p+1}, ..., y_n)$ is analytic at 'he point ξ in the complex space $\mathfrak{V}_y$.

Let $q(x)$ be a real vector function. If the transformation of coordinates $y \to x$ is $x = Ay$ then $q(x)$ becomes a function $q(Ay)$ and has components $q_1, q_1^*, \ldots, q_{2p+1}, \ldots, q_n$. We will have $q_h^*(y) = q_h(\bar{A}y)$ and thus $q_h^*(\bar{y}) = \overline{q_h(y)}$, $h \leq p$. The components q_{2p+j} are all real.

(13.11) Everything said in the present Section, except for arcs, holds also in the complex analytic domain. The cells involved, for instance in (13.5), will be referred to as *complex* n-cells (or p-cells, etc).

14. *The Weierstrass preparation theorem.* Let us consider real or complex power series $f(x_1, \ldots, x_n) = f(x)$ which are holomorphic at the origin. In this connection we shall use the following terminology which has become standard in the theory of *local rings*. We will call $f(x)$ a *unit* if $f(0) \neq 0$ and a *non-unit* if $f(0) = 0$. Units are often generically denoted by $E(x)$. A polynomial

$$x_1^p + a_1(x_2, \ldots, x_n)\, x_1^{p-1} + \ldots + a_p(x_2, \ldots, x_n)$$

where the a_h are non-units is called a *special polynomial in* x_1.

We may now state:

(14.1) The Weierstrass Preparation Theorem. *Suppose that* $f(x_1, 0, \ldots, 0) \neq 0$ *and let* p *be its least degree in* x_1. *Then*

$$f(x) = F(x_1; x_2, \ldots, x_n)\, E(x)$$

where F *is a special polynomial in* x_1 *and of degree* p *in* x_1.

For the proof see Bochner-Martin [1], p. 183.

Since the proof involves only the solution of linear equations, it introduces no irrationality and so we may complement (14.1) as follows:

(14.2) *If* $f(x)$ *is real so are the unit* E *and the special polynomial* F.

(14.3) *Factorization of special polynomials.* Let us observe that the convergent power series in $x_2, \ldots x_n$ with complex coefficients form by addition and multiplication a *ring* R with unity element in the sense of algebra. By quotient formation the ring R gives rise to a *field* F. The totality of polynomials in x_1 with coefficients in R or F generate rings $R[x_1]$, $F[x_1]$. By a well known theorem of algebra (see van der Waerden [1] p. 75) a polynomial of $R[x_1]$ is irreducible in F if and only if it is irreducible in $R[x_1]$. From this follows unique factorization in $R[x_1]$. It is also easy to verify directly:

(14.4) *If a special polynomial in x_1 is reducible in $R[x_1]$ then its factors may all be chosen special polynomials in x_1.*

(14.5) *Notation.* We shall denote by $[x]_m$, $[x;\ y]_m$, etc., a power series in the components of x or x and y, etc., with no terms of degree $< m$. Similarly for a vector whose components are of this type—the context will readily tell whether one means "vector" or "function". Exceptionally $[x;t]$, $[x;y;t]$ will stand for the same series or vectors whose coefficients are continuous and bounded functions of t (the time) in a certain range $t \geq \tau$.

§ 4. Differentiable Manifolds

15. Euclidean spaces and spheres are the simplest examples of differentiable manifolds. The formal definition runs as follows: An n-dimensional differentiable manifold M^n is a connected metric space with an open covering $\{U_a\}$ with the following properties:

(a) the U_a are n-cells;

(b) each U_a is parametrized by n real variables $x_{a_1}, \ldots, x_{a_n}$;

(c) let U, U' be two U_a with parameters x_j, x_k' and let them have a non-void intersection W. Then in W

$$x_j' = \varphi_j(x_1, \ldots, x_n) \tag{15.1}$$

where the φ_j are of class C^r, $r > 0$ — the manifold is then said to be of class C^r — and in W the Jacobian

$$J = |\partial\varphi/\partial z| \neq 0. \tag{15.2}$$

Observe that the relation (15.2) is symmetrical relative to the intersecting open sets U and U'.

If one may select the parameters such that J has a fixed sign throughout (and one may then dispose of the situation so that $J > 0$) M^n is *orientable*, otherwise it is *non-orientable*. If the φ's are analytic throughout M^n is said to be *analytic*. If M^n is compact one may suppose that $\{U_a\}$ is finite.

A differentiable M^1 is always orientable. It is a Jordan curve if it is compact and an arc otherwise.

(15.3) *An example.* The point-set of an Euclidean n-space defined by relations

$$x_h = \varphi_h(u_1, \ldots, u_p), \quad h = 1, 2, \ldots, n$$

where $\varphi(u)$ is of class C^r, $r >, 0$ for $||u|| < a$ and $\partial\varphi/\partial u$ is of rank p in that range, represents an M^p of class C^r whose basic covering consists of a single cell. If the φ_h are analytic in the range M^p is analytic.

A second covering $\{U_\beta'\}$ of M^n behaving like the basic covering $\{U_\alpha\}$ is said to cover the same M^n of class C^r if M^n is of class C^r under the joint covering $\{U_\alpha;\ U_\beta'\}$. As an example if a region Ω in an Euclidean space E^n is made of class C^r by reference to a set of coordinates x_i then the only other allowable coordinate systems x_j' for Ω are such that (15.1) and (15.2) hold.

An important concept for a differentiable M^n is the *vector field*. One understands thereby an assignment of n point functions $X_1, \ldots, X_n$ in each U, whereby with the situation as in (c) and with X_j' assigned in U' then in W

(15.4) $$X_j' = \Sigma\, \partial x_j'/\partial x_i \cdot X_i.$$

This law of transformation is also applicable to a change of allowable coordinates in a region of Euclidean space.

(15.5) *Two dimensional manifolds.* A compact M^2 may always be *triangulated*, i.e. decomposed into a finite number of "curvilinear" triangles. (A similar property holds for any M^n but is not as easily described in intuitive terms.) Be it as it may, given a triangulation of M^2 let it consist of a_0 vertices, a_1 sides and a_2 triangles. The expression

$$\chi(M^2) = a_0 - a_1 + a_2$$

is known as the *Euler-Poincaré characteristic* of M^2 and it is a topological invariant. In particular $\chi(M^2)$ is independent of the triangulation. If M^2 is orientable then

$$\chi(M^2) = 2 - 2p,\ p \geq 0,$$

and p, which is a topological invariant, is known as the *genus* of the surface. If $p = 0$ the surface is a sphere and if $p = 1$ it is a torus.

The extension of much of the above to any dimension offers no difficulty but will not be needed later. For further information the reader is referred to Lefschetz [1].

CHAPTER II

Existence Theorems. General Properties of the Solutions

The study of differential equations is initiated in the present chapter. The topics treated are the existence theorems, the dependence of the solutions on the initial conditions and the general properties of trajectories.

References: Bellman [3]; Coddington-Levinson [1]; Goursat [1], II, Ch. 19; Hurewicz [1]; Kamke [1]; Picard [1], II, Ch. 11.

§ 1. **Generalities**

1. Let t denote a real variable ranging from $-\infty$ to $+\infty$ and usually referred to as the time. We shall be primarily concerned with systems of type

$$(1.1) \qquad dx_i/dt = X_i(x_1, \ldots, x_n; t),\ i = 1, 2, \ldots, n$$

where the X_i are real valued functions of the real variables x_i and of t. In vector form (1.1) is written

$$(1.2) \qquad dx/dt = X(x; t)$$

where x, X are n-vectors. The vectors $X(x; t)$ give rise to a field $F(t)$ in the space $\mathfrak{B}_x$ of the vector x and to solve (1.2) is to find a vector x whose derivative defines a preassigned field $F(t)$.

It may be pointed out that the single differential equation

$$(1.3) \qquad \frac{d^n x}{d t^n} = f\left(x, \frac{dx}{dt}, \ldots, \frac{d^{n-1} x}{dt^{n-1}}; t\right)$$

where x is the unknown function, is equivalent to a system (1.1).

For if we set $x = x_1$ then (1.3) is equivalent to

$$(1.4)\qquad \begin{cases} \dfrac{dx_1}{dt} = x_2, \dfrac{dx_2}{dt} = x_3, \ldots, \dfrac{dx_{n-1}}{dt} = x_n, \\ \dfrac{dx_n}{dt} = f(x_1, \ldots, x_n; t), \end{cases}$$

which is of type (1.1).

(1.5) We shall have occasion to consider two spaces: the vector space $\mathfrak{B}_x$ and the product space $\mathfrak{W} = \mathfrak{B}_x \times \mathfrak{T}$, where $\mathfrak{T}$ is the time line. The spheroids in $\mathfrak{B}_x$ are written $\mathfrak{S}(x; \varrho)$ and those in $\mathfrak{W}$ $\Sigma((x; t); \varrho)$.

A solution $x(t)$ of the system (1.1) is called a *trajectory* or a *motion* or also an *integral curve*. It is thus a curve in the space $\mathfrak{B}_x$. We shall often loosely say however a "trajectory in the space $\mathfrak{W}$" to refer to the locus (curve) of the point $(x(t); t)$ of $\mathfrak{W}$ as t varies.

A system (1.2) which does not contain the time, that is, of the form

$$(1.6)\qquad dx/dt = X(x)$$

is said to be *autonomous* and, for reasons discussed more fully later, its "geometric" trajectories in the space $\mathfrak{B}_x$ are referred to as *paths*. The term that has been used by most authors, following Poincaré, is "characteristic" but as this expression covers too many things we have adopted "path" in its place.

§ 2. The Fundamental Existence Theorem

2. Let Ω be a certain region of the product space $\mathfrak{W} = \mathfrak{B}_x \times \mathfrak{T}$, and let $X(x; t)$ be an n-vector defined over Ω. Let U_λ denote a spheroid $\mathfrak{S}(Q; \lambda)$, $Q \in \mathfrak{B}_x$, of the space $\mathfrak{B}_x$ and let $R = U_\lambda \times I$, I a bounded interval of $\mathfrak{T}$, be an open set of $\mathfrak{W}$. The vector X is said to satisfy a *Lipschitz condition* in Ω, whenever given any open set R whose closure $\bar{R} \subset \Omega$, and any two points $(x; t)$, $(x'; t) \in R$ there takes place an inequality

$$\|X(x; t) - X(x'; t)\| \leq k\,\|x - x'\|,$$

where k is a constant which may depend on R.

Notice that if X does not contain t, Ω itself is a product $\Phi \times \mathfrak{T}$, where Φ is a region of $\mathfrak{B}_x$. Then the Lipschitz condition refers merely to sets U_λ such that $U_\lambda \subset \Phi$ and points x, $x' \in U_\lambda$. It takes then the form

$$||X(x) - X(x')|| \leq k\,||x - x'||,$$

where k may depend on U_λ.

Let now $P(\xi; \eta) \in \Omega$ and choose a fixed open set R_P containing P. If I_τ is the time interval $|t - \eta| < \tau$, then the set $||x - \xi|| < a$, $|t - \eta| < \tau$ is an open set $B(a, \tau)$ of $\mathfrak{W}$. We will always assume a and τ small enough to have $B \subset R_P$, and hence $\bar{B} \subset \Omega$. A set such as $B(a, \tau)$ will be conveniently referred to as a *box* of center P.

(2.1) EXISTENCE THEOREM OF CAUCHY-LIPSCHITZ. *Suppose that in the equation* (1.2) *X is continuous and satisfies a Lipschitz condition in Ω. Let $P(\xi; \eta) \in \Omega$ and let R_P be as above. Given $a > 0$ sufficiently small there may be chosen a box of center P, $B(2a, \tau(a))$, where $\tau(a) \to 0$ with a, such that through every point $(x^0; t^0) \in B(a, \tau)$ there passes a unique trajectory $x(t)$ of the system* (1.2). *This trajectory is defined over the time interval I_τ and the point $(x(t); t)$ remains in $B(2a, \tau)$ when t is in I_τ.*

Method of proof. The differential equation (1.2) together with the initial condition $x(t^0) = x^0$ is equivalent to the integral equation

$$x(t) = x^0 + \int_{t^0}^{t} X(x(t); t)\, dt. \tag{2.2}$$

This integral equation is solved by Picard's classical method of successive approximations. The process is then shown to be convergent and the solution to be unique.

3. Let k be the Lipschitz constant of R_P. Since X is continuous on the compact set $\bar{R}_P$ it has an upper bound M on that set. Define now

$$\varphi(\tau) = \frac{M}{k}(e^{k\tau} - 1) > 0.$$

Since $\varphi(\tau) \to 0$ with τ we may choose $\tau(a)$ such that $\varphi(2\,\tau(a))$

$< a$, that is to say $\tau(a) < (1/2k) \log (1 + ka/M)$. Clearly $\tau(a) \to 0$ with a.

Choose now as $(x^o; t^o)$ an arbitrary point in $B(a, \tau)$. Thus $||x^o - \xi|| < a$, $|t^o - \eta| < \tau$.

Existence of $x(t)$. Consider the sequence of vectors $x^m(t)$ defined in succession by

$$(3.1)_m \qquad x^{m+1} = x^o + \int_{t^o}^{t} X(x^m; t)\, dt,\ t \in I_\tau.$$

We will show that the sequence has a limit which satisfies (2.1).

Let us show that for $t \in I_\tau$:

$$(3.2)_m \qquad ||x^m - x^{m+1}|| \leq Mk^m |t - t^o|^{m+1}/(m+1)\,!.$$

Notice that $(3.2)_m$ implies

$$(3.3)_m \qquad ||x^m - x^{m+1}|| \leq \frac{M}{k}(2k\tau)^{m+1}/(m+1)\,!.$$

At all events from $(3.1)_o$ there follows

$$||x^o - x^1|| \leq M\,|t - t^o|,$$

so that $(3.2)_o$ holds. Suppose now that $(3.2)_p$, consequently also $(3.3)_p$ hold for every $p < m$. By repeated application of the triangle inequality we find

$$||x^o - x^{p+1}|| \leq \frac{M}{k}\left(\frac{2k\tau}{1\,!} + \frac{(2k\tau)^2}{2\,!} + \ldots + \frac{(2k\tau)^{p+1}}{(p+1)\,!}\right) < \varphi(2\tau) < a.$$

Hence for $p < m$:

$$(3.4)_p \quad ||x^{p+1} - \xi|| \leq ||x^{p+1} - x^o|| + ||x^o - \xi|| \leq \varphi(2\tau) + a < 2a.$$

Hence if $(x^o; t) \in B(a, \tau)$ then $(x^{p+1}; t) \in B(2a, \tau)$. In particular under our assumptions for $m \geq 1$ both $(x^{m-1}; t)$ and $(x^m; t) \in B(2a; \tau) \subset R_P$. We have now from $(3.1)_m$

$$||x^m - x^{m+1}|| \leq \left| \int_{t^o}^{t} ||X(x^{m-1}; t) - X(x^m; t)||\, dt \right|.$$

Since $(x^{m-1}; t)$ and $(x^m; t) \in R_P$, we may apply the Lipschitz condition to the integrand thus obtaining

$$||x^m - x^{m+1}|| \leq k \,|\int_{t^0}^{t} ||x^{m-1} - x^m||\, dt|.$$

From this and $(3.2)_{m-1}$ there follows $(3.2)_m$ by integration. Hence $(3.2)_m$, consequently also $(3.3)_m$ and $(3.4)_m$ hold for every m. In particular also every $(x^m; t) \in B\,(2a, \tau)$.

We find however more precisely from $(3.3)_m$ and $(3.4)_m$ that $\{x^m(t)\}$ is a Cauchy sequence in $\bar{U}$ where U is the open set $||x - \xi|| < \varphi\,(2\,\tau) + a$. Since $\bar{U}$ is compact the sequence has a condensation point in the set. Since it is a Cauchy sequence it has a limit and from (3.4) follows that the limit is in $||x - \xi|| \leq \varphi\,(2\,\tau) + a < 2a$ and therefore in U_{2a}. Furthermore the convergence is uniform for $t \in I_\tau$. Since $X\,(x; t)$ is continuous in $\overline{B\,(2a, \tau)}$ it follows at once that $\int_{t^0}^{t} X\,(x^m; t)\, dt$ converges uniformly to $\int_{t^0}^{t} X\,(x; t)\, dt$ for $t \in I_\tau$. Now for $t \in I_\tau$ and all m, we have

$$||x - x^0 - \int_{t^0}^{t} X\,(x; t)\, dt\,|| =$$

$$||x - x^{m+1} + \int_{t^0}^{t} X\,(x^m; t)\, dt - \int_{t^0}^{t} X\,(x; t)\, dt||$$

$$\leq ||x - x^{m+1}|| + ||\int_{t^0}^{t} X\,(x^m; t)\, dt - \int_{t^0}^{t} X\,(x; t)\, dt||.$$

Hence for $t \in I_\tau$

$$||x - x^0 - \int_{t^0}^{t} X\,(x\;\; t)\, dt||$$

can be made arbitrarily small (this uniformly in t) for m sufficiently large, and therefore it is identically zero. Thus for $t \in I_\tau$ the equation (2.2) holds. Hence $x = x\,(t)$ satisfies (1.2) and $x\,(t^0) = x^0$.

4. *Uniqueness of* $x\,(t)$. Suppose that $y\,(t)$ defines a second solution over $I_{\tau'}$: $|t - \eta| < \tau'$ such that $y\,(t^0) = x^0$. We may

suppose $\tau \leq \tau'$, and so the second solution is defined likewise over I_τ. Since $x(t)$ and $y(t)$ are differentiable they are continuous. Hence given any $\beta > 0$ there will exist a positive $\sigma < \tau$ such that $|t - t^0| < \sigma \rightarrow ||x - x^0||$, $||y - x^0|| < \beta$ and therefore $||x - y|| < 2\beta$. Now by the Lipschitz condition

$$(4.1) \qquad ||x - y|| \leq k \left| \int_{t^0}^{t} ||x - y|| \, dt \right| \leq 2\beta k |t - t^0| .$$

By repeated substitution in (4.1) we obtain

$$||x - y|| \leq 2\beta \frac{(k|t - t^0|)^m}{m!} \rightarrow 0 \text{ with } \frac{1}{m} .$$

Hence $||x - y|| = 0$ and therefore $x = y$ for $|t - t^0| < \sigma$. Thus if $x(t)$, $y(t)$ assume the same value for any $t^0 \in I_\tau$, they coincide on an interval containing t^0 and contained in I_τ.

Now consider the maximal interval I': $t' < t < t''$ containing t^0, contained in I_τ and such that $z(t) = x(t) - y(t) = 0$ on I'. If $I' \supset I_\tau$, then $x(t) = y(t)$ on I_τ, and the proof is complete. In the contrary case one of t' or t'', say t' is in I_τ. Since $z(t)$ is continuous in I_τ and is zero in I', it is also zero at t', since $t' \in \bar{I}'$. It follows from the earlier argument that $z(t)$ vanishes in an interval I'' containing t' and contained in I_τ. Thus we can augment I' by I'', and so we get a contradiction.

This completes the proof of the existence theorem.

5. *Complementary Remarks.* Of the two basic conditions in the statement of the existence theorem, the second—the Lipschitz condition—is less natural than the first, and would seem more difficult to verify. The following property will therefore be useful in this direction, and will also suffice for later applications.

(5.1) *If the partial derivatives $\partial X_i / \partial x_j$ exist and are continuous in Ω then the Lipschitz property holds.*

For we may clearly replace Ω by R_P. Applying now the mean value theorem to X_i, with $x, x' \in U_\lambda$ we find

$$X_i(x; t) - X_i(x'; t) = \sum \frac{\partial X_i(x^*; t)}{\partial x_j} (x_j - x_j') ,$$

where $x^* \in U_\lambda$ since U_λ is convex. Hence if A is an upper bound of the $|\partial X_i/\partial x_j|$ in R_P for all X_i (it exists since $\bar{R}_P$ is compact and in Ω), we find

$$||X(x;t) - X(x';t)|| \leq nA \cdot ||x - x'||,$$

which is the Lipschitz property.

(5.2) It is worth while to exhibit functions which do not satisfy a Lipschitz condition, e.g. (a) $X(x;t)$ with $X_i = 1/x_i$ for $0 < x_i < 1$; (b) $X(x;t)$ with $X_i = \sqrt{x_i}$ for $0 < x_i$.

(5.3) *Domain of continuity.* The union of all the sets such as Ω is an open set of $\mathfrak{W}$. We choose one of its components $\mathfrak{D}$ and call it *the domain of continuity of the differential equation* (1.2). Henceforth the equation is supposed to be taken together with a definite domain $\mathfrak{D}$ and all our operations will be generally restricted to that domain.

(5.4) *$\mathfrak{D}$ is an open set.*

For every point of $\mathfrak{D}$ has a connected neighborhood N which is in Ω and hence in $\mathfrak{D}$. Hence $\mathfrak{D}$ is open.

6. *Extension of the solution. Trajectories.* The existence theorem yields a solution $x(t)$ valid over a certain interval I_τ containing t^0, and such that $x(t^0) = x^0$. Let t'^0 be any point of I_τ with $x(t'^0) = x'^0$. There is a similar solution $x'(t)$ valid over an interval $I_{\tau'}$, containing t'^0 and such that $x'(t'^0) = x'^0$. Moreover by the uniqueness property $x(t) = x'(t)$ for $t \in I_\tau \cap I_{\tau'}$. We call $x'(t)$ the *continuation* of $x(t)$ to $I_{\tau'}$ and we now define $x(t)$ on $I_{\tau''} = I_\tau \cup I_{\tau'}$ by assigning to it the values $x'(t)$ for t in $I_{\tau'} - I_\tau$. We thus extend the definition of $x(t)$ to $I_{\tau''}$. By repetition of this process we thus arrive at a maximal interval I_0: $t' < t < t''$ over which $x(t)$ is defined. It may be of course that $t' = -\infty$ or $t'' = +\infty$ or both. The analogy with the classical process of analytical continuation is obvious.

If we start from $(x^0; t^0) \in \mathfrak{D}$, the set of points $(x;t)$ reached is such that from $(x^0; t^0)$ to any one of them there may be constructed a chain of boxes $B_1, \ldots, B_r$ each in a domain such as $\mathfrak{D}$ and with B_i, B_{i+1} overlapping. Since the boxes are connected so is their union and since $\mathfrak{D}$ is a component of the union of all the sets B, every $B_i \subset \mathfrak{D}$, hence $(x;t) \in \mathfrak{D}$. *Thus by continuation we never leave $\mathfrak{D}$.*

The set Γ of points $(x(t); t)$, $t \in I_\theta$, is the *trajectory in* $\mathfrak{W}$ *defined by* $x(t)$. (See 8.1a)

(6.1) *A trajectory is an arc.*

Consider the transformation $\varphi: I_\theta \to \Gamma$ defined by $\varphi(t) = (x(t); t)$, $t \in I_\theta$. It is clear that φ is continuous, and one-one. Let P be any point of Γ and let $\mathfrak{S}$ be a spheroid of Γ centered at P and on which the coordinate t varies on an interval I such that $I \subset I_\theta$. The open sets of Γ are unions of sets $\mathfrak{S}$. Since $\bar{I}$ is compact φ is topological on $\bar{I}$. Since $\mathfrak{S}$ is an open set of φI, $\mathfrak{S}$ is a union of sets $\varphi\delta$, where δ is a subinterval of I. Hence $\mathfrak{S}$, and therefore every open set of Γ, is a union of sets $\varphi\delta$, where δ is now any subinterval of I_θ. Since every $\varphi\delta$ is merely such a set for some $\mathfrak{S}$, $\varphi\delta$ is an open set of Γ. Hence the mapping φ sets up a one-one correspondence between the elements of $\{\delta\}$ and $\{\varphi\delta\}$: two aggregates of open sets on I_θ and on Γ from which their open sets can be generated by union. Hence φ sets up a one-one correspondence between the open sets of I_θ and of Γ. Thus φ is topological and Γ is an arc.

(6.2) *Critical points.* A critical point x^0 is a "point trajectory" in the space $\mathfrak{B}_x$, i.e., a point such that $X(x^0; t) = 0$ for all admissible t (all t in I_θ above). In the space $\mathfrak{W}$ this trajectory is an interval of the line $x^0 \times \mathfrak{T}$ parallel to the time axis.

§ 3. Continuity Properties

7. If $(x^0; t^0) \in \mathfrak{D}$ then we may enclose it in a box $B(a, \tau)$. It will be recalled that for any $t, t^0 \in I_\tau$ and $x^0 \in U_a$ there is a solution

$$(7.1) \qquad x(t) = x^0 + (x^1 - x^0) + \dots$$

where under the conditions just stated the series is uniformly convergent. Let us consider $x(t)$ as a function of t, x^0, t^0, and write it accordingly $x(t; x^0; t^0)$. Then:

(7.2) *If* $t, t^0 \in I_\tau$ *and* $x^0 \in U_a$, *then* $x(t; x^0; t^0)$ *is a continuous function of* $(t; x^0; t^0)$.

Owing to the uniform convergence it is only necessary to prove that $x^{m+1} - x^m$ is continuous over the range under consideration, or in the last analysis that x^m has this property. This is trivial for x^0 so we assume it for x^m and prove it for

x^{m+1}. In the proof of the existence theorem it has been shown that under the conditions under consideration all the x^m are confined to a certain box $B' \subset \mathfrak{D}$ in whose points $X(x; t)$ is continuous and bounded. It follows that in $(3.1)_m$ the integrand is continuous and bounded under our conditions. Since the integral is thus a continuous function of $(t; x^0; t^0)$, the same holds for x^{m+1} and (7.2) follows.

We now remove the restrictions on t, x^0, t^0, save of course that only points $(x, t) \in \mathfrak{D}$ are to be considered and we will prove continuity under the same conditions. More precisely:

(7.3) *If* $(x^0; t^0) \in \mathfrak{D}$ *and* $x(t; x^0; t^0)$ *is the solution such that* $x(t^0; x^0; t^0) = x^0$ *then* $x(t; x^0; t^0)$ *is continuous in* $(t; x^0; t^0)$.

Consider the trajectory Γ_0 through $(x^0; t^0)$. Upon combining successive approximations with the process of continuation we find that since the closed arc of Γ_0 from $(x^0; t^0)$ to $(x; t)$ is compact, it may be covered by a finite number of boxes such as B, say $B_1, \ldots, B_r$, where the first contains $(x^0; t^0)$, the last $(x; t)$ and consecutive boxes overlap. More precisely if $B_i = I_{\tau_i} \times U_i$ then $I_{\tau_i}, I_{\tau_{i+1}}$ overlap and so we may choose a value t^i in their common part. We will now keep $(x^0; t^0)$ fixed and consider a new variable initial point $(x'; t')$ in B_1. The trajectory Γ through this point meets the set $t = t^1$ at a certain point $(x^1; t^1)$ whose coordinate x^1 is a continuous function of $(x'; t')$ by (7.2). It follows that we can choose a sphere $\Sigma((x^0; t^0); \varrho) \subset B_1$ with ϱ so small that when $(x'; t') \in \Sigma$ then $(x^1; t^1) \in B_2$. Now for $(x^1; t^1) \in B_2$ and $t \in I_{\tau_2}$ the solution $x(t)$ such that $x(t^1) = x^1$, is a continuous function of $(x^1; t^1)$ and hence of $(t; x'; t')$ for $t \in I_{\tau_2}$ and $(x'; t') \in \Sigma$. Confining now $(x'; t')$ to Σ, Γ will contain the point $(x^1; t^1) \in B_2$, and so it will contain a point $(x^2; t^2)$, where $x^2 = x(t^2)$. The same continuity argument shows that when ϱ is sufficiently small $(x^2; t^2) \in B_3$ and we may now proceed as before. After a finite number of steps we shall find a ϱ so small that for $(x'; t') \in \Sigma$, $x(t; x'; t')$ will remain in B_r and be continuous in $(t; x'; t')$ for $t \in I_{\tau_r}$ and $(x'; t') \in \Sigma$. This implies (7.3).

8. Several noteworthy consequences may be drawn from the preceding results. The notations being the same corresponding to $M(x^0; t^0)$ we may choose ε, δ such that the sets $\mathfrak{S}(x^0; \varepsilon) \times t^0$, $\Sigma(M, \delta) \subset B_1$. Consider now $x(t; x'; t')$, $(x'; t') \in \Sigma$. By the ex-

istence theorem the function is defined for every $t \in I_{\tau_1}$ and hence for $t = t^0$. Let $x(t; x'; t') = x''(t)$, so that $x(t^0; x'; t') = x''(t^0)$. By (7.3) $x''(t)$ is a continuous function of $(x'; t')$. Therefore δ may be chosen such that $(x''(t^0); t^0) \in \mathfrak{S}(x^0; \varepsilon) \times t^0$. In other words any trajectory passing near enough to $(x^0; t^0)$ in $\mathfrak{D}$ will cross $\mathfrak{S}(x^0; \varepsilon) \times t^0$; of course the converse will hold if $\delta \geq \varepsilon$. Or explicitly:

(8.1) *Corresponding to any $M(x^0; t^0) \in \mathfrak{D}$ and any $\varepsilon > 0$ there exists a $\delta > 0$ such that a trajectory crossing $\Sigma(M, \delta)$ intersects the hyperplane $t = t^0$ within $\mathfrak{S}(x^0; \varepsilon)$ (i.e., it meets $\mathfrak{S}(x^0; \varepsilon) \times t^0$).*

(8.1a) Referring to (6) and in its notations let γ denote the part of the trajectory Γ say for $t \geq t^0 \in I_\theta$ and let $G = \bar{\gamma} - \gamma$. If $G \neq \infty$ let it contain the point $Q(x''; t'')$ not in the boundary of $\mathfrak{D}$. Since γ passes arbitrarily near Q by (8.1) $x(t)$ must cross t'', which is ruled out since t'' is an end-point of I_θ. Thus as t tends to one of the endpoints of I_θ the trajectory Γ can only tend to infinity or to the boundary of the domain $\mathfrak{D}$.

An extension of the continuity theorem in a new direction is the following. Suppose that X is in fact a function $X(x; t; y)$ which is continuous and bounded in a region Ω of $\mathfrak{W} \times \mathfrak{V}_y$. Suppose moreover that with respect to Ω the Lipschitz condition is still fulfilled in the same form as earlier. The *domain of continuity* Δ is defined as before as a component of the union of all the sets Ω. We have now the following stronger result which extends (7.3) and may understandably be stated in the brief form:

(8.2) *The solution is continuous in $(t; x^0; t^0; y)$ when $(x^0; t^0; y)$ ranges over Δ.*

The proof may be related to (7.3) by a well known device. Enlarge (1.2) by adding the differential equation

$$dy/dt = 0 \tag{8.3}$$

so that (1.2) and (8.3) form a system such as (1.2) for the vector $(x; y)$. Owing to the special form of (8.3) the Lipschitz condition still suffices to prove the existence theorem and hence all its corollaries, including among them (7.3), which in the present instance becomes (8.2).

Consider again the trajectory Γ_0 and an arc MN of Γ_0, where

$(x^0; t^0)$ are the coordinates of M and $(x^1; t^1)$ are those of N. Introduce the two sets

$$S_0 = \bar{\mathfrak{S}}(x_0; \varrho_0) \times t^0,\ S_1 = \bar{\mathfrak{S}}(x_1; \varrho_1) \times t_1,$$

where ϱ_0, ϱ_1 are so chosen that both sets are in $\mathfrak{D}$. Let $M' \in S_0$ and let Γ': $x(t; M')$ be the trajectory through M'. It is clear from the argument proving (7.3) that for ϱ_0 small enough $x(t; M')$ may be extended throughout the whole closed interval $\bar{I}$: $t^0 \leq t \leq t^1$. Consider a mapping φ of the compact product $\bar{I} \times \bar{S}_0$ into $\mathfrak{D}$ sending $\bar{I} \times M$ into MN and $\bar{I} \times M'$ into $M'N'$. This mapping φ is one-one, since $(x'; t')$ and $(x''; t'')$ certainly have different images if $t' \neq t''$ or $x' \neq x''$ (otherwise distinct trajectories would meet). Since $\bar{I} \times \bar{S}_0$ is compact φ is topological.

Take now a small n-cell E_0^n containing M and contained in S_0. Applying φ merely to $\bar{I} \times E_0^n$ and denoting the hyperplanes $t = t^i$ by π_i we have:

(8.4) Structural Theorem. *Given a sufficiently small n-cell E_0^n in π_0 containing M, a trajectory through a point M' of E_0^n intersects π_1 in a single point N' such that $M' \to N'$ defines a topological mapping θ of E_0^n onto a similar $E_1^n \subset \pi_1$ containing N and of course $N = \theta M$. Moreover let λ be the closed arc $M'N'$ of the trajectory through M', and let $\lambda(t)$ be the point of λ corresponding to any $t \in \bar{I}, I$: $t^0 \leq t \leq t^1$. Then $(M', t) \to \lambda(t)$ defines a topological mapping φ of the cylinder $\bar{I} \times E_0^n$ such that $\varphi(\bar{I} \times M) = MN$, $\varphi(\bar{I} \times M') = \lambda$.*

(8.5) *General solution.* This concept may now be introduced with reasonable clarity. Let $\mathfrak{B}_c$ be n-*d*imensional and $f(t; c)$ a function such that: (a) for c in a certain region Λ of $\mathfrak{B}_c$, f is a solution of (1.2) in the domain $\mathfrak{D}$; (b) if $M_0 = f(t^0; c^0)$, $M = f(t^0; c)$, where c, c^0 are in Λ, there are n-cells E^n in $\mathfrak{B}_x \times t^0$ and ε^n in Λ, such that for $c \in \Lambda$ the correspondence $c \to M$ is a topological mapping φ of ε^n onto E^n, such that $\varphi c^0 = M_0$. In other words, c may be chosen in ε^n so as to yield any solution with its initial value at t^0 in E^n. The function $f(t; c)$ is known as a *general* solution. By (8.4) the concept of general solution is manifestly independent of the particular point M_0 chosen on the trajectory $f(t; c^0)$, i.e., it does not depend upon t^0 but solely upon the trajectory itself.

§ 4. Differentiability Properties

9. These properties are embodied in the following very comprehensive proposition:

$(9.1)_r$. *Let* $X(x; t; y)$ *and related designations be as before, and let* X *be of class* C^r *in* x, y *and* t *in a certain region* Ω *of the space* $\mathfrak{V}_x \times \mathfrak{V}_y \times \mathfrak{T}$. *Then the solution* $x(t; t^0; x^0; y)$ *of* (1.2) *such that* $x(t^0; t^0; x^0; y) = x^0$, $(t^0; x^0; y) \in \Omega$, *is of class* C^r *in* $t^0; x^0; y$, *and of class* C^{r+1} *in* t.

In the proof we follow more or less Gronwall [1] and Sansone [1], Vol. I, p. 27.

We shall require the following:

(9.2) Gronwall's Lemma. *Let* $f(t)$ *be a scalar function such that*

$$(9.2a) \qquad 0 \leqq f(t) \leqq \lambda + \int_{t^0}^{t} (\mu f(t') + \nu)\, dt',$$

where λ, μ, ν *are positive constants,* $t \geqq t^0$, *and* $f(t)$ *is continuous in the range* $t^0 \leqq t \leqq t^1$. *If* $t^1 - t^0 = T$ *then*

$$(9.2b) \qquad f(t) \leqq (\nu T + \lambda)\, e^{\mu T}, \quad t^0 \leqq t \leqq t^1 .$$

(Gronwall [1]).

Set $f(t) = e^{\mu(t-t_0)} g(t)$, so that $g(t)$ is continuous in the same interval as $f(t)$. If $\gamma = \sup g(t)$ in $t^0 \leqq t \leqq t^1$ then $g(t)$ attains the value γ for some $t^* : t^0 \leqq t^* \leqq t^1$. Now (9.2a) yields:

$$\gamma\, e^{\mu(t^* - t^0)} \leqq \lambda + \int_{t^0}^{t^*} (e^{\mu(t' - t^0)} \mu \gamma + \nu)\, dt'$$
$$= \lambda + \nu (t^* - t^0) + \gamma \{e^{\mu(t^* - t^0)} - 1\}.$$

From this follows $\gamma \leqq \lambda + \nu T$ and this implies (9.2b).

An immediate consequence is:

(9.3) *Let* f, λ, ν *be n-vectors and* μ *an* $n \times n$ *matrix, where* λ, μ, ν *are constant and* f *has the same continuity property as before. If*

$$(9.3a) \qquad f(t) = \lambda + \int_{t^0}^{t} (\mu f(t') + \nu)\, dt'$$

then also

$$\|f(t)\| \leqq (\|\nu\| T + \|\lambda\|) e^{n\|\mu\| T}. \tag{9.3b}$$

For (9.3a) yields

$$0 \leqq \|f(t)\| \leqq \|\lambda\| + \int_{t^o}^{t} (n \|\mu\| \cdot \|f(t')\| + \|\nu\|) dt'$$

from which (9.3b) follows by a direct application of the lemma.

Proof of (9.1). Property $(9.1)_o$ is merely (8.2) as regards t^o, x^o, y, and obvious as regards t, so let us consider $(9.1)_1$. We proceed first with regard to y. It is to be proved that the vector $\partial x/\partial y_k$ exists and is continuous in y. It will be assumed that all functional values are taken in Ω and for the present t will remain in a finite range $|t - t^o| \leq T$. For any $f(y)$ let $\Delta f = f(y + \Delta y) - f(y)$, where $\Delta y = (0, \ldots, 0, \Delta y_k, 0, \ldots, 0)$. We start from the integral equation

$$x = x^o + \int_{t^o}^{t} X(x; t'; y)\, dt'. \tag{2.2}$$

Owing to the continuity of x in y, $\Delta x \to 0$ with Δy, and for Δy small $(x + \Delta x; t; y + \Delta y)$ remains in Ω. Thus (2.2) yields for Δy_k small

$$\frac{\Delta x}{\Delta y_k} = \int_{t^o}^{t} \left\{ \left(\frac{\partial X}{\partial x} + \varepsilon_1 \right) \frac{\Delta x}{\Delta y_k} + \left(\frac{\partial X}{\partial y_k} + \varepsilon_2 \right) \right\} dt', \tag{9.4}$$

where it is sufficient to know that ε_1 and $\varepsilon_2 \to 0$ with Δy_k uniformly in t for $|t - t^o| \leq T$. Moreover

$$\frac{\partial X}{\partial x} = \left(\frac{\partial X_r}{\partial x_s} \right),$$

the Jacobian matrix of the vector X as to the vector x.

Consider now the system

$$dz/dt = \partial X/\partial x \cdot z + \partial X/\partial y_k, \tag{9.5}$$

with the initial condition $z(t^o) = 0$, it being understood that x stands for the solution $x(t)$ of (1.2). Since the system is linear with continuous coefficients (by the class C^1 hypothesis for X)

it has a unique solution and this solution is continuous in t. Hence $||z||$ has an upper bound ξ in $|t - t^o| \leq T$.

The solution of (9.5) under consideration satisfies

$$z = \int_{t^o}^{t} \left(\frac{\partial X}{\partial x} z(t') + \frac{\partial X}{\partial y_k} \right) dt' . \tag{9.6}$$

Upon subtracting (9.6) from (9.4) and setting $u(t) = \dfrac{\Delta x}{\Delta y_k} - z$ we find for $u(t)$ the relation

$$u(t) = \int_{t^o}^{t} \left\{ \left(\frac{\partial X}{\partial y} + \varepsilon_1 \right) u + (\varepsilon_1 z + \varepsilon_2) \right\} dt' . \tag{9.7}$$

Owing to the continuity of $\partial X/\partial y$ and boundedness of ε_1 the norm of their sum has an upper bound μ in $|t - t^o| \leq T$. Clearly also $||\varepsilon_1 z + \varepsilon_2|| < \eta$, where $\eta \to 0$ with Δy_k. Hence (9.7) yields in view of (9.3):

$$||u(t)|| \leqq \eta T e^{n\mu T}$$

which $\to 0$ with Δy_k. This shows that $\Delta x/\Delta y_k \to z$ as $\Delta y_k \to 0$. Hence $\partial x/\partial y_k$ exists, is continuous in t in $|t - t^o| \leq T$, and is the solution of the differential equation (9.5) which is zero for $t = t^o$. The extension step by step from the limited to a larger t interval is an immediate consequence of the fact that $\partial x/\partial y_k$ satisfies (9.5). Thus $\partial x/\partial y_k$ exists throughout Ω and satisfies (9.5).

Let us regard now x^o as a parameter. The same reasoning will apply with the following minor modifications. Instead of (9.4) we will have

$$\frac{\Delta x}{\Delta x_j^o} = e_j + \int_{t^o}^{t} \left(\frac{\partial X}{\partial x} \frac{\Delta x}{\Delta x_j^o} + \varepsilon \right) dt' \tag{9.4'}$$

where e_j is the unit vector with components δ_{hj}. The system (9.5) is now replaced by

$$dz/dt = \partial X/\partial x \cdot z , \tag{9.5'}$$

the desired solution being this time such that $z(t^o) = e_j$. Instead of (9.6) we have now

$$z = e_j + \int_{t^o}^{t} \frac{\partial X}{\partial x} z \cdot dt' . \tag{9.6'}$$

The same argument as for x^0 holds for t^0 also. The initial step yields however a slight variant. We have

$$\frac{\Delta x}{\Delta t^0} = X(x^0; t; y) + \varepsilon_2 + \int_{t^0}^{t} \left(\frac{\partial X}{\partial x} + \varepsilon_1 \right) \frac{\Delta x}{\Delta t^0} \, dt',$$

where ε_1, $\varepsilon_2 \to 0$ with Δt^0. The equation in z is (9.5′) but the solution to be envisaged is such that $z(t^0) = X(x^0; t^0; y)$. The rest of the argument goes through with insignificant modifications.

One proceeds from class C^1 to class C^2 by replacing the given equation by (9.5) or its parallels such as (9.5′), etc., down to class C^r. The class as to t follows by direct application of the rules for derivatives. Similarly for mixed derivatives. This completes the proof of (9.1).

(9.8) *Remark.* It is a trivial observation that even under the mere conditions of the fundamental existence theorem dx/dt exists and is continuous along a trajectory. It implies however the non-trivial geometrical facts that in a real system a trajectory possesses a continuously turning tangent and that its arc length ds is defined.

§ 5. Analyticity Properties

10. The argument in deriving the continuity properties of $x(t)$ rests essentially upon the following three propositions:

(a) A continuous function of a continuous function is continuous.

(b) If $f(x; t)$ is continuous in $(x; t)$ over a suitable range so is $\int_{t^0}^{t} f(x; t) \, dt$.

(c) A uniformly convergent series of continuous functions is continuous.

These three propositions made it possible to "transfer" continuity from the approximations $x^m(t)$ to the solution $x(t)$ itself. Since these three properties hold also with "continuous" replaced by "analytic" or "holomorphic" the same transfer will operate for analyticity or holomorphism. It will be necessary however to distinguish carefully between analyticity as to x^0, as to t, t^0, or as to all three. The difference arises from the fact that we may wish to consider not only $X(x; t)$ analytic in $(x; t)$

but also merely in x alone, or in certain additional parameters that may be present in X.

Suppose $X(x; t)$ in (1.2) analytic in x and continuous in $(x; t)$ in a certain region Ω. Since the partial derivatives exist and are continuous in Ω the Lipschitz condition holds in Ω (5.1) and so the existence theorem is applicable.

We now proceed as before replacing *continuity* by *analyticity* wherever need be and obtain first an analogue Δ of $\mathfrak{D}$ which we call *domain of analyticity*. Then we have the following properties, which we merely state since the proofs are unmodified. The space $\mathfrak{V}_x$ and the values of the variables and functions may be real or complex unless otherwise restricted.

(10.1) *If $X(x; t)$ is analytic in both variables and Δ is the domain of analyticity then the solution $x(t; x^0; t^0)$ such that $(x(t); t) \in \Delta$, $x(t^0; x^0; t^0) = x^0$ is analytic in all three arguments* (see (7.3)).

(10.2) *If $X(x; t)$ is merely continuous in t (t real) then $x(t; x^0; t^0)$ is merely analytic in x^0* (see (7.3)).

(10.3) *If $X(x; t; y)$ is analytic in y also and Δ is defined accordingly as in* (7), *then $x(t; x^0; t^0; y)$ is analytic in all arguments in the case* (10.1) *and in $(x^0; y)$ alone in the case* (10.2).

11. (11.1) Poincaré's Expansion Theorem. *Let the differential equation with t real and the domain of analyticity Δ:*

(11.1a) $$dx/dt = X(x; t; y),$$

possess for $y = 0$ a solution $\xi(t)$ on the closed interval $\bar{I}: t^0 \leq t \leq t^1$ such that for fixed t the $X_i(x; t; y)$ may be expanded in power series of the $(x_i - \xi_i(t))$ and of the y_j uniformly convergent in some range

$$||x - \xi(t)|| < \alpha, \ ||y|| < \alpha, \ t \in \bar{I}.$$

Then setting $\xi(t^0) = \xi^0$, the equation (11.1a) *has a solution $\xi(t; x^0; y)$ such that $\xi(t; \xi^0; 0) = \xi(t)$ and that for fixed t the $\xi_i(t; x^0; y)$ may be expanded in power series of the $(x_i^0 - \xi_i^0)$ and the y_j convergent for $t \in \bar{I}$ and $(x^0; y)$ in a certain region*

$$||x^0 - \xi^0|| < \beta, \ ||y|| < \beta.$$

In particular if $\xi(t; \xi^0; y) = \xi(t; y)$ then the latter is a solution *such that* $\xi(t; 0) = \xi(t)$ *and that* $\xi(t; y)$ *may be expanded in power series of the* y_j *convergent in a region* $\|y\| < \beta$.

This theorem has been utilized by Poincaré, Picard and others in many fruitful ways in questions of approximation, likewise in the search for solutions with special properties (periodicity among others) neighboring specified solutions.

If we make the change of variable $x - \xi(t) = z(t)$, then $z(t)$ satisfies the differential equation

$$dz/dt = q(z; t; y) \tag{11.2}$$

which behaves like (11.1a) save that now $\xi(t)$ is replaced by 0 and that, for $t \in \bar{I}$, q_i may be expanded in a power series in the z_i, y_j uniformly convergent in $T(\alpha)$: $\|z\| < \alpha$, $\|y\| < \alpha$. We merely need to prove now for (11.2) the analogue of (11.1) with $\xi(t) = \xi^0 = 0$, and the existence of a solution $\xi(t; z^0; y)$ possessing a series expansion in powers of the z_i^0, y_j valid for $t \in \bar{I}$ and $(z^0; y)$ in a certain $T(\beta)$.

Let (11.2) be amplified by

$$dy/dt = 0\,. \tag{11.3}$$

Thus (11.2), (11.3) form a system in the unknown $(y; z)$ with the right hand sides holomorphic in $T(\alpha)$ for every $t \in \bar{I}$. The associated domain of analyticity Δ_1 in $\mathfrak{B}_y \times \mathfrak{B}_z \times \mathfrak{T}$ contains the product $T(\alpha) \times \bar{I}$.

By the mechanism of proof of (7.3) as paraphrased for (10.2) there is a salution $\xi(t; y^0; z^0)$ such that $\xi(t; 0; 0) = 0$, $y = y^0$ valid for every $t \in \bar{I}$ and analytic in the initial values y^0, z^0 when they are in a certain box B of center $(0, 0)$. Hence the solution $\xi(t; y^0; z^0)$ may be expanded in a power series of the y_i^0, z_j^0 valid in a region $T(\beta) \subset B$. Since $y = y^0$, this is the required property of ξ and (11.1) is proved.

§ 6. Equations of Higher Order

12. We refer particularly to the single equation (1.3). Since it is equivalent to (1.4) the extension of the preceding results is immediate. It may be observed that the Lipschitz condition is equivalent here to:

$$|f(x_1, \ldots, x_n; t) - f(x_1', \ldots, x_n'; t)| < k \sum |x_h - x_h'|,$$

and this is the form in which it is to be utilized in connection with the existence theorem. With this interpretation of the Lipschitz condition we may state the following proposition which follows from those already proved.

(12.1) *The symbols and conditions being as in* (2.1) *there exists a unique solution* $x(t)$ *of* (1.3) *valid in a certain interval* $|t - t^0| < \tau$ *and such that*

$$\left(\frac{d^p x}{dt^p}\right)_{t = t^0} = x^0_{p+1}, \quad p \leq n - 1.$$

This solution is a continuous function of t, t^0 *and the* x^0_{p+1}; *it is also of class* C^r, *after the manner of* (7.3) *or* (9). *Moreover if the coefficients* a_i *are analytic under conditions similar to those of* (10.1) *then the solution has the appropriate analyticity properties.*

§ 7. Autonomous Systems

13. Consider an autonomous system, assumed of class C^1 at least,

$$(13.1) \qquad dx/dt = X(x).$$

If $\mathfrak{T}$ denotes the space of t (real line) and S is a region of $\mathfrak{B}_x$ in which the partial derivatives $\partial X_i/\partial x_j$ exist and are continuous, then by (5.1) we may take $\Omega = S \times \mathfrak{T}$, in the proof of the theorem of existence.

Let $\mathfrak{D}$ be a component of the set of all the points of $\mathfrak{B}_x$ which have a neighborhood such as S. The domain of continuity is now by definition a set $\mathfrak{D} \times \mathfrak{T}$. However the significant set is actually $\mathfrak{D}$ and it will be called the *characteristic domain of continuity* of (13.1), or merely the *domain* wherever the meaning is otherwise clear. A similar characteristic domain of analyticity $\Delta \subset \mathfrak{D}$ may be introduced when X is analytical.

Beyond this point the simplifications in the various corollaries to (2.1) offer no difficulties and are left to the reader.

Suppose that $x(t)$ is a solution of (13.1). If we make the change of variable $t' = t - \tau$, the form of (13.1) is unchanged. Hence

$x(t-\tau)$ is a solution whatever τ. Thus from a single solution, one may derive here a whole family of solutions depending upon an arbitrary parameter τ.

Consider the trajectory Γ_τ corresponding to $x(t-\tau)$ in the space $\mathfrak{V}_x \times \mathfrak{T}$ and let γ be its projection in $\mathfrak{V}_x$. Γ_τ is the set of points $(x(t-\tau); t) \in \mathfrak{V}_x \times \mathfrak{T}$, and so γ is the set of points $x(t-\tau)$ in $\mathfrak{V}_x$. We have called γ a *path* of (13.1). This path is independent of τ. Suppose in fact that $\Gamma_{\tau'}$ has the projection γ'. Thus γ' is the set of all points $x(t-\tau')$. Now if M_t, M'_t are the points of γ, γ' corresponding to the value t, evidently $M_{t+\tau-\tau'} = M'_t$ and so $\gamma = \gamma'$. In other words γ, γ' are merely different parametrizations of the same locus.

The paths are all contained in the domain $\mathfrak{D}$ and in this domain through each point $M(x^0)$ there passes a single path γ. For the paths are the projections of the trajectories Γ_τ through the points $(x^0; \tau)$ for all τ. If Γ_0: $x(t)$ is the trajectory through $(x^0; 0)$, i.e. the solution such that $x(0) = x^0$, then Γ_τ: $x(t-\tau)$ is a trajectory and hence *the* trajectory through $(x^0; \tau)$ (since it is unique). Thus all the Γ_τ are merely the trajectories corresponding to all the solutions $x(t-\tau)$, and as we have seen the corresponding path is unique.

(13.2) *Critical points*. Referring to (6.2) a *critical* point of (13.1) is a point (= a vector) x^0 of the domain $\mathfrak{D}$ which annuls $X(x)$, i.e. a solution of $X(x) = 0$ in $\mathfrak{D}$. Evidently $x(t) = x^0$ is a path. Unless otherwise stated the term path will be reserved in the sequel for those which are not critical points.

(13.3) *If x^0 is a critical point then no trajectory other than $x(t) = x^0$ may reach $(x^0; t^0)$ by the process of extension.*

Since $(x^0; t^o) \in \mathfrak{V}_x \times \mathfrak{T}$ if a second trajectory $x^*(t)$ could reach it there would be two trajectories through the point, which is ruled out.

As a corollary:

(13.4) *If $x(t)$ tends to a critical point x^0, then $t \to +\infty$ or $t \to -\infty$.*

14. Side by side with the system (13.1) we may consider the system

$$(14.1) \quad dx_1/X_1(x_1, \ldots, x_n) = \ldots = dx_n/X_n(x_1, \ldots, x_n)$$

which may be interpreted in the following way. Let us agree to consider also a single point as a differentiable curve. We may then say that (14.1) requires to find differentiable curves λ such that at each point of λ the vector dx is collinear with $X(x)$, or is zero wherever $X(x) = 0$.

It is clear that the path γ through any point M of $\mathfrak{D}$ is a suitable λ.

Suppose first M critical. Then γ and λ both reduce to M and so they coincide. Suppose now M non-critical. At the point M one of the components of X, say $X_h \neq 0$. Since X_h is continuous and $M \in \mathfrak{D}$ there is a sphere $\mathfrak{S}(M, \varrho)$ in $\mathfrak{D}$ such that $X_h \neq 0$ and is bounded away from zero in the sphere. As a consequence, in the sphere the system (14.1) is equivalent to

$$(14.2) \qquad dx_i/dx_h = q_i(x) = X_i(x)/X_h(x),\ i \neq h\,.$$

This system is of the form (1.2) with x_h in place of t. Owing to our assumptions (see 13) the $\partial q_i/\partial x_j$ exist and are continuous in $\mathfrak{S}(M, \varrho)$ and so the existence theorem (2.1) may be applied within the sphere and with x_h as the independent variable. It implies that there is a unique differentiable arc μ in $\mathfrak{S}(M, \varrho)$ containing M and along which (14.2) holds, i.e. along which (14.1) holds. Now the arc μ' of the path γ through M, situated in $\mathfrak{S}(M, \varrho)$, is a differentiable arc such as μ, and therefore $\mu' = \mu$. Thus

(14.3) *The solution of* (14.1) *at any point M of $\mathfrak{D}$ is the path through M.*

15. *Closed paths.* A path γ is said to be closed whenever it is a Jordan curve. Thus γ is compact and every point of γ is reached from a given point M in a finite time. It follows then (argument of (13.1)) that the solution $x(t)$ corresponding to γ is periodic. Conversely if $x(t)$ is periodic then its path γ is closed. Let τ be the period. Thus $x(0) = x(\tau)$ and $x(0) \neq x(\theta)$, $0 < \theta < \tau$, else θ would be the period and not τ. Since as t varies from 0 to τ, limits excluded, $x(t)$ describes an arc of γ, γ is a Jordan curve. Thus

(15.1) *A n.a.s.c. for a path γ to be closed is that its solution $x(t)$ be periodic.*

16. We shall now prove for paths certain results in close analogy with those of (8) for trajectories. The basic properties are (16.5, 17.4, 17.6). The first two state roughly that an arc of path may be imbedded in a "tube" of paths. This property is very convenient in the topological applications. The difficulties in the proof arise from the presence of closed paths.

Let M be a non-critical point, γ the path through M, E^{n-1} a differentiable $(n-1)$-cell through M, represented by

$$(16.1) \qquad x_i = \varphi_i (u_1, \ldots, u_{n-1}), \ ||u|| < a,$$

or in vector notation

$$(16.2) \qquad x = \varphi (u),$$

where $\varphi (0) = x^0$, (16.1) is of class C^r, $r \geq 1$, and the Jacobian matrix of the functions φ_i is of rank $(n-1)$ on E^{n-1}. The latter is said to be *transverse* to γ at M whenever γ is not tangent to E^{n-1} at M. This means in substance that the differential vector dx along γ at M is linearly independent of those along E^{n-1} at M. In other words its components, which are $X_i (x^0)\, dt$, are not of the form

$$\Sigma\, \partial x_i / \partial u_j \cdot du_j \,.$$

In the last analysis the vectors $X (x^0)$ and

$$\partial x / \partial u_j]_{x^0;\, u=0}$$

are linearly independent, which is equivalent to the condition that the determinant

$$(16.3) \qquad D (x; u) = \left| \frac{\partial x_i}{\partial u_1}, \ldots, \frac{\partial x_i}{\partial u_{n-1}}, X_i \right| \neq 0$$

for $x = x^0$, $u = 0$. Since D is continuous it will remain $\neq 0$ for u and $x - x^0$ sufficiently small. Hence if we choose a in (16.1) small enough the path γ' through any point M' of E^{n-1} will likewise not be tangent to the cell at M'. Under these conditions then the cell E^{n-1} is transverse to all the paths that cross it.

The notations remaining the same let us associate explicitly with γ the solution $x (t; M')$ such that $x (0; M') = x'^0 = M'$.

This solution satisfies the relation

$$(16.4) \qquad -x + \varphi(u) + \int_0^t X(x(t'; \varphi(u)))\,dt' = 0.$$

Consider this as a system of n relations for the determination of $u_1, \ldots, u_{n-1}, t$ in terms of x. For the values zero of u; $x - x^0$; t up to sign the Jacobian as to u; t is precisely $D(0; 0)$ and hence $\neq 0$. Hence (16.4) represents a transformation $\mathfrak{V}_u \times \mathfrak{T} \to \mathfrak{V}_x$ which is regular in the neighborhood of $u = 0, t = 0$ mapping that point into x^0. It defines therefore a topological mapping f of a suitable neighborhood $||u|| < \beta$, $|t| < \tau$ of the origin in $\mathfrak{V}_u \times \mathfrak{T}$ onto a neighborhood of x^0 in $\mathfrak{V}_x$. If $\alpha < \beta$ and I_τ is the time interval $|t| < \tau$, f maps topologically the cylinder $E^{n-1} \times I_\tau$ into a neighborhood C of x^0 in $\mathfrak{V}_x$ in such manner that: (a) $f(u \times 0) = M'$ and in particular $f(0 \times 0) = M$; (b) $f(u \times I_\tau)$ is mapped onto the arc of γ' described in the time $-\tau < t < \tau$.

It is convenient to introduce here the following "cylindrical" terminology. The image $C = f(E^{n-1} \times I_\tau)$ will be referred to as an *open path cylinder* and the arcs $f(u \times I_\tau)$, $f(0 \times I_\tau)$ as the *generators* and the *axis* of the cylinder. When the dimension is two we shall say "path rectangle." We note the following two properties:

(16.5) *Every path passing sufficiently near to M must have an arc in the path cylinder C and hence it must cross the cell E^{n-1} transverse to γ at M.*

Since C is a neighborhood of M, if the path γ' passes sufficiently near to M it will meet C at some point P and therefore γ' will contain the generator through P.

(16.6) *If γ' through $M' \in E^{n-1}$ followed beyond M' returns to a point M'' of E^{n-1} then it must take at least a time τ to do so.*

For f maps $u \times I_\tau$ topologically on $\mathfrak{V}_x$, and to return to E^{n-1} it must take at least the time τ that it takes to describe the generator through M' beyond M'.

17. The situation remaining the same let γ be followed beyond M up to a point $N(x^1) \neq M$ such that, between M and N, γ is an arc, and let E_1^{n-1} be transverse to γ at N and so small that

it does not meet E^{n-1}. Let the cell E^{n-1} have the parametric representation similar to (16.2):

$$x = \psi(v), \; \|v\| < \beta, \; \psi(0) = x^1 . \tag{17.1}$$

Let t^1 be the time at which N is reached beyond M. As before since E_1^{n-1} is transverse to γ at N:

$$\left| \frac{\partial \psi_i}{\partial v_1}, \ldots, \frac{\partial \psi_i}{\partial v_{n-1}}, X_1 \right|_{(v = 0, x^1, t^1)} \neq 0 . \tag{17.2}$$

If γ' is followed beyond M' it will first reach E_1^{n-1} at a point N' at time t where t and $v(N')$ are determined from the relation

$$-\psi(v) + x(u) + \int_0^t X(x(u; t'); t')\, dt' = 0 . \tag{17.3}$$

The Jacobian of this system as to t, v reduces to within sign, for $t = t^1$, $v = 0$, $u = 0$, to (17.2). Hence (17.3) has a unique solution $v(u)$, $t(u)$ for a small enough, such that $v(0) = 0$, $t(0) = t^1$. Furthermore the mapping $\varphi: u \to v(u)$ is one-one and continuous. Since φ^{-1} is manifestly of the same nature as φ, it is also continuous for v small, and so φ is topological for a small. Thus $M' \to N'$ defines a topological mapping, namely φ, of E_1^{n-1}, and in fact of $\bar{E}_1^{n-1}$ for a small enough, onto a closed subcell of E^{n-1}.

Consider now a segment $\lambda: 0 \leq s \leq 1$, and the transformation $F: E^{n-1} \times \lambda \to \mathfrak{V}_x$ defined in the following manner. Let the arc $M'N'$ of γ' be described in time t''. Then F sends $M' \times \lambda$ into $M'N'$ in such manner that the point $M' \times s$ is imaged into the point of γ' reached at time st''. In particular $F(M' \times 0) = M'$, $F(M' \times 1) = N'$. It is clear also that F is a mapping. We shall show that if the constant a is sufficiently small then F is also *univalent*: it images distinct points into distinct points.

Suppose in fact that F images two distinct points $(M' \times s')$ and $(M'' \times s'')$ into the same point P. Thus P is on both $M'N'$ and $M''N''$. This can only arise if γ' followed say beyond M' passes through M''. Now we may assume a so small that (16.6) becomes applicable. Thus it will require a time at least τ to describe the arc $M'M''$ of γ'. Moreover if F fails to be univalent however small one chooses α, the preceding situation must arise for some M' arbitrarily near M. Hence by an obvious limiting process γ must return to M after a time t^2 where $\tau \leq t^2 \leq t^1$. Since MN is an arc this is impossible. Hence F is univalent.

Thus F is a one-one mapping of the compact set $\bar{E}^{n-1} \times \lambda$ onto a subset C of $\mathfrak{B}_x$. As a consequence F is *topological*. The image set C of the closed cylinder $\bar{E}^{n-1} \times \lambda$ is called a *closed path cylinder*. The arc MN is referred to as the *axis* of the cylinder, the transverse cells E^{n-1}, E_1^{n-1} as its *bases*, the images $F(M' \times \lambda)$ as its *generators*. We have therefore proved:

(17.4) *Let γ be any path, MN an arc of γ, E^{n-1} and E_1^{n-1} differentiable cells transverse to γ at M and N. Then there exists a closed path cylinder whose axis is MN and whose bases are in the two given cells.*

When $n = 2$ the cylinder is called a rectangle and the cells are arcs but otherwise nothing is changed.

That $M' \to N'$ defines a topological mapping φ of E^{n-1} into a subcell $E_1'^{n-1}$ of E_1^{n-1}, holds only because MN is an arc. Suppose that $N = M$ so that γ is closed. The time of description still being t^1 one may choose an integer k such that if M_i is reached at time it^1/k, $i \leq k$, then the paths M_iM_{i+1} are all arcs. Here of course $M_k = M$. Choose now E_i^{n-1} transverse to γ at M_i and $E_k^{n-1} = E^{n-1}$. Then if M' is sufficiently near M on E^{n-1} the path γ' through M' will meet E_i^{n-1} in M_i', with $M_k' = N'$ on E^{n-1}. By the earlier argument for E^{n-1} sufficiently small $M' \to M_1'$ defines a topological mapping f_1 of E^{n-1} onto a subcell $E_1'^{n-1}$ of E_1^{n-1}. Similarly for E_1^{n-1}, hence again for E^{n-1} sufficiently small $M_1' \to M_2'$ defines a topological mapping f_2 of $E_1'^{n-1}$ onto a subcell $E_2'^{n-1}$ of E_2^{n-1}, etc. We thus obtain for E^{n-1} small enough obvious topological mappings f_i and $f = f_k \ldots f_1$ will be a topological mapping of a subcell of E^{n-1} onto one of E^{n-1}. This is the result which we had in view.

It has just been assumed that N follows M on γ. By changing t into $-t$ the same results would be obtained when N precedes γ. We may therefore state:

(17.5) *Let N follow [precede] M on the path γ and let E^{n-1}, E_1^{n-1} be transverse to γ at M, N. There can be chosen a subcell E'^{n-1} of E^{n-1} such that if $M' \in E'^{n-1}$ then the path γ' through M' followed forward [backward] from M' will first meet E_1^{n-1}*

in a point N' and $M' \to N'$ defines a topological mapping f of E'^{n-1} into a subcell of E_1^{n-1}. If γ is closed $N = M$ and $E_1^{n-1} = E^{n-1}$.

Supposing again γ closed the mapping F will be topological on $E^{n-1} \times \mu, \mu: 0 \leq s < 1$, and continuous on $\bar{E}^{n-1} \times \bar{\mu}$. The set $F(\bar{E}^{n-1} \times \mu)$ is called a *path-ring* with obvious definitions of generators and base. We have here:

(17.6) *If γ is a closed path there is a path-ring of axis γ and whose base is in any given cell E^{n-1} transverse to γ.*

18. Suppose now for the sake of simplicity that our differential equation is analytic. Let also the transverse cell E^{n-1} to the closed path γ at the point M be analytic and parametrized by (16.2). Let the point $M'(u)$ of E^{n-1}, with u small, be the initial point of the path γ' at time zero. Let finally γ be described once in time 2π. Since the solution $x(t)$ corresponding to γ' is analytic in the initial values and in t it will also be analytic in u and t. It will thus be of the form

$$x = \varphi(u; t) \tag{18.1}$$

where φ is analytic in u and t. As t varies from zero on, the point $x(t)$ varies from M' and reaches again E^{n-1} at a point M_1' and time t^1. This point coincides with a point represented in the system (16.1) by $(u'; 0)$. Thus we have the relation

$$-\varphi(u; t^1) + \varphi(u'; 0) = 0\,. \tag{18.2}$$

The Jacobian of the left side relative to $u; t^1$ is merely $D(u; t^1)$ and $D(0; 2\pi) = D(M, 0) \neq 0$. Hence one may solve (18.2) for $u; t^1$ in terms of u'. In particular $u = \psi(u')$ where ψ is analytic. But this relation is manifestly reciprocal and is inverted by traversing γ' in the reverse direction from M_1' to M'. Hence $u' = \omega(u)$ where ω is analytic. Thus:

(18.3) *If a path γ' near to the closed path γ crosses the cell E^{n-1} in consecutive points M', M_1', the transformation $\omega: M' \to M_1'$ is analytic.*

It may be observed that generally ω transforms a certain neighborhood U of M on E^{n-1} into another neighborhood V of the same point. The only thing however that matters regarding ω is the manner in which it operates very close to the point M.

19. *Integral of a system.* Let us still consider an autonomous system (13.1). Let $f(x)$ be a scalar function which is continuously differentiable in a region Ω of $\mathfrak{V}_x$ where (13.1) is valid. We will say that $f(x)$ is an *integral* of the system (13.1) if $df/dt = 0$ along any path of (13.1) in the region Ω, or equivalently if

$$\Sigma X_h \, \partial f/\partial x_h = 0 \tag{19.1}$$

in Ω. Then along any path $f(x) = C$. Let us suppose that f contains x_n, and let x^0 be any point where $\partial f/\partial x_n \neq 0$. By the implicit function theorem one may solve $f = C$ in a suitable neighborhood U of x^0 for $x_n - x_n{}^0$ in terms of the $x_h - x_h{}^0$, $h < n$, and C as

$$x_n - x_n{}^0 = \varphi(x_1 - x_1{}^0, \ldots, x_{n-1} - x_{n-1}{}^0; C), \quad \varphi(0; C) = 0 .$$

Substituting in (13.1) there results a system in $n - 1$ variables. Thus the existence of any integral of a system (13.1) enables one to lower the dimension of the system by one unit.

If the system (13.1) had $n - 1$ integrals (functionally independent) its solution would be reduced to a quadrature. As an example the system in two variables

$$dx_1/dt = X_1(x_1, x_2), \quad dx_2/dt = X_2(x_1, x_2)$$

with X_1, X_2 homogeneous and of the same degree has the integral x_2/x_1, and hence it can be solved by a quadrature.

CHAPTER III

Linear Systems

Linear systems, which we take up in the present chapter, offer the simplest application of the general theorems of the preceding chapter. They are also important as a preparation to the treatment of non-linear systems in the following chapters. Often indeed most known information regarding a non-linear system is embodied in its "first approximation" which is linear.

Reference: Goursat [1], II, Ch. 16.

§ 1. Various Types of Linear Systems

1. By *linear* systems are meant those of the vector form

$$dx/dt = A(t)\,x + b(t)\,. \tag{1.1}$$

They include the highly important type with x a scalar variable

$$d^n x/dt^n + a_1(t)\, d^{n-1} x/dt^{n-1} + \dots + a_{n+1}(t) = 0\,. \tag{1.2}$$

The special systems (1.1) with $b = 0$, or (1.2) with $a_{n+1} = 0$ are known as *homogeneous*. That is to say they are the systems

$$dx/dt = A(t)\,x, \tag{1.3}$$

x a vector,

$$d^n x/dt^n + a_1(t)\, d^{n-1}x/dt^{n-1} + \dots + a_n(t)\,x = 0\,, \tag{1.4}$$

x a scalar variable.

(1.5) *Variation equations.* Consider the general vector system

$$dx/dt = X(x;\,t) \tag{1.6}$$

or with components separated:

$$dx_i/dt = X_i(x_1, \ldots, x_n; t) \,. \tag{1.7}$$

Suppose that we have obtained a solution $x(t)$ representing a trajectory over $U \times I$; U: $\|x - x^0\| < a$; I: $t^1 < t < t^2$ and that it may be extended to one over $U_1 \times I_1$ where $\bar{U} \subset U_1$, $\bar{I} \subset I_1$. Consider now possible trajectories *very near* the given one, in the sense that any one of them is of the form $x(t) + \xi(t)$ over $U_1 \times I_1$, and that the norm $\|\xi(t)\|$ is so small that the products of the ξ_i may be neglected. If we assume that the partial derivatives $\partial X_i/\partial x_j$ exist and are continuous for $(x(t); t) \in U \times I$, then we have from (1.7) after neglecting the products of the ξ_i:

$$d\xi_i/dt = \Sigma\, \partial X_i/\partial x_j \cdot \xi_j \,,$$

or in vector and matrix form

$$d\xi/dt = \partial X/\partial x \cdot \xi \,. \tag{1.8}$$

The system (1.8) is known as the *system of variation equations* of (1.6) or of (1.7). Similarly if we have

$$d^n x/dt^n = F(x, dx/dt, \ldots, d^{n-1} x/dt^{n-1}, t) \,, \tag{1.9}$$

x a scalar variable and $x^0(t)$ is a solution, then the consideration of the "neighboring" solutions $x^0(t) + \xi(t)$, where the squares and products of ξ and its derivatives are neglected, leads to a variation equation, obtained thus: if we set

$$\left(\frac{d^k x}{dt^k}\right)_{x = x^0} = y_{k+1} \,, \quad \left(\frac{d^0 x}{dt^0}\right)_{x = x^0} = x^0 = y_1 \,,$$

then the equation is

$$d^n\xi/dt^n - \Sigma\, \partial F/\partial y_k \cdot d^{k-1}\xi/dt^{k-1} = 0$$

where $d^0\xi/dt^0 = \xi$.

The common feature of the variation systems is that they are *linear homogeneous*. Thus frequently even if the original system is beyond our scope, once a particular solution is known the passage to the variation equations may serve to obtain important information regarding the relation of the solution in question to the neighboring solutions.

§ 2. Homogeneous Systems

2. Consider the system (1.3), and let R be the interior of the set of all points t where $A(t)$ is continuous. Since the $\partial X_i/\partial x_j$ of (II, 5.1) are here the $a_{ij}(t)$ and the components of R are intervals I, the domain of (1.3) is of the form $\mathfrak{V}_x \times I$. Hence:

(2.1) *If* $(x^0; t^o) \in \mathfrak{V}_x \times I$ *then there passes a unique trajectory through* $(x^0; t^0)$. *If* $A(t)$ *is continuous for* $t > \tau$ *then there passes a unique trajectory through* $(x^0; t^0)$ *whatever* $t^0 > \tau$ *and whatever* $x^o \in \mathfrak{V}_x$ (II, 6).

(2.2) *If* A *is analytic at* t^0 *then the solution* $x(x'; t)$ *considered as a function of the initial vector* x' *is analytic at* $(x^0; t^0)$ (II, 10.1).

(2.3) *If* A *is a function of the vector* y *and is analytic at* $(y^0; t^0)$ *then the solution is analytic in* $(x'; t; y)$ *at* $(x^0; t^0; y^0)$ (II, 10.3).

(2.4) *If* t^1, t^2 *are the end points of* I, *then the boundary of the domain* $\mathfrak{V} \times I$ *is* $\mathfrak{V} \times t^1 \cup \mathfrak{V} \times t^2$. *If* $t^0 \in I$, *whatever* x^0, *then any trajectory reaching* $(x^0; t^0)$ *may be extended over the whole interval* I.

The boundary part of the statement is obvious. Regarding the rest applying (II, 9.3) to (II, 2.2) shows that the solution $x(t)$ is bounded on any segment contained in I and hence finite at every point of I. It follows then from (II, 8.1a) that the trajectory can only tend to the values $t = t^1, t^2$, and hence may be extended to the whole of I.

3. (3.1) *The solutions of* (1.3) *make up an* n*-dimensional vector space.*

If $x(t)$, $y(t)$ are two solutions and k, l scalars then $kx + ly$ is likewise a solution since

$$d/dt \cdot (kx + ly) = kAx + lAy = A(kx + ly) .$$

Therefore the solutions make up a vector space $\mathfrak{W}$. Let $t^0 \in I$, where the domain of (1.3) is $\mathfrak{D} = \mathfrak{V}_x \times I$ and let $x^1, \ldots, x^n$ be n independent points of $\mathfrak{V}_x$. According to the existence theorem there is a solution $x^h(t)$ such that $x^h(t^0) = x^h$. To prove (3.1) it is sufficient to show that $\{x^h(t)\}$ is a maximal set of linearly independent solutions, or that:

(a) *the* $x^h(t)$ *are linearly independent;*

(b) *they span all the solutions, i.e. every solution is of the form* $\Sigma k_h x^h(t)$.

If the $x^h(t)$ are linearly dependent there must exist a non-trivial identical relation

$$\Sigma k_h x^h(t) = 0$$

valid for all $t \in I$. Hence for $t = t^o$:

$$\Sigma k_h x^h(t^o) = \Sigma k_h x^h = 0$$

which violates the assumption that the x^h are linearly independent. This proves (a).

If $x(t)$ is any solution, by (2.5) $x(t^o)$ exists whatever $t^o \in I$. Since x^o is in $\mathfrak{V}_x$ and $\{x^h\}$ is a base for the space we will have

$$x^o = \Sigma k_h x^h .$$

It follows that $\Sigma k_h x^h(t)$ is a solution of (1.3) taking likewise the value x^o for $t = t^o$. Since by the existence theorem there is only one such solution we must have

$$x(t) = \Sigma k_h x^h(t) \tag{3.2}$$

and this is (b). Thus (3.1) is proved.

A maximal linearly independent set of solutions $\{x^h(t)\}$ for (1.3) is known as a *base* for (1.3). Our argument yields also readily:

(3.3) *A n.a.s.c. in order that a set of n solutions* $\{x^h(t)\}$ *be a base is that every solution* $x(t)$ *may be represented uniquely in the form* (3.2) *where the* k_h *are scalars.*

(3.4) *A n.a.s.c. in order that* $\{x^h(t)\}$ *be a base is that if* $x_{ih}(t)$ *are the coordinates of* $x^h(t)$, *then for some* $t^o \in I$, *and hence for every* $t \in I$ *the determinant*

$$D(t) = |x_{ih}(t)| \neq 0 . \tag{3.4a}$$

Suppose $\{x^h(t)\}$ is a base. If $x(t)$ is any non-trivial solution we will have a non-trivial relation (3.2) and hence for $t = t^o$:

$$x_i(t^o) = \Sigma k_h x_{ih}(t^o) . \tag{3.5}$$

Thus (3.5) considered as a system of n equations of the first degree in the unknowns k_h, does have a solution. Furthermore the solution is unique. For regardless of $x(t)$, if we are given $x(t^0) = x^0$, then corresponding to any solution $(k_1, \ldots, k_n)$ of (3.5) the function $\Sigma\, k_h\, x^h(t)$ will be a solution of (1.3) taking the value x^0 for $t = t^0$. Since the function is unique (II, 2.1) so is the set $(k_1, \ldots, k_n)$. It follows that $D(t^0) \neq 0$. Since t^0 is any $t \in I$, (3.4a) holds and so the condition of (3.4) is necessary.

Suppose now that $D(t^0) \neq 0$ for some $t^0 \in I$. By (2.5) $x(t^0)$ is defined. Under our assumption (3.5) will then have a unique solution $(k_1, \ldots, k_n)$ and so $\Sigma\, k_h\, x^h(t)$ will be a solution assuming the same value $x(t^0)$ for $t = t^0$ as $x(t)$. Since such a solution is unique there subsists a relation (3.2). Thus $\{x^h(t)\}$ spans all the solutions and since it consists of n vectors it is a base. By the necessity of the condition (3.4a) will then hold for every $t \in I$. This proves (3.4).

(3.6) The transformation $x^0 = x(t^0) \rightarrow k$ resulting from (3.5) is a topological mapping $\mathfrak{B}_x \rightarrow \mathfrak{B}_k$. Hence $\Sigma\, k_h\, x^h(t)$ is a general solution.

(3.7) *A noteworthy property of the variation equation.* Consider an equation (1.6) with X analytic in x and a domain of analyticity $\Delta = \Delta_1 \times \mathfrak{T}$, where $\mathfrak{T}$ is the real t line and Δ_1 a region of $\mathfrak{B}_x$. Suppose also that we have a solution $x(c; t)$, $c = (c_1, \ldots, c_r)$ analytic in c for c in a certain region Λ of $\mathfrak{B}_c$, and all t, and with a Jacobian matrix J of rank r in Λ. In particular for a certain $c^0 \in \Lambda$, $x(t; c^0) = x(t)$, the solution of (1.7) whose variation equation is (1.8). Then:

(3.8) *The r functions $\partial x(c^0; t)/\partial c_i$ are linearly independent solutions of the variation equation* (1.8).

Linear independence is an immediate consequence of the fact that the Jacobian matrix J is of rank r, so that we only need to prove that the functions in question are solutions. Let $\Delta c = (0, \ldots, 0, \Delta c_i, 0, \ldots)$. Since $x(t; c^0 + \Delta c)$ is a solution for $c^0 + \Delta c \in \Lambda$ if we substitute for the x_h in (1.7) the expansions

$$x_h(t; c^0 + \Delta c) = x_h(t; c^0) + \Delta c_i/1! \cdot \partial x_h(t; c^0)/\partial c_i^0 + \ldots$$

the result must be an identity in Δc_i. If we identify the powers of Δc_i on both sides we find that the vector whose components are $\partial x_h(t;c^0)/\partial c_i^0$ is a solution of the variation equation. This proves (3.8).

Remark. The argument could be extended without particular difficulty to the non-analytical case.

4. Side by side with (1.3) it is interesting to consider the associated matrix equation

$$dX/dt = AX. \tag{4.1}$$

Written out explicitly it takes the form

$$dx_{ih}(t)/dt = \Sigma\, a_{ij}(t)\, x_{jh}(t). \tag{4.2}$$

It shows that the vector x^h whose coordinates are the elements $x_{jh}(t)$ of the hth column of X, is a solution of (1.3). Thus the columns of X give rise to n solutions of the differential equation (1.3). Further developments rest upon the important relation

$$|X(t)| = |X(t^0)| \exp\left(\int_{t^0}^{t} (\text{trace } A)\, dt\right), \tag{4.3}$$

for all $t^0, t \in I$. The proof is as follows: By a well known rule for the derivative of a determinant we find

$$\frac{d\,|X(t)|}{dt} = \Sigma_i \begin{vmatrix} \cdots \\ \dfrac{dx_{ik}}{dt} \\ \cdots \end{vmatrix} = \Sigma_i \begin{vmatrix} \cdots\cdots \\ \Sigma_j\, a_{ij}\, x_{jk} \\ \cdots\cdots \end{vmatrix} = \Sigma_{i,j} \begin{vmatrix} \cdots\cdots \\ a_{ij}\, x_{jk} \\ \cdots\cdots \end{vmatrix},$$

where the rows unwritten in each determinant are as in X itself. In the last determinant there are proportional rows unless $i = j$. Hence terms with $i = j$ are the only terms which do not vanish and they have the values $a_{ii}|X|$. Hence

$$d\,|X|/dt = |X| \text{ trace } A$$

from which (4.3) follows by integration.

An immediate consequence of (4.3) is

(4.4) *If* $|X(t)| \neq 0$ *for some* $t \in I$ *then it is* $\neq 0$ *for all* $t \in I$.

A matrix-solution of (4.1) such that $|X(t)| \neq 0$ is called a *non-singular solution.*

(4.5) *If* x^h *is the solution of the differential equation* (1.3) *represented by the* h^{th} *column of the matrix solution* X *of* (4.1) *then a n.a.s.c. in order that* $\{x^h(t)\}$ *be a base for* (1.3) *is that* X *be a non-singular solution of* (4.1).

This is an immediate consequence of (3.4).

Remark. It is clear now that (4.4) is likewise implicit in (3.4). However the basic relation (4.3) is important for its own sake; furthermore there is some interest in having a direct proof of the type given here, rather than one such as for (3.4), based upon the existence theorem.

Owing to our habitual identification of vectors with one-column matrices we have naturally associated with (1.3) the matrix equation (4.1) with X as a right multiplier. We could equally identify vectors with one-row matrices and we would then associate in place of (1.3), (4.1) the equations

$$dx/dt = xB\ ,\ dX/dt = XB$$

and everything said so far would hold with rows and columns interchanged.

(4.6) *Remark.* If X is a non-singular solution of (4.1) then $X \cdot a$, a an arbitrary vector, is the general solution of (1.3).

5. *Adjoint systems.* The special vector and matrix equations

(5.1) $$dy/dt = -yA$$

(5.2) $$dY/dt = -YA$$

where A is the same as in (1.3) and (4.1) are said to be *adjoint* to (1.3) and (4.1). This time the row-vectors making up Y represent solutions of (5.1). Since

(5.1)′ $$dY'/dt = (-A')\,Y'\ ,$$

(1.3)′ $$dX'/dt = -X'(-A')\ ,$$

(1.3)′ may be considered as the adjoint system to (5.1)′. This brings out the symmetry in the situation. If the matrix A is *skew-symmetric* then $A = -A'$ ($a_{ii} = 0$, $a_{ij} = -a_{ji}$) and (4.1) is the same as (5.1)′, (5.2) the same as (1.3)′. All the systems considered are then said to be *self-adjoint.*

From (4.1) and (5.2) there follows:

$$(5.3) \qquad Y\frac{dX}{dt} + \frac{dY}{dt}X = 0$$

and therefore

$$(5.4) \qquad YX = C\,,$$

where C is a constant matrix. Let Y be a solution of (5.2) such that $|Y| \neq 0$ and set $X = Y^{-1}C$. Then (5.4) holds, and hence also (5.3). Consequently

$$(5.5) \qquad Y\frac{dX}{dt} = YAX\,.$$

Since $|Y| \neq 0$, Y^{-1} exists and so from (5.5) follows (4.1). Consequently every solution of (4.1) is of the form $Y^{-1}C$. This proves:

(5.6) *If Y is a non-singular solution of the adjoint equation to* (4.2) *then every solution X of* (4.1) *is represented by $Y^{-1}C$ where C is an arbitrary constant matrix. The non-singular solutions correspond to $|C| \neq 0$.*

Taking $C = E$ we have $YX = E$, and so X^{-1} is a special solution of (5.2). Since $Y^{-1} = X$, we recognize in (5.6) proposition (3.3) in another formulation.

For obvious reasons of symmetry the same holds with (4.1) replaced by (5.2), Y by X and $Y^{-1}C$ by CX^{-1}.

In the applications the important associated types are really (1.3) and (5.1). For this reason it will be worth while to be a little more explicit regarding the relation of a system (1.3) to its adjoint. If we write (1.3) as

$$(5.7) \qquad dx_i/dt = \Sigma\, a_{ij}\, x_j\,,$$

then the adjoint (5.1) assumes the form

$$(5.8) \qquad dy_j/dt = -\Sigma\, a_{ij}\, y_i\,.$$

If $(x_1, \ldots, x_n)$, $(y_1, \ldots, y_n)$ are any two solutions of the respective systems then clearly

$$\Sigma\,(x_i\, dy_i + y_i\, dx_i) = 0$$

and so

(5.9) $$\Sigma\, x_i\, y_i = C\,,$$

a constant. This is the analogue of (5.4), and could in fact be deduced from it. If we have a base $\{y^j\}$, $y^j = (y_{j1}, \ldots, y_{jn})$, for the solutions of the adjoint to (1.3) then $|y_{ji}| \neq 0$. From (5.9) follows also

(5.10) $$\Sigma\, y_{ji}\, x_i = C_j\,,$$

a system of equations of the first degree which may be solved for the x_i. The solution thus obtained is in terms of n arbitrary constants, the C_j, and is the *general* solution of (5.7). We may thus state:

(5.11) *Given a linear homogeneous system of differential equations* (5.7) *if we have n linearly independent solutions of the adjoint system* (5.8), *then the complete solution of the system* (5.7) *itself is reduced to the solution of an algebraic system of n linear equations in n unknowns.*

6. *The linear homogeneous differential equation of order n.* Since (1.4) is reducible to a system (1.3) its properties may be deduced from those of (1.3). For convenience in the applications we shall express them directly. For the present and to the end of the present section the functions and variables are scalars.

For the sake of expediency we introduce the customary operator $D = d/dt$. Then (1.4) takes the form

(6.1) $$D^n x + a_1(t)\, D^{n-1} x + \ldots + a_n(t)\, x = 0\,,$$

or with the conventions $D^0 = 1$, $a_0 = 1$:

(6.2) $$\Sigma\, a_k(t)\, D^{n-k} x = 0\,.$$

Writing $x = x_1$, the equivalent system (5.7) is:

(6.3) $$\begin{cases} Dx_1 = x_2 \\ \cdot\ \cdot\ \cdot\ \cdot\ \cdot\ \cdot \\ Dx_{n-1} = x_n \\ Dx_n = -\, a_n\, x_1 - a_{n-1}\, x_2 - \ldots - a_1\, x_n\,. \end{cases}$$

This yields in particular:

(6.4) $$x_k = D^{k-1}\, x\,.$$

The domain of (6.3) is of the form $\mathfrak{V}_x \times I$, where I is a component of the interior of the set of points of the real t line at which all the $a_i(t)$ are continuous.

Let us examine the linear dependence of the solutions of (6.1) at first directly, i.e. without referring to (6.3). A set of solutions $\{\xi^i(t)\}$ is said to be *linearly dependent* whenever there can be found real scalars C_i not all zero such that

$$(6.5) \qquad \sum C_i \xi^i(t) = 0\,.$$

When this holds we also have

$$(6.6)_k \qquad \sum C_i D^k \xi^i(t) = 0\,,$$

for all k. Of course since by (6.1) the $D^n \xi^i$ are linearly dependent upon $\xi^i, \ldots, D^{n-1}\xi^i$, (6.6) need only be considered for $k < n$. It is clear that the $\xi^i(t)$ will be linearly dependent whenever $(6.6)_0, \ldots, (6.6)_{n-1}$ have a solution in the C_i not all zero, and hence certainly if the matrix

$$(6.7) \qquad (D^k \xi^i),\ (k = 0, \ldots, n-1;\ i = 1, 2, \ldots, r)$$

has more than n columns. Hence there cannot be more than n linearly independent solutions. Given on the other hand n solutions, a n.a.s.c. for their linear dependence is that the determinant called *Wronskian*:

$$\Delta(\xi^1, \ldots, \xi^n) = \begin{vmatrix} \xi^1 & \ldots & \xi^n \\ D\xi^1 & \ldots & \\ \vdots & & \\ D^{n-1}\xi^1 & \ldots & \end{vmatrix} = 0\,.$$

We have shown (3.1) that there are n linearly independent solutions and no more. A set of n linearly independent solutions $\xi^1, \ldots, \xi^n$ is known as a *base* for the solutions. Any other ξ is given by a relation

$$\xi(t) = \sum c_i \xi^i(t)\,.$$

If we think of $\xi(t)$ as a function $\xi(t; c_1, \ldots, c_n)$ then it is described as a *general* solution of (6.1).

Coupling what precedes with the properties of the Wronskian we have:

(6.8) *A n.a.s.c. in order that* $\{\xi^1, \ldots, \xi^n\}$ *be a base for the solutions of* (6.1) *is that the Wronskian* $\Delta(\xi^1, \ldots, \xi^n) \neq 0$ *for some* $t^0 \in I$, *and hence for every* $t \in I$.

(6.9) *Given* n *linearly independent functions* $\xi^1, \ldots, \xi^n$ *each* n *times differentiable on an interval* $I: t_1 < t < t_2$, *they satisfy the differential equation*

$$\Delta(x; \xi^1, \ldots, \xi^n) = 0 \tag{6.9a}$$

and this equation is unique to within a factor in t.

If we replace x by ξ^i, Δ acquires two identical rows and so vanishes. Hence ξ^i satisfies (6.9a). Unicity will follow if we can prove more generally:

(6.10) *If the two equations*

$$D^n x + a_1(t) D^{n-1} x + \ldots + a_n(t) x = 0 \tag{6.10a}$$

$$D^n x + b_1(t) D^{n-1} x + \ldots + b_n(t) x = 0 \tag{6.10b}$$

are satisfied over the interval I *by the same set of* n *linearly independent functions then they are identical.*

For otherwise the difference is of order $< n$, not identically zero, yet with n linearly independent solutions which is ruled out. Therefore (6.10) holds and so does (6.9).

We also prove:

$$\Delta(t) = \Delta(t^0) \exp\left(-\int_{t_o}^{t} a_1(t)\, dt\right). \tag{6.11}$$

In fact by differentiation

$$d\Delta/dt = \begin{vmatrix} \xi^i \\ D\xi^i \\ \vdots \\ D^{n-2}\xi^i \\ D^n \xi^i \end{vmatrix} = -\begin{vmatrix} \xi^i \\ \\ \vdots \\ D^{n-2}\xi^i \\ \sum a_k D^{n-k} \xi^i \end{vmatrix} = -a_1 \Delta,$$

from which to (6.11) is but a step.

It is not difficult to see that in what precedes the Wronskian matrix plays the role of the previous matrix X. In fact referring to the system (6.3) associated with, and equivalent to (6.1), if we set $D^{k-1} \xi^i(t) = x_{ki}(t)$ then

$$\Delta(\xi^1, \ldots, \xi^n) = (x_{ki}(t)) = X.$$

To $\xi^i(t)$ there corresponds now the vector solution $x^i(t)$ of (6.3) and clearly the correspondence $\xi^i(t) \leftrightarrow x^i(t)$ is one-one and preserves linear dependence. As a consequence the properties just obtained for (6.1) could be obtained directly from those of (6.3).

7. We will now discuss adjoints for the type (6.1). First the adjoint of (6.3) is

$$\text{(7.1)} \qquad \begin{cases} Dy_1 = a_n y_n \\ Dy_2 = -y_1 + a_{n-1} y_n \\ \cdots\cdots\cdots \\ Dy_n = -y_{n-1} + a_1 y_n . \end{cases}$$

If we differentiate the $(k+1)$st relation k times and set $y_n = y$ we find by addition

$$\text{(7.2)} \qquad D^n y - D^{n-1}(a_1 y) + \ldots + (-1)^n a_n y = 0,$$

and we now define (6.1), (7.2) as *adjoint* to one another. To derive the analogue of (5.9) we must solve (7.1) for the y_k. We first have

$$D^i y_{k+i} = -D^{i-1} y_{k+i-1} + D^{i-1}(a_{n-k-i+1} y)$$

and this yields

$$y_k = \sum_{h=0}^{n-k} (-1)^h D^h (a_{n-k-h} y).$$

Substituting finally in (5.9) we obtain as the analogue of (5.9):

$$\text{(7.3)} \qquad \sum_{h,k} (-1)^h D^{k-1} x \, D^h (a_{n-k-h} y) = C.$$

If we possess a base $\{\eta^i\}$ for the solutions of the adjoint equation (7.2) to (6.1) then the *general* solution $x(t)$ of (6.1) may be obtained by solving the linear (algebraic) system

$$(7.4) \quad \sum_{h,k} (-1)^h D^{k-1} x \, D^h (a_{n-k-h} \eta^i) = C_i \quad (i = 1, 2, \ldots, n)$$

for x.

The relation just obtained for adjoints may also be derived directly by means of integration by parts. For any two functions $x(t)$, $y(t)$ with a suitable number of derivatives we have:

$$y D^k x = D(yD^{k-1} x) - DyD^{k-1} x$$
$$\cdots\cdots\cdots\cdots$$
$$D^h yD^{k-h} x = D(D^h yD^{k-h-1} x) - D^{h+1} yD^{k-h-1} x$$
$$\cdots\cdots\cdots\cdots$$
$$D^{k-1} yDx = D(D^{k-1} yx) - D^k y \cdot x .$$

From this follows

$$yD^k x + (-1)^{k-1} xD^k y = D \sum (-1)^h D^h yD^{k-h-1} x .$$

Replacing k by $n - k$ and y by $a_k y$ we find:

$$y \cdot a_k D^{n-k} x + (-1)^{n-k-1} xD^{n-k} (a_k y)$$
$$= D \sum (-1)^h D^h (a_k y) \, D^{n-k-h-1} x .$$

Hence if x is a solution of (6.1) and y a solution of its adjoint,

$$D \sum_{h,k} (-1)^h D^h (a_k y) \, D^{n-k-h-1} x = 0 ,$$

from which with a rearrangement of indices it is but a step to (7.3).

Green's formula. Generally speaking let $L(x)$, $M(x)$ be two linear differential operators of same order

$$L(x) = \sum a_k(t) \, D^{n-k} x , \quad M(x) = \sum b_k(t) \, D^{n-k} x .$$

We shall say that L, M are adjoint to one another if

$$(7.5) \qquad yL(x) - xM(y) = dF/dt$$

where F is a polynomial in x, y and their derivatives of orders $\leq n - 1$ with coefficients functions of t. It is assumed throughout that all the functions of t under consideration are continuous together with their derivatives as far as required on some interval I. If $t^1, t^2 \in I$ we will then have as a consequence of (7.5):

(7.6) $$\int_{t^1}^{t^2} (yL(x) - xM(y))\, dt = F\Big]_{t^1}^{t^2},$$

a result known generally as *Green's formula.* It is particularly useful wherever one deals with preassigned "boundary" conditions at t^1, t^2.

(7.7) *Example.* If we have

$$L(x) = M(x) = d/dt\,(p\, dx/dt) + qx\,,$$

then

$$yL(x) - xL(y) = d/dt\,(p\,(y\, dx/dt - x\, dy/dt))$$

and so L is self-adjoint. We have then

$$\int_{t^1}^{t^2} (yL(x) - xL(y))\, dt = \Big[p\,(y\, dx/dt - x\, dy/dt)\Big]_{t^1}^{t^2}.$$

Thus the integral at the left depends only on the values of the functions and their derivatives at the end points t^1, t^2.

§ 3. Non-Homogeneous Systems

8. We return now to vector variables. Consider first the system

(8.1) $$dx/dt = A(t)\, x + b(t)\,,$$

where x and b are vectors and A is a matrix. If x, x' are any two solutions, then $x' - x$ satisfies the associated homogeneous system (1.3). Hence if one has any special solution of (8.1) every other is obtained by adding to it a suitable "complementary function": a solution of the associated homogeneous equation.

To find a solution of (8.1) we have recourse to the adjoint of (1.3):

$$dy/dt = -yA(t)\,.$$

Let $Y(t)$ be the solution of the associated matrix equation and more particularly the one such that $Y(t_0) = E$. Now

(8.2) $$Y\frac{dx}{dt} + \frac{dY}{dt}x = \frac{d\,(Yx)}{dt} = Yb\,.$$

Hence

$$x = Y^{-1} (x (t_0) + \int_{t_0}^{t} Yb\, dt) . \tag{8.3}$$

Similarly the matrix equation

$$dX/dt = A (t) X + B (t) \tag{8.4}$$

has the general solution

$$X = Y^{-1} (X (t_0) + \int_{t_0}^{t} YB\, dt) . \tag{8.5}$$

The same method may be applied to the solution of the non-homogeneous "scalar" equation

$$\Sigma\, a_k D^{n-k} x = b (t) . \tag{8.6}$$

If y satisfies the adjoint equation (7.2) to the related homogeneous equation then (7.5) yields this time

$$by = D \sum_{h,\, k} (-1)^h D^h (a_k y) D^{n-k-h-1} x . \tag{8.7}$$

If we possess a base $\{\eta^i\}$ for the solutions of (7.2), the related relations (8.7) duly integrated yield

$$\int^{t} b (t)\, \eta^i (t)\, dt = \Sigma (-1)^h D^h (a_k \eta^i) D^{n-k-h-1} x . \tag{8.8}$$

This is a linear (algebraic) system in $x, Dx, \ldots, D^{n-1} x$, whose determinant is $\Delta (\eta^1, \ldots, \eta^n) \neq 0$, and so it may be solved for x.

Here again if $b = 0$ then the equation becomes

$$\Sigma\, a_k D^{n-k} z = 0 . \tag{8.9}$$

Hence $x = x' + z$, where z is the general solution of (8.9), x' is a solution of (8.6), and x is the general solution of (8.6).

§ 4. Linear Systems with Constant Coefficients

9. Suppose that in

$$dx/dt = Ax , \tag{9.1}$$

A is a constant. The associated matrix equation

$$(9.2) \qquad dX/dt = AX$$

has already been treated (I, 11) and found to have the solution

$$(9.3) \qquad X = e^{At}$$

which is non-singular (I, 10.11). By (4.5) if $x^h = (x_{1h}, \ldots, x_{nh})$, the set $\{x^h\}$ is a base for (9.1), and the solution is thus complete.

It is of interest to examine the form of the solution. We first observe that if we apply the transformation $x = Py$, $|P| \neq 0$, then (9.1) is replaced by

$$(9.4) \qquad dy/dt = By,\ B = P^{-1}AP.$$

We choose P so that B is in normal form. In place of (9.2) here

$$(9.5) \qquad dY/dt = BY,$$

with the non-singular solution e^{Bt}. Notice now that $e^{At} = Pe^{Bt}P^{-1}$ (I, 10.9), hence we merely need to find the form of e^{Bt}. Now if $B = \operatorname{diag}(B_1, \ldots, B_r)$ then $e^{Bt} = \operatorname{diag}(e^{B_1t}, \ldots, e^{B_rt})$, (I, 10.4). If

$$(9.6) \qquad B_h = C(\lambda_h) = \begin{pmatrix} \lambda_h & & & \\ 1 & \lambda_h & & \\ & \ddots & \ddots & \\ & & 1 & \lambda_h \end{pmatrix}$$

the same calculation as in (I, 10.6) will yield if ϱ_h is the order of B_h:

$$(9.7) \qquad e^{B_ht} = e^{\lambda_ht} \begin{pmatrix} 1 & & & & 0 \\ \frac{t}{1!} & 1 & & & \\ \frac{t^2}{2!} & \frac{t}{1!} & 1 & & \\ \cdots & \cdots & & \ddots & \\ \cdots & \cdots & & & \\ \frac{t^{\varrho_h-1}}{(\varrho_h-1)!} & \cdots & \cdots & \frac{t}{1!} & 1 \end{pmatrix},$$

and e^{Bt} is made up of these diagonal blocks.

Special case: $A \sim \text{diag}(\lambda_1, \ldots, \lambda_n)$. Then clearly $Y = \text{diag}(e^{\lambda_1 t}, \ldots, e^{\lambda_n t})$ and so there is a base $\{y^h\}$, where

$$(9.8) \qquad y^h = (\delta_{hj} e^{\lambda_j t}) = (0, \ldots, 0, e^{\lambda_h t}, 0, \ldots, 0),$$

so that y^h has a single coordinate $\neq 0$ in the hth place. The general solution of (9.4) is

$$(9.9) \qquad y = (C_1 e^{\lambda_1 t}, C_2 e^{\lambda_2 t}, \ldots, C_n e^{\lambda_n t}).$$

The general solution for x is $x = Py$ or:

$$x_i = \Sigma\, p_{ij} C_j e^{\lambda_j t}, \quad (i = 1, 2, \ldots, n).$$

In point of fact it is not particularly difficult to describe the solutions in the general case. We will begin with the y^i. Set $\sigma_h = \varrho_1 + \cdots + \varrho_{h-1}$. Let also y_{mk} denote as usual the coordinates of y^k. Then we have

$$y_{mk} = \frac{t^{m-k}}{(m-k)!} e^{\lambda_j t}, \ \sigma_h < k \leq m \leq \sigma_{h+1},$$

and all other coordinates y_{mk} are zero. Hence x^k will have its coordinates x_{ik} of the form

$$x_{ik} = \varphi_{ik}(t)\, e^{\lambda_j t}$$

where φ_{ik} is a polynomial in t of degree $\leq \varrho_h - 1$. The general solution x will then have its coordinates x_i given by

$$(9.10) \qquad x_i = \Sigma\, C_k \varphi_{ik}(t)\, e^{\lambda_j t},$$

where the C_k are arbitrary constants. If B_h is of order one, the φ_{ik}, with $k = \sigma_{h+1}$, are all equal to φ_{1k} and the corresponding terms in the general solution are $\varphi_{1k} C_k e^{\lambda_j t}$.

(9.11) Let us suppose explicitly that A is real and that the characteristic roots are $\lambda_1, \bar{\lambda}_1, \ldots, \lambda_r, \bar{\lambda}_r, \lambda_{2r+1}, \ldots, \lambda_n$ where the λ_{2r+i} are real. Then (9.1) may be reduced to the normal form with the λ's forming a system $(\lambda_1, \bar{\lambda}_1), \ldots, (\lambda_r, \bar{\lambda}_r), \lambda_{2r+1}, \ldots, \lambda_n$ with r conjugate pairs and the rest real. The coordinates may likewise be chosen such that for real points they form sets $(y_1, \bar{y}_1, \ldots, y_r, \bar{y}_r, y_{2r+1}, \ldots, y_n)$ with the last $n - 2r$ real. The modifications in (9.10) are quite simple. We merely note that the general real solution when the roots are distinct is

$$x_k = \sum_{i=1}^{r} (c_i e^{\lambda_i t} + \bar{c}_i e^{\bar{\lambda}_i t}) + \sum_{i=2r+1}^{n} c_i e^{\lambda_i t}, \tag{9.12}$$

where the c_i, $i > 2r$, are real.

10. Consider now a non-homogeneous system (8.1) where A is a constant matrix. Since X, Y in (5.4) are any two non-singular solutions of (4.1) and (5.2) we may assume $C = E$, and here $X = e^{At}$, hence $Y = e^{-At}$, $Y^{-1} = e^{At}$. Thus the solution (8.3) assumes now the form

$$x = \int_{t_0}^{t} e^{A(t-u)} b(u)\, du + e^{At} \cdot x(t_0). \tag{10.1}$$

As an application consider the equation

$$d^2x/dt^2 + x = b(t), \tag{10.2}$$

or equivalently

$$dx/dt = y, \; dy/dt = -x + b(t). \tag{10.3}$$

We have here

$$Y = Y^{-1} = \begin{pmatrix} \sin t, & \cos t \\ \cos t, & -\sin t \end{pmatrix}$$

and therefore

$$\{x, y\} = \int_0^t b(u) \{\sin(t-u), \cos(t-u)\}\, du$$

or explicitly

$$x = \int_0^t b(u) \sin(t-u)\, du; \quad \frac{dx}{dt} = \int_0^t b(u) \cos(t-u)\, du. \tag{10.4}$$

The general solution, is

$$x = C \cos(t-\alpha) + \int_0^t b(u) \sin(t-u)\, du, \tag{10.5}$$

with C, α as arbitrary constants. This is the well known form of the general solution of (10.2).

§ 5. Linear Systems with Periodic Coefficients: Theory of Floquet

11. Consider again our basic system (1.3), or

$$(11.1)\qquad dx/dt = Ax\,.$$

We suppose t real and A real, continuous for all t, and with the real non-zero period ω. Thus $A\,(t+\omega) = A\,(t)$. We will allow however for convenience *complex* solutions of (11.1), and so complex scalars. The domain of (11.1) is $\mathfrak{B}_x \times \mathfrak{T}$.

Let $X = (x_{ij})$, $|X| \neq 0$, be a non-singular solution of

$$(11.2)\qquad DX = dX/dt = AX\,.$$

Thus if $x^j = (x_{1j}, \ldots, x_{nj})$, then $\{x^j\}$ is a base for the solutions of (11.1). Evidently $X\,(t+\omega)$ satisfies (11.2) for the value $t+\omega$ of t and as we know $|X\,(t+\omega)| \neq 0$ also. Now

$$DX\,(t+\omega) = A\,(t+\omega)\,X\,(t+\omega) = A\,(t)\,X\,(t+\omega)\,.$$

Hence $X\,(t+\omega)$ is a non-singular solution of (11.2) with non zero determinant, for the value t itself. From this follows

$$(11.3)\qquad X\,(t+\omega) = X\,(t)\cdot C$$

where C is constant. Since $|X\,(t)| \neq 0$, $X^{-1}\,(t)$ exists, $C = X^{-1}\,(t)\,X\,(t+\omega)$ and $|C| \neq 0$. In fact by (4.3):

$$(11.4)\qquad |C| = e^{\int_o^\omega (\text{trace } A)\,dt}$$

If we replace $\{x^h\}$ by another base the effect is to replace X by XP, $|P| \neq 0$, and hence C by $P^{-1}X^{-1}\,(t)\;X\,(t+\omega)\,P = P^{-1}\,CP \sim C$. Since P is an arbitrary non-singular matrix and the choice of base is essentially immaterial, we may assume X such that C is a matrix in normal form: $C = \text{diag}\,(C_1, \ldots, C_r)$, where C_i is a block $C\,(\lambda)$. In particular if the characteristic roots $\mu_1, \ldots, \mu_n$ of C are all distinct $C = \text{diag}\;(\mu_1, \ldots, \mu_n)$. The μ_j are known as the *characteristic exponents* of the system (11.1).

Since $|C| \neq 0$, likewise $|C_i| \neq 0$. Hence (I, 10.10) there exists a matrix B_i of the same order as C_i such that $e^{\omega B_i} = C_i$ and if $B = \text{diag}\,(B_1, \ldots, B_r)$, then $e^{\omega B} = C$.

Notice that if the characteristic roots are all distinct, or if they are not and still $C = \text{diag}\,(\mu_1, \ldots, \mu_n)$ then choosing for λ_i any determination of $1/\omega \log \mu_i$, we may take $B = \text{diag}\,(\lambda_1, \ldots, \lambda_n)$. If the μ_i are all distinct, the same will be true as regards the λ_i.

12. Having chosen B consider the matrix

$$Z(t) = e^{tB} \cdot X^{-1}. \tag{12.1}$$

If we recall that B and its power series commute we find

$$\begin{aligned} Z(t+\omega) &= e^{(t+\omega)B} \cdot (X^{-1}(t+\omega)) = e^{\omega B} \cdot e^{tB}(X(t)\,C)^{-1} \\ &= e^{tB} \cdot e^{\omega B} \cdot C^{-1} \cdot X^{-1}(t) \\ &= e^{tB} X^{-1}(t) = Z(t). \end{aligned}$$

In other words $Z(t)$ has the period ω. Moreover by (I, 10.11):

$$|Z(t)| = |e^{tB}| \cdot |X^{-1}(t)| \neq 0 \tag{12.2}$$

for all values of t. Setting $X^* = ZX = e^{tB}$ then

$$dX^*/dt = BX^*. \tag{12.3}$$

The relations between the coordinates assume the form:

$$x_{ih}^* = \Sigma\, z_{ij}(t)\, x_{jh}.$$

Here x_{ih}^* are the coordinates of the vector x^{*h} referred to the coordinate system $(x_1^*, \ldots, x_n^*)$ deduced from the initial system by the linear transformation

$$x_i^* = \Sigma\, z_{ij}(t)\, x_j, \tag{12.4}$$

which is non-singular owing to (12.2). The relation (12.3) yields

$$dx_{ih}^*/dt = \Sigma\, b_{ij}\, x_{jh}^*.$$

Thus $x^{*h} = (x_{1h}^*, \ldots, x_{nh}^*)$ is a solution of the linear homogeneous system with constant coefficients,

(12.5) $$dx_i^*/dt = \Sigma\, b_{ij}\, x_j^* .$$

Since $|X^*| = |x_{ij}^*| = |Z| \cdot |X| \neq 0$, the set $\{x^{*h}\}$ is a base for (12.5). The elements of this base may be written

$$x_{ih}^* = \Sigma\, \varphi_{jh}\,(t)\, e^{\lambda_j t}$$

where the λ_j are the characteristic roots of B, and the φ's are polynomials. Whenever the λ_j are all distinct then the φ's are constant. It follows that the initial base $\{x^h\,(t)\}$ for (11.1) is of the form

$$x_{ih} = \psi_{jh}\,(t)\, e^{\lambda_j t}$$

where the ψ's are polynomials in t with coefficients periodic and of period ω, or else if the λ_j are all distinct mod $2\pi/\omega$, merely periodic functions of t.

We note then the following properties:

(12.6) THEOREM. *By a transformation of variables* (12.4) *the periodic system* (11.1) *may be reduced to a linear homogeneous system with constant coefficients.*

(12.7) *There is a base* $\{x^h\}$ *for the solutions of the periodic system* (11.1) *whose* n *elements are of the form*

$$x^h = \{\psi_{1h}\,(t)\, e^{\lambda_1 t}, \ldots, \psi_{nh}\,(t)\, e^{\lambda_n t}\},$$

where the ψ*'s are as above.*

(12.8) *When the characteristic exponents* μ_i *are all distinct mod* $2\pi/\omega$ *the solutions are of the form* $\Sigma\, \psi_j\,(t)\, e^{\lambda_j t}$, *where* ψ_j *is periodic and of period* ω.

(12.9) *Real solutions.* All that is required is to obtain a real base, and a process for the purpose has been developed in (9).

(12.10) *Remark.* Generally the determination of the characteristic exponents is difficult. For it necessitates the determination of a base for the solutions of (11.1) and following the elements of that base, for instance by analytical continuation, throughout a whole period ω. And no one has ever viewed analytical continuation as a practical procedure.

CHAPTER IV

Stability

This short chapter is devoted to the fundamental concept of stability. Various types of "point" and "trajectory" stabilities are discussed and carefully defined and certain transformations which preserve stability properties are also discussed.

References: Bellman [3]; Dykhman [1]; Krassovskii [1]; Lefschetz [2]; Levinson [3]; Liapunov [1]; Malkin [1, 2, 3, 7]; Perron [3]; Persidski [1, 2]; Poincaré [4].

§ 1. Historical Considerations

1. Historically, stability seems to have been first discussed by Lagrange in connection with the equilibrium of conservative systems. We merely recall that the state of a conservative system depends upon a certain real vector x, its derivative $x'(t)$ and two real continuous functions,

the *kinetic energy* $K(x; x')$ which is positive save that $K = 0$ when and only when $x' = 0$,

the *potential energy* $V(x)$, known only up to an arbitrary constant,

which satisfy the law of conservation of energy

$$K + V = \text{const.}$$

If K and V are of class C^1 the positions of equilibrium may be defined as those where

$$\partial V / \partial x_i = 0, \; i = 1, 2, \ldots, n$$

where n is the dimension of x. However, without referring to differentiability one may define equilibrium as an extremum of V, a definition which will be ample for our purpose.

The following proposition first proved (later) by Dirichlet, was formulated by Lagrange:

(1.1) THEOREM. *Whenever in a certain position of the system the potential energy V is a minimum, the position is an equilibrium and that equilibrium is stable.*

Behind this theorem there lies the following definition of stability and it is this stability that we shall prove:

(1.2) *Corresponding to any $\varepsilon > 0$ there is an $\eta > 0$ such that if $\varrho = ||x|| + ||x'|| < \eta$ at the beginning of the motion then $\varrho < \varepsilon$ ever after.*

We may as well assume that the position of equilibrium is $x = 0$ and that $V(0) = 0$. Since V is continuous there is a $\sigma > 0$ such that in $\mathfrak{S}(\sigma): ||x|| < \sigma$, we have $V(x) > 0$ save at $x = 0$ where $V = 0$. Take now any $\varepsilon < \sigma$. Since V is positive on the compact set $||x|| = \varepsilon$, it will have a positive lower bound μ on the set. Similarly K is positive on the product of the sets $||x|| \leqq \varepsilon$ and $||x'|| = \varepsilon$ which is compact, and so it has a positive lower bound ν on this product.

Take now the system to an initial position (x_0, x_0') where at all events $||x_0|| < \varepsilon$ and denote by K_0, V_0 the corresponding values of K, V. Then

$$K = K_o + V_o - V. \tag{1.3}$$

Now K_0, $K_0 + V_0 \geqq 0$ and vanish only when $x_0 = x_0' = 0$. Since they are continuous there is an $\eta < \varepsilon$ such that if $||x_0|| + ||x_0'|| < \eta$ then $K_0 + V_0$ and $K_0 < \inf(\mu, \nu)$. By (1.3) then $V < \mu$, $K < \nu$, hence $||x|| + ||x'|| < 2\varepsilon$. This proves Lagrange's theorem.

Observe now that if K, V are of class C^1 the motion is governed by Lagrange's equations

$$d/dt \cdot \partial K/\partial x_i' - \partial K/\partial x_i = - \partial V/\partial x_i, \; i = 1, 2, \ldots, n \,.$$

The system will have the origin as critical point and stability asserts an evident property regarding the trajectories passing near the critical point. This suggests then a stability concept for differential equations and this is the concept which we shall now discuss.

§ 2. Stability of Critical Points

2. Consider our usual real system

$$dx/dt = X(x; t) \tag{2.1}$$

and suppose that it satisfies the conditions of the basic existence theorem (II, 2.1) over a domain $I \times R$, where I is the ray $t \geqq \tau$ and R is a connected open set of the vector space $\mathfrak{B}_x$. Let $A(x_o)$ be a critical point of the system: $X(x_o; t) = 0$ for $t \geqq \tau$. We may choose A as origin and it will be readily verified that this does not affect the definitions. We have then $X(0: t) = 0$, $t \geqq \tau$.

There are several stability concepts basically due to Liapunov, and to be described presently. Two additional such concepts are given in § 7.

We designate by t_o any time $\geqq \tau$ and by $x(t)$ the solution of (2.1) such that $x(t_o) = x_o$. The number $\varepsilon > 0$ will always be such that $||x|| < \varepsilon$ is in the set R.

Stability. Given any ε and t_o there corresponds to them an $\eta(\varepsilon, t_o) > 0$ such that if $||x_o|| < \eta$ then $||x(t)|| < \varepsilon$ for all $t \geqq t_o$. The stability is *uniform* if one may choose for η a function $\eta(\varepsilon)$ of ε alone (η independent of t_o).

Instability. Conditional stability. Instability corresponds to this: given any ε, however small and arbitrary t_o, for some $||x_o|| < \varepsilon$ we have $||x(t_1)|| = \varepsilon$ for some $t_1 > t_o$. We have conditional stability whenever this situation holds only for some but not all x_o such that $||x_o|| < \varepsilon$.

Asymptotic stability. The origin is asymptotically stable, when it is stable and in addition for some ε, t_o, and $0 < \xi$, $||x_o|| < \eta(\varepsilon, t_o)$, there corresponds a $T(x_o, t_o, \xi)$ such that $||x(t)|| < \xi$ for $t > t_o + T$. The asymptotic stability is *uniform* whenever it is uniformly stable, and moreover one may choose for T a function $T(\xi)$ of ξ, alone (T independent of x_o and t_o).

Remark. In an autonomous system t_o may be shifted to zero, and so stability is always uniform.

(2.2) One could also define stability in terms of suitable neighborhoods instead of the "ε, η" spheroids. Thus the formulation for stability of a critical point would run as follows: given $t_o \geqq \tau$ and any neighborhood U of A there is another $V(U, t_o) \subset U$ such that if $x(t_o) \in V$ then for $t \geqq t_o$, $x(t) \in U$. The equivalence with the definition already given is obvious, but at times the new formulation is more convenient.

(2.3) *Positive and negative stability.* The preceding definitions refer to the behavior of the trajectories for t large. It may be indicated as "positive stability, . . .". The same questions may be raised regarding the behavior of the trajectories for t small, i.e., $(-t)$ large, and they will lead to the concepts of *negative* stability, ...

§ 3. Stability in Linear Homogeneous Systems

3. Take a system

$$dx/dt = P(t)\, x \tag{3.1}$$

where the matrix $P(t)$ is continuous and bounded for $t \geq \tau$. If $x(t)$ is a solution so is $Cx(t)$. If $x(t)$ satisfies the stability condition relative to ε, η then $Cx(t)$ satisfies it relative to $C\varepsilon$ and $C\eta$, and conversely. Hence

(3.2) *The earlier stability condition is equivalent for* (3.1) *to the following: there exists a positive* $\zeta(t_0) < 1$ *such that if* $\|x(t_0)\| < \zeta$ *then* $\|x(t)\| < 1$ *for all* $t \geq t_0$. *Uniform stability corresponds to the possibility of choosing* ζ *independent of* t_0.

A more interesting property however is the following:

(3.3) *Stability at the origin for* (3.1) *is equivalent to the boundedness of all solutions for* t *large.*

Suppose that we have stability with $\varepsilon, t_0, \eta(\varepsilon, t_0)$ as before. If $x(t)$ is any solution choose $C > 0$ so that $\|Cx(t_0)\| < \eta$. Then $\|Cx(t)\| < \varepsilon$ for $t \geq t_0$ and hence $\|x(t)\| < M = \varepsilon/C$ for t large.

Conversely suppose that all the solutions are bounded for t large. Let $X(t)$ be a non-singular solution of the matrix equation associated with (3.1). Thus every solution assumes the form $x(t) = X(t)\, X^{-1}(t_0)\, x_0$. It is clear that for $t \geq t_0$ and a certain positive function $\mu(t)$

$$\|X(t)\, X^{-1}(t_0)\| \leq \mu(t_0)\,.$$

Hence if we take $\zeta(t_0) = \dfrac{1}{\mu(t_0)}$ then $\|x_0\| < \zeta$ will imply

$||x(t)|| < 1$ for $t \geq t_0$. Hence the system (3.1) is stable at the origin and (3.3) is proved.

It is easily seen that uniform stability under (3.3) corresponds to this: if $||x(t_0)|| < 1$ then $||x(t)|| < M$ for $t \geq t_0$, where M does not depend on t_0.

4. *An Example.* Consider a real homogeneous system with constant coefficients

$$dx/dt = Ax .$$

Suppose that the characteristic roots of the matrix A are distinct and that none are pure complex. Apply the linear transformation reducing A to the real normal form. As a consequence the given system is replaced by a similar one which for the real points is

$$dx_j/dt = \lambda_j x_j;\ d\bar{x}_j/dt = \bar{\lambda}_j \bar{x}_j,\ i = 1, 2, \ldots, r$$

$$dx_{2r+h}/dt = \lambda_{2r+h}\, x_{2r+h}$$

where λ_{2r+h} and x_{2r+h} are real. The real general solution is

$$x = (C_1 e^{\lambda_1 t}, \bar{C}_1 e^{\bar{\lambda}_1 t}, \ldots, C_{2r+1} e^{\lambda_{2r+1} t}, \ldots) .$$

Since $e^{\lambda t} \to 0\ [\to +\infty]$ for $t \to +\infty$ if λ has negative [positive] real part we can state this: Let k be the number of λ_s with negative real parts. Making all the constants C except those corresponding to such λ_s equal to zero there is obtained a k dimensional linear family of solutions which are asymptotically stable at the origin and the family is maximal. Thus we have

asymptotic stability for $k = n$;
conditional asymptotic stability for $0 < k < n$;
instability for $k = 0$.

We shall find that this type of behavior is quite general.

§ 4. Uniformly Regular Transformations

5. It is evident that a linear transformation of coordinates does not affect the stability properties. A broader type of transformation having the same property is introduced here.

Let T be a mapping: $y = f(x; t)$ which maps origin into origin and is topological for each $t \geq \tau$ between two neighborhoods $U(t)$ of the origin in $\mathfrak{B}_x$ and $V(t)$ of the origin in $\mathfrak{B}_y$, where all the $U(t)$, $t \geq \tau$, contain a fixed neighborhood U of the origin in $\mathfrak{B}_x$ and all the $V(t)$ contain a fixed neighborhood V of the origin in $\mathfrak{B}_y$.

The mapping T is said to be *uniformly regular at the origin* whenever given any $\varrho > 0$ there is a corresponding $\varrho_1 > 0$ such that if $x \in U$, $||x|| \geq \varrho$ and $t \geq \tau$, then $||Tx|| \geq \varrho_1$ and likewise for T^{-1}.

(5.1) *A mapping T uniformly regular at the origin preserves the stability properties in both directions.*

It is sufficient to prove that ordinary, uniform or asymptotic stability at the origin in $\mathfrak{B}_y$ implies the same for $\mathfrak{B}_x$. The argument for uniform stability applies to ordinary stability also, so we confine our attention to the former.

Let the origin in $\mathfrak{B}_y$ be uniformly stable, and let $\mathfrak{S}(\varepsilon)$, $\mathfrak{S}_1(\varepsilon)$ denote: $||x|| < \varepsilon$, $||y|| < \varepsilon$ in $\mathfrak{B}_x$, $\mathfrak{B}_y$. We must show that given any $\varepsilon > 0$ there is an $\eta > 0$ such that $x \in \mathfrak{S}(\eta)$ at any time $t \geq \tau$ implies $x \in \mathfrak{S}(\varepsilon)$ ever after. We may suppose ε so chosen that $\mathfrak{S}(2\varepsilon) \subset U$. Now corresponding to ε there is an $\varepsilon_1(\varepsilon) > 0$ such that $||T\,x|| \geq \varepsilon_1$ for $x \in U - \mathfrak{S}(\varepsilon)$ and all $t \geq \tau$, and we suppose ε_1 so chosen that $\mathfrak{S}_1(2\,\varepsilon_1) \subset V$. Hence for all $t \geq \tau$ and $y \in \mathfrak{S}_1(\varepsilon_1)$, $T^{-1}y$ is not in $U - \mathfrak{S}(\varepsilon)$. Since the origin goes into the origin and $T^{-1}\mathfrak{S}_1(\varepsilon_1)$ like $\mathfrak{S}_1(\varepsilon_1)$ is connected, and since also $T^{-1}\mathfrak{S}_1(\varepsilon_1)$ meets $\mathfrak{S}(\varepsilon)$, it cannot have points outside U. Hence for all $t \geq \tau$: $T^{-1}y \in \mathfrak{S}(\varepsilon)$. The uniform stability assumption for $\mathfrak{B}_y$ implies that there is an $\eta_1(\varepsilon_1) > 0$ such that if $y \in \mathfrak{S}_1(\eta_1)$ at time $t \geq \tau$ then $y \in \mathfrak{S}_1(\varepsilon_1)$ ever after. Owing to the restriction on the mapping T there is an $\eta > 0$ such that $y \in V - \mathfrak{S}_1(\eta_1)$ at any time $t \geq \tau$ implies $||T^{-1}y|| \geq \eta$ ever after. By the same reasoning as before if $x \in \mathfrak{S}(\eta)$ at any time $t \geq \tau$ then $Tx \in \mathfrak{S}_1(\eta_1)$ at the same time, which implies $Tx \in \mathfrak{S}_1(\varepsilon_1)$, hence $x \in \mathfrak{S}(\varepsilon)$ ever after. Hence the origin in $\mathfrak{B}_x$ is uniformly stable.

Let now the origin be asymptotically stable in $\mathfrak{B}_y$. To show that the same holds in $\mathfrak{B}_x$ we must show that if $\varepsilon_1 > 0$ there is an $\varepsilon(\varepsilon_1) > 0$ with ε_1 such that $T^{-1}\mathfrak{S}_1(\varepsilon_1) \subset \mathfrak{S}(\varepsilon)$ for all $t \geq \tau$. If this is false there is an $\alpha > 0$ such that $T^{-1}\mathfrak{S}_1(\varepsilon_1)$ has points outside $\mathfrak{S}(\alpha)$ for all $\varepsilon_1 > 0$. By hypothesis there is an $\eta_1 > 0$ such that if $t \geq \tau$ and $x \in U - \mathfrak{S}(\alpha)$ then $||Tx|| \geq \eta_1$. We may suppose $\mathfrak{S}(2\alpha) \subset U$ and it follows as above that $T^{-1}\mathfrak{S}_1(\eta_1) \subset \mathfrak{S}(\alpha)$.

This contradiction proves the assertion as to asymptotic stability, and completes the proof of (5.1).

6. (6.1) *The following property of the mapping T suffices to deduce uniform stability of the origin in $\mathfrak{V}_x$ from the same in $\mathfrak{V}_y$: when $||x|| = \varrho$ then $0 < \varrho_1 \leq ||Tx|| \leq \varrho_2$ for all $t \geq \tau$ and conversely.*

Assume uniform stability in $\mathfrak{V}_y$. Then given any $\varepsilon > 0$ there exist $\varepsilon_1, \varepsilon_2 > 0$ such that $\varepsilon_1 \leq ||Tx|| \leq \varepsilon_2$ for $||x|| = \varepsilon$ and all $t \geq \tau$. The assumption as to $\mathfrak{V}_y$ implies that there is an $\eta_1(\varepsilon_1) > 0$, then by the restriction on T that there is an $\eta > 0$ such that if $||Tx|| = \eta$ then $0 < \alpha_1 \leq ||Tx|| \leq \alpha_2 < \eta_1$ for all $t \geq \tau$. It is clear that η together with ε establishes uniform stability of the origin in $\mathfrak{V}_x$. A similar proof holds with $\mathfrak{V}_x$ and $\mathfrak{V}_y$ interchanged.

7. The verification of uniform regularity is generally by no means simple. The following property characterizes a type of transformation which will cover many future requirements.

(7.1) *Let T be given by a relation $y = A(t)\,x + a(x;t)$ where for $t \geq \tau$, A^{-1} exists and A and A^{-1} are continuous and bounded, and where furthermore the $a_i(x;t)$ are power series in the x_j beginning with terms of degree at least two, with coefficients continuous and uniformly bounded in t for $t \geq \tau$, and convergent in a fixed spheroid $||x|| < B$. Then T is uniformly regular at the origin.*

We find at once from the implicit function theorem that T^{-1} is given by a relation of the same type as T:

$$x = A^{-1}(t)\,y + b(y;t)\,. \tag{7.2}$$

Suppose first that T is linear. Thus T and T^{-1} are given by

$$y = A(t)\,x,\quad x = A^{-1}(t)\,y\,.$$

Remembering that $||x|| = \sup\{|x_h|\}$ and similarly for $||y||$, and setting $||A||, ||A^{-1}|| < \alpha$, we find

$$||y|| = \sup|\Sigma\, a_{ij}\, x_j| \leq n\alpha\, ||x||\,.$$

Similarly $||x|| \leq n\alpha\, ||y||$. Hence

$$\frac{1}{n\alpha} \leq \frac{||x||}{||y||} \leq n\alpha\,.$$

Hence $||y|| \geq \varepsilon$ implies $||x|| \geq \varepsilon/n\alpha$ and *vice versa*. Hence T is uniformly regular at the origin.

Consider now a general T of the type under discussion. Let $r = ||x||$ and set $u = x/r$. Then

$$y = r\{A(t)\,u + ra^*\}$$

where a^* is bounded. Therefore given any small positive ξ, say $< 1/2$, we may choose an $R > 0$ such that if $r < R$ then $||y|| \leq (1 + \xi)\,||Ax||$. Moreover we may choose R so that it be similarly related to T^{-1}. Hence assuming $||x||, ||y|| < R$, we will have this time

$$\frac{1}{na(1+\xi)} \leq \frac{||x||}{||y||} \leq na(1+\xi)$$

and the conclusion is the same as before. This proves (7.1).

§ 5. Stability of Trajectories

8. Let C be a class of trajectories say of the system (2.1) and let Γ_0: $x^0(t)$ be an element of C. We say that Γ_0 is *stable relative to* C whenever given $\varepsilon > 0$ and $t_0 \geq \tau$, there exists an $\eta(\varepsilon) > 0$ such that if Γ_1: $x^1(t) \in C$ and $||x^1(t_0) - x^0(t_0)|| < \eta$, then $||x^1(t) - x^0(t)|| < \varepsilon$ for all $t \geq t_0$. If $x^1(t) \to x^0(t)$ as $t \to +\infty$ then the stability is said to be *asymptotic*.

If the only class for which Γ_0 is stable is the empty one then Γ_0 is said to be *unstable*. If Γ_0 is stable [asymptotically] relative to the class of *all* trajectories, it is said to be [asymptotically] stable. A trajectory which is neither stable nor unstable is known as *conditionally stable*. It is then stable [asymptotically] relative to a true subclass of the class of all trajectories.

9. *Orbital stability*. Often the type of stability defined proves too stringent, or at all events one may only establish a weaker type, orbital stability. Roughly speaking it asserts that if a trajectory Γ_1 passes closely enough to Γ_0 at a certain time, it remains quite close to it ever after. More explicitly Γ_0 is orbitally stable relative to a class C of trajectories whenever given $\varepsilon > 0$ there is an $\eta(\varepsilon)$ and a $\tau(\varepsilon)$ such that if Γ_1 passes at time τ through $\mathfrak{S}(\Gamma_0, \eta)$ (nearer than η to Γ_0) then Γ_1 remains within $\mathfrak{S}(\Gamma_0, \varepsilon)$ for $t > \tau$. The other types of orbital stability are defined in the obvious way.

The following example illustrates the difference between ordinary and orbital stability. Let $\{\Gamma_\lambda\}$ be a family of closed trajectories depending continuously upon a parameter λ. That is to say the solution $x(t, \lambda)$ corresponding to λ is a continuous function of λ. Then orbital stability is here a consequence of continuity as to λ. Let, however, the period $T(\lambda)$ of $x(t, \lambda)$ vary with λ, a case easily realized. Then no matter how small ν, it is not possible to maintain $||x(t, \lambda_0 + \nu) - x(t, \lambda_0)||$ arbitrarily small since in time it will come arbitrarily near to $||x(t^0 + kT(\lambda_0), \lambda_0) - x(t^0, \lambda_0)||$, where k is any preassigned number between 0 and 1.

§ 6. Stability of Mappings

10. There is a parallel concept of stability for a mapping F of a space R into itself and a fixed point P of F. The stability is relative to the iterates $F, F^2, \ldots$ of F. Thus P is said to be stable relative to F whenever for every neighborhood U of P in R there is another $V \subset U$ such that if a point $Q \in V$ then for n above a certain N the points $F^n Q \in U$. If no such V exists then P is unstable. If there is a subspace S of R such that $P \in S$, that $FS \subset S$ and that P is stable relative to F confined to S, then P is conditionally stable. As an example let a Euclidean plane referred to coordinates x, y undergo the similitude $F: x' = ax, y' = by$ with the origin as fixed point. Then the origin is:

stable if $|a| < 1, |b| < 1$;
unstable if $|a| > 1, |b| > 1$;
conditionally stable if $|a| < 1, |b| > 1$ or the other way around.

The stable sets in the third case are respectively the x and the y axes.

§ 7. Further Definitions of Stability (Antosiewicz)

11. Returning to the system (2.1) suppose that $X(0; t) = 0$ for $t \geq \tau$ so that the origin is a critical point. We now state two refinements of the basic definitions of asymptotic stability as given in (2.1).

(11.1) We shall say that the origin is equi-asymptotically stable (Massera [2]) if it is stable in Liapunov's sense and if, given any

$t_0 \geqq \tau$, there is an $\eta(t_0) > 0$ such that whenever $\|x_0\| < \eta$ then $x(t) \to 0$ as $t \to +\infty$ uniformly in x_0.

(11.2) We shall say that the origin is exponentially asymptotically stable (Malkin [1]) if there is a $\lambda > 0$ and, given any $\varepsilon > 0$, a $\delta(\varepsilon) > 0$ such that whenever $\|x_0\| < \delta$, $t_0 \geqq \tau$ then $\|x(t)\| \leqq \varepsilon \exp[-\lambda(t - t_0)]$ for all $t \geqq t_0$.

Exponential asymptotic stability is obviously the strongest of all the types of stability that we have introduced. It is clear, moreover, that equi-asymptotic stability is implied by uniform asymptotic stability and implies, in turn, ordinary asymptotic stability. However, it does not in general imply uniform (non-asymptotic) stability, not even in the simplest case of a single first order linear equation.

Linear homogeneous systems have the following interesting property:

(11.3) THEOREM. *If the origin of a system*

$$\frac{dx}{dt} = P(t)\,x \tag{11.4}$$

where $P(t)$ is continuous for $t \geqq \tau$, is asymptotically stable then it is equi-asymptotically stable; if it is uniformly asymptotically stable then it is exponentially asymptotically stable.

The first part of the assertion follows trivially from the fact that any solution of (11.4) has the form

$$x(t) = X(t)\,X^{-1}(t_0)\,x_0 \tag{11.6}$$

where $X(t)$ is a (non-singular) solution of the matrix equation associated with the system (11.4).

As to the second part, recall the remark made at the end of the proof of (3.3) which implies that $\|X(t)\,X^{-1}(t_0)\| < M$ for $t \geqq t_0 \geqq \tau$ where M is a constant independent of t_0. By the uniform asymptotic stability there is a $T > \tau$ such that $\|X(t)X^{-1}(t_0)\| < 1/2$ for $t \geqq t_0 + T$. Hence we find that $\|X(t)\,X^{-1}(t_0) < 2^{-m}M$ for $t \geqq t_0 + mT$ where m is a positive integer, and this in turn implies that $\|X(t)X^{-1}(t_0)\| < 2M \exp[-\lambda(t - t_0)]$ for all $t \geqq t_0 \geqq \tau$ where $\lambda = \log(2/T)$. The proof is now complete.

In the case of non-linear systems, uniform asymptotic stability is not, in general, equivalent with exponential asymptotic stability as is shown by the single equation $dx/dt = -x^3$.

CHAPTER V

The Differential Equation

$$dx/dt = Px + q(x; t)$$

$$(P \text{ a constant matrix; } q(0; t) = 0)$$

Consider a general system

$$(\alpha) \qquad dx/dt = X(x; t)$$

with a critical point at O. Upon taking O as the origin we will have $X(0; t) = 0$ say for $t \geqq \tau$. It may be possible to write

$$(\beta) \qquad X(x; t) = X_1(x; t) + X_2(x; t)$$

where in some sense for x small and $t \geqq \tau$, X_2 is small in comparison with X_1. For instance X could be a power series in x with coefficients bounded in t beginning with terms of degree m and X_1 would consist of the terms of degree m of the series. The system

$$(\gamma) \qquad dx/dt = X_1(x; t)$$

is known as the *first approximation* to the system (α).

One may well expect that the behavior of the solutions of the first approximation near the origin would tell a great deal regarding the same for the given system. However in this general form very little is known concerning this situation. Ample information is available only when X_1 is linear. In this chapter and the next we study the behavior at the origin of a so called *quasi-linear* system

$$(\delta) \qquad dx/dt = Px + q(x; t)$$

where x, q are n-vectors, P is a constant *non-critical* matrix (no characteristic root zero or pure complex), $q(0; t) = 0$ and q is, in some sense, small in comparison with x. The first approximation

$$(\varepsilon) \qquad dx/dt = Px$$

is linear with constant coefficients, and so it is as simple as possible. We will be especially interested in the stability behavior. In the very general case to be discussed the stability of the given system will be governed by that of (ε).

The stability pattern to be encountered was set by Poincaré and Liapunov. However they dealt primarily with analytic systems. Analyticity restrictions were removed by various authors and notably by Perron. The treatment of the non-analytical case which we give rests upon Bellman [2] with an interesting improvement due to Levinson [3].

The first two sections take up the non-analytical case. In §§ 3, 4 we discuss the analytical case which is continued in the next chapter. In particular § 4 is devoted to a fundamental expansion theorem due to Liapunov, which leads directly to appropriate stability properties.

References: Bellman [3]; Coddington-Levinson [1]; Lefschetz [2]; Liapunov [1]; Picard [1]; Poincaré [1, 3, 4].

§ 1. General Remarks

1. We consider then a system

$$(1.1) \qquad dx/dt = Px + q(x; t)$$

where x and q are n-vectors, P is a constant matrix, $q(0; t) = 0$ for $t \geqq \tau$, so that the origin is a critical point, and q is continuous in a set $\Omega(A, \tau) : ||x|| < A, t \geqq \tau$ of $\mathfrak{V}_x \times \mathfrak{T}$. We also designate by $\Omega(A)$ the region $||x|| < A$ of $\mathfrak{V}_x$. To comply in part with the exigencies of the existence theorem (II, 2.1) $q(x; t)$ is assumed to satisfy a Lipschitz condition *uniformly* in t for $t \geqq \tau$ in $\Omega(A)$. That is to say for any pair of points x, x' of $\Omega(A)$ and any $t \geqq \tau$ we have

$$(1.2) \qquad ||q(x; t) - q(x'; t)|| \leqq a\,||x - x'||$$

where a is a positive constant. This implies furthermore for $x' = 0$

$$(1.3) \qquad \|q(x;t)\| \leq a\,\|x\|,\ x \in \Omega(A)\,.$$

We may affirm then that the trajectories exist and are unique in $\Omega(A, \tau)$.

At times one will impose upon q other "smallness" conditions. They are best expressed in the well known O, o notations. Let us recall that:

$f(z) = O(z)$ means that there is a positive a such that $|f(z)/z| < a$ for z sufficiently small;

$f(z) = o(z)$ means that $f(z)/z \to 0$ with z.

A matrix whose characteristic roots all have negative real parts is known as *stable*.

2. Our primary concern will be the nature and stability properties of the solutions in the vicinity of the origin. Here an important role will be played by the first approximation

$$(2.1) \qquad dx/dt = Px\,.$$

Roughly speaking its stability properties will be those of the solution of the full system (1.1). A good deal will then depend upon the nature of the characteristic roots $\lambda_1, \ldots, \lambda_n$ of the matrix P. The coordinates are assumed so chosen that the matrix P is in normal form in no sense a restriction. It may mean that there are pairs of complex coordinates x_h, x_h^*, the real points corresponding to $x_h^* = \bar{x}_h$. Accordingly there are also pairs of components q_h, q_h^* of q behaving in accordance with (I, 13.10).

§ 2. The General Non-Analytic System

3. We will refer to the system (1.1) satisfying the Lipschitz condition (1.2) as a *basic* system. Our first proposition, in whose proof we follow Levinson [3] is:

(3.1) STABILITY THEOREM. *Let a basic system satisfy* (1.3) *with a constant* α *which* $\to 0$ *with* A. (*Thus* $q\,(x; t) = o(\|x\|)$ *uniformly in* t *for* $t \geqq \tau$.) *Then if* P *is stable the system is asymptotically stable at the origin.*

4. As regards the theorem it is clear that one may decrease A and augment τ. Keeping both fixed for the present let ξ be

such that $||\xi|| < A$ and let $x(t)$ be the solution such that $x(\tau) = \xi$. We will allow t to vary beyond τ only so that for $\tau \leq t' \leq t$ we have $||x(t')|| \leq A$. Under the circumstances according to (III, 10.1)

$$(4.1) \quad x(t) = Y(t - \tau) x(\tau) + \int_\tau^t Y(t - t') q(x(t'); t') dt'$$

where $t \geq \tau$ and $Y(t) = e^{Pt}$. Since P is in normal form every term of $Y(t)$ is of the form $\beta\, t^r e^{\lambda_h t}$. Let $0 < \lambda < \inf\{Re(-\lambda_h)\}$. Thus

$$\beta\, t^r e^{\lambda_h t} = (\beta t^r e^{(\lambda_h + \lambda)t})\, e^{-\lambda t}.$$

Since the parenthesis is bounded there is a $\gamma > 0$ such that

$$(4.2) \qquad\qquad ||Y(t)|| < \gamma e^{-\lambda t}.$$

We choose now A small enough to have $\alpha < \lambda/\gamma$. Thus if $\sigma = \alpha\gamma$ then $0 < \sigma < \lambda$. Now from (1.3) and (4.2) follows:

$$(4.3) \quad ||x(t)|| \leq \gamma e^{-\lambda(t-\tau)} ||x(\tau)|| + \gamma\alpha e^{-\lambda t} \int_\tau^t e^{\lambda t'} \cdot ||x(t')|| dt'.$$

Hence if $u(t) = e^{\lambda t} \cdot x(t)$ we find

$$||u(t)|| \leq \gamma ||u(\tau)|| + \sigma \int_\tau^t ||u(t')|| dt'.$$

Applying now (II, 9.3) we obtain

$$||u(t)|| \leq \gamma ||u(\tau)|| \cdot e^{\sigma(t-\tau)}$$

and hence

$$(4.4) \qquad\qquad ||x(t)|| \leq \gamma ||x(\tau)|| e^{-(\lambda-\sigma)(t-\tau)}.$$

Let now $0 < \varepsilon < A$ and let $0 < \eta < \varepsilon/\gamma$, ε. From (4.4) follows that if $||x(\tau)|| < \eta$, then $||x(t)|| < \varepsilon < A$ for all $t \geq \tau$. Thus the range of t is $\tau \leq t < +\infty$. Moreover the origin is stable.

From the same relation (4.4) follows then that $x(t) \to 0$ as $t \to +\infty$ and so the stability is asymptotic, thus proving our theorem.

5. While the proof of the stability theorem has been direct enough it does not embody any constructive indication regarding

the solution $x(t)$ near the origin. Let us view a tentative method of successive approximation. Let $x^0(t)$ be the solution of the first approximation (2.1) such that $x^0(\tau) = \xi$, a preassigned vector. Then define recursively

$$x^{n+1}(t) = x^0(t) + \int_\tau^t Y(t - t')\, q(x^n(t'); t')\, dt'. \tag{5.1}$$

We shall now prove with Bellman [3]:

(5.2) THEOREM. *Let the characteristic roots of a basic system all have negative real parts. Suppose also that uniformly in t for* $t \geq \tau$ *and* $||x||, ||x'|| \leq A$:

$$||q(x; t) - q(x'; t)|| = O(||x||) \cdot ||x - x'||. \tag{5.2a}$$

Then for $||\xi||$ *small enough the successive approximations converge to a solution* $x(t)$ *of the system such that* $x(\tau) = \xi$. *This solution is thus a general solution.*

It is clear that (5.2a) is stronger in a suitable $\Omega(A)$ than a mere Lipschitz condition. Furthermore it implies

(5.2b) *For* $||x||$ *small enough* $||q(x; t)|| \leq \delta\, ||x||$,

where $\delta \to 0$ *with* $||x||$ *and is independent of t for* $t \geq \tau$.

This may also be expressed more simply as:

(5.2c) $q(x; t) = o(||x||)$, *uniformly in t for* $t \geq \tau$.

6. Passing to the proof of (5.2) let $0 < \mu < \lambda < \inf\{Re(-\lambda_h)\}$. The general solution of the first approximation (2.1) is a linear combination of the vectors represented by the columns of e^{Pt}. The terms of these vectors are each of the form

$$Bt^r e^{\lambda_h t} = B(t^r e^{(\lambda_h + \mu)t})\, e^{-\mu t}.$$

Since $Re(\lambda_h + \mu) < 0$, the parenthesis is bounded. Since $B \to 0$ with ξ there is a positive constant β depending solely on $||\xi||$ and $\to 0$ with ξ, such that

$$||x^0(t)|| < \beta e^{-\mu t}, \quad t \geq \tau. \tag{6.1}$$

We propose to prove that for every m

$$(6.2)_m \qquad ||x^m(t)|| < 2\beta e^{-\mu t}, \ t \geqq \tau.$$

Since $(6.2)_o$ holds we assume $(6.2)_m$ and prove $(6.2)_{m+1}$. We find at once from (5.2b):

$$(6.3)_m \qquad ||q(x^m(t); t)|| < 2\beta\delta e^{-\mu t}, \ t \geqq \tau.$$

Then remembering (4.2), (5.1) and $(6.3)_m$ we have in succession:

$$||x^{m+1}(t)|| \leqq ||x^o(t)|| + \int_\tau^t ||Y(t-t')|| \cdot ||q((x^m(t'); t')|| dt'$$

$$\leqq \beta e^{-\mu t} + 2n\beta\gamma\delta \int_\tau^t e^{-\lambda(t-t')} \cdot e^{-\mu t'} dt'$$

$$\leqq \beta e^{-\mu t}(1 + 2n\gamma\delta/(\lambda - \mu)) < 2\beta e^{-\mu t},$$

for A, hence δ small enough. This proves $(6.3)_m$ for all m.

We must now show that $x^m(t)$ converges to a solution of (1.1). We have from (5.2a):

$$||q(x^{m+1}; t) - q(x^m; t)|| \leqq \epsilon ||x^{m+1} - x^m||$$

where $\epsilon \to 0$ with β and hence with ξ. Thus·

$$||x^{m+1}(t) - x^m(t)|| \leqq \int_\tau^t ||Y(t-t'|| \cdot ||\{q(x^m(t'); t')$$

$$-q(x^{m-1}(t'); t')\}|| dt'$$

$$< \epsilon\gamma e^{-\lambda t} \int_\tau^t e^{\lambda t'} ||x^m(t') - x^{m-1}(t')|| dt'.$$

Hence in succession

$$\max ||x^{m+1}(t) - x^m(t)|| < \left(\frac{n\epsilon\gamma}{\lambda}\right) \max ||x^m(t) - x^{m-1}(t)||,$$

$$||x^{m+1}(t) - x^m(t)|| < \left(\frac{n\epsilon\gamma}{\lambda}\right)^m \max ||x^1(t) - x^o(t)||.$$

Upon choosing ξ small enough one may assume ϵ so small that $n\epsilon\gamma/\lambda < 1$. Hence the series $\sum (x^{m+1} - x^m)$ will then converge uniformly for $\tau \leqq t$. Hence $x^m(t) \to x(t)$, a solution of (1.1) as $n \to +\infty$. This completes the proof of Theorem (5.2).

7. The result just obtained may be generalized. Following again Bellman [3] we have:

(7.1) Generalized Stability Theorem. *Let the basic system* (1.1) *possess k characteristic roots $\lambda_1, \ldots, \lambda_k$ with negative real parts and the rest with positive real parts, and let it satisfy again uniformly in t for $t \geqq \tau$, and $||x||, ||x'|| < A$:*

$$||q(x;t) - q(x';t)|| = ||x - x'|| \cdot o(||x|| + ||x'||).$$

Then a suitable process of successive approximation yields a solution depending continuously on k parameters and which is asymptotically stable at the origin.

Here $P = \text{diag}(P_1, P_2)$ where P_1 is of order k and has the characteristic roots $\lambda_1, \ldots, \lambda_k$. Correspondingly $Y(t) = \text{diag}(Y_1(t), Y_2(t))$ where Y_1 is of order k. Let

$$Z_1(t) = \begin{pmatrix} Y_1, & 0 \\ 0, & 0 \end{pmatrix}, \quad Z_2(t) = \begin{pmatrix} 0, & 0 \\ 0, & Y_2 \end{pmatrix}$$

where Z_1, Z_2 are of order n. Thus $Y = Z_1 + Z_2$.

Suppose now that $x(t)$ satisfies

$$(7.2) \qquad x(t) = x^0(t) + \int_\tau^t Z_1(t - t')\, q(x(t'); t')\, dt' - \int_t^{+\infty} Z_2(t - t')\, q(x(t'); t')\, dt',$$

where $x^0(t)$ is any solution of the first approximation (2.1). If it is shown that for τ large enough $x(t)$ remains in $\Omega(A, \tau)$ then $||q||$ will be bounded. Since the second integral has components which are sums of terms of the form

$$a \int_t^{+\infty} (t - t')^r e^{\lambda(t-t')} q_j(x(t'); t')\, dt'$$

where $Re\, \lambda > 0$, it is convergent. Hence $x(t)$ represents then a function which satisfies (1.1).

We will set up a system of successive approximations $x^n(t)$ based on (7.2) which will be in $\Omega(A, \tau)$ throughout and thus will converge to an appropriate solution of (1.1).

The first k columns of $Z_1(t)$ represent solutions $x^{01}, \ldots, x^{0k}$ of (2.1) which $\to 0$ as $t \to +\infty$ and we take

$$(7.3) \qquad x^o(t) = d_1 x^{o1} + \cdots + d_k x^{ok}.$$

Thus $x^o(t) \to 0$ as $t \to +\infty$.

Notice also that if in the notations of (6)

$$(7.4) \qquad ||x(t)|| \leqq 2\beta e^{-\mu t},\ t \geqq \tau$$

then also

$$(7.5) \qquad ||q(x(t);t)|| < 2\beta\delta e^{-\mu t},\ t \geqq \tau.$$

Since the λ_h, $h \leqq k$ have negative real parts and the λ_{k+h} have positive real parts we have in fact as for (4.2):

$$(7.6) \qquad ||Z_1(t)|| < \gamma_1 e^{-\lambda t},\ t > 0,\ \lambda > 0,$$

$$(7.7) \qquad ||Z_2(t)|| < \gamma_2 e^{\varrho t},\ t < 0,\ \varrho > 0.$$

Our successive approximation reads now

$$(7.8) \qquad x^{m+1}(t) = x^o(t) + \int_\tau^t Z_1(t-t')\, q(x^m(t');t')\, dt' - \int_t^{+\infty} Z_2(t-t')\, q(x^m(t');t')\, dt'.$$

The treatment is now the same as in (6) and details may be omitted. One shows first that if the d_h are small $(7.4)_m$ holds: i.e. if it is satisfied by x^m then $(7.4)_{m+1}$ holds. Hence if one chooses the d_h small, thus making $(7.4)_o$ hold, (7.4) will be true for all m. Then the series $\sum(x^{m+1} - x^m)$ is shown to converge uniformly. The fact that the second integral in (7.8) has an infinite upper limit calls for a special argument in proving that $x(t) = \lim x^m(t)$ solves (1.1). This is done as follows. Take some positive $t_1 > t$. Then

$$(7.9) \qquad x^{m+1}(t) - x^o(t) - \int_\tau^t Z_1(t-t')\, q(x^m(t');t')\, dt' + \int_t^{t_1} Z_2(t-t')\, q(x^m(t');t')\, dt' = -\int_{t_1}^{+\infty} Z_2(t-t')\, q(x^m(t');t')dt'.$$

Now

$$
\begin{aligned}
&\left\| \int_{t_1}^{+\infty} Z_2(t-t')q(x^m(t');t')dt' \right\| \\
&< 2n\beta\,\delta\,\gamma_2 e^{\varrho t} \int_{t_1}^{+\infty} e^{-\varrho t'} e^{-\mu t'}\,dt' \\
&< \frac{2n\beta\,\delta\,\gamma_2}{\varrho+\mu} e^{-\mu t_1} = \varepsilon\, e^{-\mu t_1}.
\end{aligned}
$$

Since every term at the left in (7.9) is continuous one may replace x^m, x^{m+1} by their common limit $x(t)$ and the left hand side tends to the corresponding limit. By the result just obtained then

$$
\Big\| x(t) - x^0(t) - \int_{\tau}^{t} Z_1(t-t')\,q(x(t');t')\,dt' + \int_{t}^{t_1} Z_2(t-t')\,q(x(t');t')\,dt' \Big\| < \varepsilon\, e^{-\mu t_1}.
$$

As $t_1 \to +\infty$ this inequality shows that $x(t) = \lim x^m(t)$ satisfies (7.2) and hence also (1.1). Since the solution thus obtained satisfies (7.4) and depends upon the k arbitrary parameters d_h, it is asymptotically stable relative to the origin and so (7.1) is proved.

When the system is autonomous one may give somewhat more precision to our stability theorems. Stated in full we have:

(7.10) STABILITY THEOREM FOR AUTONOMOUS SYSTEMS. *Let the system* (1.1) *be autonomous*: $q = q(x)$, *where* q *still satisfies* (5.2a). *Then if there are exactly* k *characteristic roots with negative [positive] real parts there is a family of paths depending continuously on* k *parameters which is asymptotically stable [which is unstable]. In particular if all the characteristic roots have negative [positive] real parts the system is asymptotically stable [is unstable].*

This theorem will serve in a sense as a pattern for many cases later.

In view of (7.1) it is only necessary to dispose of the case of k roots with positive real parts. Let $t' = -t$. Then the system

$$dx/dt' = -Px - q(x)$$

falls under (7.1) with exactly k characteristic roots with negative real parts. Hence it has a solution $x(a_1, \ldots, a_k, t')$ depending continuously upon the parameter $a = (a_1, \ldots, a_k)$ and which is asymptotically stable. Thus along the associated path Γ_a the point $M(x(t'))$ tends to the origin as $t' \to +\infty$. Since Γ_a is a path of (1.1) also, but as such described in the opposite sense, as $t \to +\infty$ the point M leaves the origin. Hence the origin is unstable and (7.10) follows.

§ 3. Analytic Systems: Generalities

8. Analytic systems of the general type of the present chapter have been investigated by Poincaré [1], [3], Picard [1], III, Ch. 1, and Dulac [1], by an indirect method introduced by Poincaré and based on certain partial differential equations. This procedure will be described in (VI, § 1). Liapunov [1], on the other hand attacked the problem directly and obtained explicit series for the solutions near a critical point. This is what we shall do in the present and next sections.

By one or the other procedure one may treat at practically no cost, as we shall do, complex systems: complex vector x, real time.

We shall use here the notations of (I, 14.5): $[x]_m$ denotes a vector whose components are power series in those of x beginning with terms of degree at least m. We assume furthermore that the series represent holomorphic functions in the set $\Omega(A)$ and the series as well as their coefficients are uniformly bounded in the set for $t \geq \tau$.

Consider then a system

$$(8.1) \qquad dx/dt = Px + q(x; t), \; q = [x]_2,$$

with x complex and t real. As before everything will revolve around a comparison of the solutions of (8.1) with the general solution of the first approximation.

$$(8.2) \qquad du/dt = Pu.$$

If $\{u^h(t)\}$ is a base for the solutions of (8.2) the general solution of (8.2) is

$$(8.3) \qquad u(t) = \Sigma\, a_h u^h(t)$$

where the a_h are constants.

Regarding the characteristic roots $\lambda_1, \ldots, \lambda_n$ of the matrix P two assumptions will have to be made. The first is the one to which we are already accustomed:

(8.4) *The characteristic roots all have negative real parts.*

The second and new assumption, really imposed by the mechanism governing the construction of the series, is:

(8.5) *The distinct characteristic roots satisfy no relation with integers* m_h

$$\lambda_j = \Sigma\, m_h \lambda_h,\ m_h \geqq 0,\ \Sigma\, m_h > 1\,.$$

A set of numbers $\{\lambda_h\}$ having the preceding two properties will be referred to as *well behaved.*

§ 4. The Expansion Theorem of Liapunov

9. We may now state the basic theorem to be proved and which is due to Liapunov ([1], p. 246). In the statement u will stand for the general solution (8.3) of the first approximation, a for the vector of the parameters a_j in (8.3), N for the number of distinct λ_j, and as before $\Omega(A)$ for the set $||x|| \leqq A$.

(9.1) THEOREM. *Let the system* (8.1) *be as described, with the set of characteristic numbers* $\{\lambda_j\}$ *well behaved. Then the system has a general solution given by a series*

$$(9.1\text{a}) \qquad x(t) = \sum_{m=1}^{+\infty} Z^m(t; a)\,,$$

where the $Z^m(t; a)$ *have the following properties:*

(a) $Z^1(t; a) = u(t)$;

(b) $Z^m(t; a) = \displaystyle\sum_{m_1 + \ldots + m_N = m} X^{m_1, \ldots, m_N}(t; a) \exp i \sum_{m=1}^{N} m_j \lambda_j\,,$

where

(c) *the* $X^{m_1, \ldots, m_N}$ *are vectors whose components are polynomials in* t *whose coefficients are forms of degree* m *in the* a_j*, the coefficients of these forms being continuous and bounded functions of* t *for* $t \geqq \tau$;

(d) *there exist positive numbers* ϱ, τ *such that the series* (9.1a) *converges absolutely and uniformly to a point in* $\Omega(A)$ *for* $\|a\| \leq \varrho$ *and* $t \geq \tau$.

If we write (m) to denote the partition of m into N summands $m_1, \ldots, m_N$, and accordingly denote by $X^{(m)}$ the vector $X^{m_1, \ldots, m_N}$, (9.1a) assumes the simpler form

$$x(t) = \sum_{m} \sum_{(m)} X^{(m)}(t; a) \exp t \sum m_j \lambda_j. \tag{9.1b}$$

The method of proof, essentially due to Liapunov, is as follows. A series of vectors

$$x(t; \varepsilon) = \varepsilon x^1(t) + \varepsilon^2 x^2(t) + \ldots \tag{9.2}$$

is formally substituted in (8.1) and the powers of ε are identified. This yields a recurrent collection of systems for the x^m. Suitable solutions put together yield a series

$$x(t) = x^1(t) + x^2(t) + \ldots \tag{9.3}$$

which is shown to converge to an actual solution of (8.1). The role of ε is thus merely to assign a weight to the x^m. Liapunov dispensed with ε altogether and identified directly the terms of equal weight.

Before proceeding with the proof it will be convenient to discuss a preliminary lemma due to Liapunov. For the present all letters designate ordinary complex scalars and not vectors.

Consider the scalar function of a scalar variable u

$$f(u) = (1 - u/A)^{-n} - 1 - nu/A = \binom{n+1}{2}\left(\frac{u}{A}\right)^2 + \ldots,$$

where A is a positive number, and in the power series let u be replaced by the formal power series in ε

$$y = \varepsilon y_1 + \varepsilon^2 y_2 + \ldots.$$

The result ordered as to powers of ε is

$$\varphi(y) = \varepsilon^2 \varphi_2(y_1) + \varepsilon^3 \varphi_3(y_1, y_2) + \ldots,$$

where the φ_h are polynomials. The substitution $y_m \to k^m y_m$, k any scalar, yields the same result as $\varepsilon \to k\,\varepsilon$. Hence, the homogeneity relation

$$(9.4)\quad \varphi_m(ky_1, k^2y_2, \ldots, k^{m-1}y_{m-1}) = k^m \varphi_m(y_1, \ldots, y_{m-1}).$$

Choose now $z_1, z_2, \ldots,$ as follows: z_1 is arbitrary, and

$$z_m = M\,\varphi_m(z_1, \ldots, z_{m-1}),\ m > 1,$$

where M is a fixed positive constant. Then

(9.5) LEMMA. *There is a positive constant $\gamma(M, A)$ depending solely upon M and A such that if $|z_1| < \gamma$ then the series*

$$z = z_1 + z_2 + \ldots$$

is absolutely convergent. (Liapunov [1] p. 219).

Consider the relation

$$(9.6)\qquad F(z_1, z) = -z + z_1 + Mf(z) = 0.$$

Since $F(0, 0) = 0$, $(\partial F/\partial z)_{0,0} = -1$, by the implicit function theorem for analytic functions there exists a unique function $z^*(z_1)$ holomorphic in z_1 at the origin, satisfying (9.6) and such that $z^*(0) = 0$. About $z_1 = 0$ this function may be represented by a power series

$$z^*(z_1) = z^*_1 z_1 + \ldots + z^*_p z_1^p + \ldots$$

whose radius of convergency $\gamma(M, A) > 0$. To prove the lemma it is sufficient to show that

$$(9.7)_m\qquad z_m = z^*_m z_1^m.$$

We shall prove $(9.7)_m$ by induction. Clearly $(9.7)_1$ holds. Suppose that $(9.7)_h$ holds for $h < m$. Then

$$z^*_m z_1^m = z_1^m M\,\varphi_m(z^*_1, \ldots, z^*_{m-1})$$

$= M\,\varphi_m(z^*_1 z_1, \ldots, z^*_{m-1} z_1^{m-1})$, (by the homogenity property)

$= M\,\varphi_m(z_1, \ldots, z_{m-1})$ (by induction)

$= z_m$ (by definition).

Thus $(9.7)_m$ holds. Hence

$$z^*(z_1) = z = z_1 + z_2 + \dots$$

converges absolutely when $|z_1| < \gamma(M, A)$, and this is the lemma.

10. Before plunging into the main body of the proof let us first observe that if (9.1) actually does represent a solution of (8.1) then it must be a general solution since it is a power series in the n independent parameters $a_1, \dots, a_n$. Thus it is not necessary to give further attention to the "general solution" part of the theorem.

Substituting now the series (9.2) in (8.1), the identification of like powers of ε yields the recurrent system:

$$(10.1)_1 \qquad dx^1/dt = Px^1,$$

$$(10.1)_m \quad dx^m/dt = Px^m + r^m(t; x^1, \dots, x^{m-1}), \; m > 1,$$

where the components of r^m are polynomials in those of the x^p, $p < m$, with coefficients continuous and uniformly bounded in t for $t \geq \tau$.

The solution does give rise to a process of successive approximations. It is to be proved that Σx^m converges and that its limit is a solution of (8.1).

Since the choice of coordinates is manifestly immaterial we shall assume them such that P is in normal form. We observe then that the terms of $Y = e^{Pt}$ are all of the form $\beta t^r e^{\lambda_h t}$.

It is clear that the components r_h^m of r^m are polynomials in those of the x^{m-j} with coefficients which are bounded functions of t for $t \geq \tau$. Furthermore if one makes in $x(t; \varepsilon)$ the substitution $x^s(t) \to k^s x^s(t)$ the result is the same as replacing ε by $k\varepsilon$. Hence the important homogeneity relation similar to (9.4):

$$(10.2) \quad r^m(t; kx^1, \dots, k^{m-1}x^{m-1}) = k^m r^m(t; x^1, \dots, x^{m-1}).$$

Since $(10.1)_1$ is merely (8.2), its general solution is $x_1 = u(t)$ given by (8.3). We thus have explicitly

$$(10.3) \quad x_h^1 = (a_{h0} + a_{h1}t + \dots + a_{hs_h} t^{s_h}) e^{\lambda_h t}, \; s_h < n,$$

where the a_{hj} are the same as some of the a_h of (8.3), merely numbered differently. Thus x_h^1 is a product $X_h^{(1)} e^{\lambda_h t}$, $X_h^{(1)}$ as in (9.1c). More generally:

(10.4)$_m$ *The system* (10.1)$_m$ *has a solution of the form*

$$x^m = \sum_{(m)} X^{(m)} (t; a) \exp t \sum m_j \lambda_j \tag{10.4a$_m$}$$

where $X^{(m)}$ *is as described in theorem* (9.1).

Since (10.4)$_1$ holds we assume (10.4)$_p$ for every $p < m$ and prove (10.4)$_m$. Since the x_h^p, $p < m$, as well as the terms of Y all have an $e^{\lambda_h t}$ as factor and the λ_h all have negative real parts, (10.1)$_m$, $m > 1$, has a solution

$$x^m = \int_{+\infty}^{t} Y(t - t') \, r^m (t'; x^1(t'), \ldots, x^{m-1}(t')) \, dt' . \tag{10.5$_m$}$$

Since (10.4)$_p$, $p < m$, holds we have

$$\begin{aligned} & r^m (t; x^1(t), \ldots, x^{m-1}(t)) \\ & = \sum R^{(m)} (t; X^{(1)}, \ldots, X^{(m-1)}) \exp t \sum m_j \lambda_j \end{aligned} \tag{10.6$_m$}$$

where the component $R_h^{(m)}$ of $R^{(m)}$ behaves like r_h^m, with the $X^{(p)}$ (for all partitions of p into ν summands) in place of x^p. In particular $R^{(m)}$ satisfies a homogeneity relation such as (10.2), its weight being m, and p that of $X_s^{(p)}$. The coefficients of these polynomials are continuous and bounded functions of t for $t \geq \tau$.

The terms of $Y(t)$ in the row h are all of the form $y_{hk}(t) e^{\lambda_k t}$ where $y_{hk} = a t^r$. Hence by (10.5)$_m$ and (10.6)$_m$:

$$\begin{aligned} x_h^m = \sum e^{\lambda_k t} \int_{+\infty}^{t} & y_{hk}(t - t') R_k^{(m)}(t'; a) \\ & \times \exp t' (-\lambda_k + \sum m_j \lambda_j) \cdot dt' . \end{aligned} \tag{10.7$_m$}$$

By assumption (8.5) the last parenthesis is never zero. Hence integration yields for x_h^m and hence for x^m a relation (10.4a)$_m$. Thus (10.4)$_m$ holds for all m.

11. We have disposed so far of the formal part of the argument, for we have obtained, formally, a series

$$x = x^1 + x^2 + \ldots \tag{11.1}$$

which satisfies (9.1abc). There remains to prove (9.1d): for suitable ϱ, τ and $||a|| \leq \varrho$, $t \geq \tau$, (11.1) converges absolutely and uniformly to a point of Ω_A. Once that is accomplished, in view of the integral representation (10.5), one shows at once that $x(t)$ given by (11.1) is a solution of (8.1) and Liapunov's theorem (9.1) will have been proved. Thus everything is now reduced to proving (9.1d).

Certain estimates regarding the $|x_h{}^m|$ will have to be made based on (10.4), ..., (10.7). One meets here with the difficulty that as $t \to +\infty$ so do the $x_h{}^{(m)}(a; t)$ and $y_{hk}(\pm t)$ but the exponentials say in (10.4) all $\to 0$. For this reason it will be necessary to "borrow" as it were from the exponentials suitable factors cancelling out the growth of the $X_h{}^{(m)}$ and y_{hk}.

Let $\lambda > 0$ be so chosen that $2\lambda < \inf \{Re(-\lambda_h)\}$ and let $\mu_h = \lambda_h + 2\lambda$.

If we set

$$X^{*(m)}(t; a) = e^{-m\lambda t} X^{(m)}(t; a)\,, \tag{11.2}$$

we will have in place of (10.4) and (10.6):

$$x^m = e^{-m\lambda t} \sum X^{*(m)}(t; a) \exp t \sum m_j \mu_j\,, \tag{11.3}$$

$$\begin{aligned} &r^m(t; x^1(t), \ldots, x^{m-1}(t)) \\ &= e^{-m\lambda t} \sum R^{(m)}(t; X^{*(1)}, \ldots, X^{*(m-1)}) \exp t \sum m_j \mu_j\,. \end{aligned} \tag{11.4}$$

The integral expression (10.7) becomes thus

$$\begin{aligned} x_h{}^m = \sum \int_{+\infty}^{t} e^{(\mu_k - 2\lambda)(t - t')} y_{hk}(t - t') R_k{}^{(m)}(t'; X^{*(1)}, \ldots) \\ \times \exp t'(-m\lambda + \sum m_j \mu_j) \cdot dt'\,. \end{aligned}$$

This last expression may also be written

$$\begin{aligned} x_h{}^m = \sum e^{(\mu_k - 3\lambda) t} \int_{+\infty}^{t} e^{-\lambda(t' - t)} y_{hk}(t - t') \\ \times R_k{}^{(m)}(t'; X^{*(1)}, \ldots) \exp t'\{(3 - m)\lambda - \mu_k + \sum m_j \mu_j\} \cdot dt'\,. \end{aligned} \tag{11.5}$$

12. We shall now proceed with our estimates. Since the $q(x; t)$ (cf. (8.1)) are uniformly bounded in $\Omega(A)$, the $q_i(x; t)$ have an

upper bound M on the set. Thus the $q_i(x; t)$ have on $\Omega(A)$ the common majorante

$$Q(x) = M\{\Pi\left(1 - \frac{x_h}{A}\right)^{-1} - 1 - \frac{1}{A}\Sigma x_h\}.$$

Notice now that in (10.3) the powers of t are all $< n$. Let us assume at all events $\tau > 0$, hence $t > 0$ and define

$$z_h^1 = \varrho\, nt^n\, e^{(\mu_h - 2\lambda)t} = e^{-\lambda t} Z^{(1)}(t)\, e^{\mu_h t},$$

$$Z_h^{(1)}(t) = \varrho\, nt^n e^{-\lambda t}, \varrho > 0.$$

We also notice that for $\tau \geqq 1$, which we assume henceforth,

$$|x_h^1| < |z_h^1|.$$

We will set $\zeta_1 = \sup|z_h^1|$ and form $\zeta_2, \zeta_3, \ldots$, by the recurrence $\zeta_m = M\varphi_m(\zeta_1, \ldots, \zeta_{m-1})$ where φ_m is as in lemma (9.5). According to the lemma

$$\zeta = \zeta_1 + \zeta_2 + \cdots$$

converges provided that $\zeta_1 \leqq \gamma(M, A)$ where γ is the same as in the lemma. The convergence is assured for $\varrho\, n\, \tau^n < \gamma(M, A)$, or for $\varrho < \varrho_0 = \gamma(M, A)/n\,\tau^n$.

Now $y_{hk}(t)$ is of the form $\alpha\, t^r$, $r < n$. Hence

$$d/dt\,(y_{hk}(t)\, e^{-\lambda t}) = y_{hk}(t)\, e^{-\lambda t}(r - \lambda t)/t.$$

The maximum of $y_{hk}(t)\, e^{-\lambda t}$ occurs thus for $t = r/\lambda$ and is $\alpha(r/\lambda)^r e^{-r} < \alpha(n/\lambda)^n$. Let η/n^2 be the largest of these numbers. Thus

$$|y_{hk}(t)\, e^{-\lambda t}| < \eta/n^2.$$

Similarly $Z_h^{(1)}(t)$ reaches its maximum for $t = n/\lambda$ and we will choose $\tau > n/\lambda$. Thus $Z_h^{(1)}(t)$ is monotone decreasing for $t \geqq \tau$ and $\to 0$ as $t \to +\infty$.

We proceed now to form the analogues of the r_h^m with q replaced by Q and the x^p by the z^p, and designate the result by $[r_h^m(z^1, \ldots, z^{m-1})]$, defining z^m inductively by

$$z_h^m = [r_h^m(z^1, \ldots, z^{m-1})].$$

The components z_h^m of z^m are thus all equal. We have now (11.4) with expressions $Z^{(m)}(t)$ and $[R_h^{(m)}(Z^{(1)}, \ldots, Z^{(m-1)})]$ taking the place of $X_h^{*(m)}$ and $R_h^{(m)}(X^{*(1)}, \ldots)$, and find at once from the definitions that

$$|X_h^{*(m)}| < Z_h^{(m)}; \ |R_h^{(m)}(X^{*(1)}, \ldots, X^{*(m-1)})| < [R_h^{(m)}(Z^{(1)}, \ldots, Z^{(m-1)})].$$

We also note that in the sequence $Z^{(1)}, Z^{(2)}, \ldots$, the components of any $Z^{(p)}$ are linear combinations with positive coefficients of the components of its predecessors. Since the components $Z_h^{(1)}$ are monotone decreasing for $t \geqq \tau$, this holds for all the components just mentioned and hence also for $[R_h^{(m)}]$.

We shall now show that for τ above a certain value

$$(12.1)_m \qquad |x_h^m| < e^{-(m-\varepsilon)\lambda t}\zeta_m,$$

where $0 < \varepsilon < 1$. We first extend the symbol $\ll$ to a comparison of coefficients of series in powers of the $e^{\mu_h t}$. Thus (11.5) yields

$$(12.2) \qquad x_h^m \ll \sum_{(m)} e^{(\mu_k - 3\lambda)t} \cdot \eta[R_h^{(m)}(Z^{(1)}, \ldots Z^{(m-1)})] \times \int_{+\infty}^{t} \exp t' \{(3 - m)\lambda - \mu_k + \Sigma m_j \mu_j\} \, dt',$$

where $\ll$ is as in (I, 12) and we have written $[R_h^{(m)}]$ in place of $[R_k^{(m)}]$ since all these components are equal.

Let $\mu' = \inf\{Re - \mu_h\}$, $\mu'' = \sup\{Re - \mu_h\}$. Thus

$$Re\{(3 - m)\lambda - \mu_h + \Sigma m_j \mu_j\} \leqq 3\lambda + \mu'' - (\lambda + \mu')m < -1$$

say for $m > m_0$. Hence assuming that $(12.1)_p$ holds for $p \leqq m_0$, we will have from (12.2) for $m > m_0$:

$$(12.3) \quad |x_h^m| < |e^{-m\lambda t} \cdot \eta \cdot \Sigma[R_h^{(m)}(Z^{(1)}, \ldots, Z^{(m-1)})] \exp t \Sigma m_j \mu_j| = |e^{-m\lambda t} \cdot \eta \cdot [r_h^m(z^1, \ldots, z^{m-1})]| \, |e^{-m\lambda t} \cdot \eta \cdot z_h^m|.$$

Upon replacing z_h^1 by ζ_1, $[r^m]$ reduces to $M\varphi_m(\zeta_1, \ldots, \zeta_{m-1}) = \zeta_m$. Hence $|z_h^m| < \zeta_m$.

Take now $\tau > \dfrac{\log \eta}{\varepsilon\lambda}$. Then $e^{-\varepsilon\lambda t} \cdot \eta < 1$ for $t \geqq \tau$. As a consequence we find from (12.3) that $(12.1)_m$ holds for $m > m_0$ and thus for every m if it holds up to and including m_0.

13. Returning to (11.3) the expressions $X^* \to 0$ as $t \to +\infty$. The number of such expressions for $m \leq m_0$ is finite. Hence for τ large enough and $t \geq \tau$, $m \leq m_0$, we will have

$$\Sigma\, |X_h{}^{*(m)}| < \zeta_m .$$

Since $|\exp t \Sigma\, m_h\, \mu_h| < 1$, $(12.1)_m$ will hold for $m \leq m_0$ also and hence for every m. Since the series $\zeta = \zeta_1 + \zeta_2 + \ldots$ is convergent the series

$$x(t) = x^1 + x^2 + \ldots$$

converges absolutely and uniformly for $t \geq \tau$ and $||a|| \leq \varrho$. On the other hand we have $\zeta_m < \zeta$ and so by $(12.1)_m$:

$$||x^m|| < ne^{-(m-\varepsilon)\lambda t} \cdot \zeta$$

and therefore

$$(13.1) \qquad ||x|| < \frac{n\,\zeta\, e^{-(1-\varepsilon)\lambda t}}{1 - e^{-\lambda t}}.$$

This last expression $\to 0$ as $t \to +\infty$. Hence for τ sufficiently large and $t \geq \tau$ we will have $x \in \Omega(A)$. This completes the proof of Liapunov's theorem.

14. *Complements.* (14.1) *Generalization of the Liapunov expansion theorem.* Let $\{\lambda_1, \ldots, \lambda_k\}$ be a well behaved set of characteristic roots, each being repeated as often as its multiplicity. One may choose now $a_{k+1} = \ldots = a_n = 0$ and also everywhere take the m_k as a partition of m into ν^* parts, where ν^* is the number of distinct roots in the set $\{\lambda_1, \ldots, \lambda_k\}$. The same argument as before yields this time a solution $x(t; a_1, \ldots, a_k)$ depending analytically on k parameters.

(14.2) *Case of autonomous system with distinct characteristic roots with all real parts negative.* All the difficulties in the proof disappear then automatically. The complicated compensation process of (11) becomes unnecessary since the $r_h{}^m, \ldots$ do not contain the time explicitly. A base for the first approximation is now made up of the vectors

$$u^h = (0, \ldots, 0, e^{\lambda_h t}, 0, \ldots, 0)$$

and the general solution of the first approximation is

$$u = (a_1 e^{\lambda_1 t}, \ldots, a_n e^{\lambda_n t}) .$$

One verifies then that

$$r_h{}^m = \Sigma \, \varrho_h{}^{(m)} \, u_1{}^{m_1} \ldots u_n{}^{m_n} .$$

In view of (8.5) the integration (10.5) may be carried out and yields the terms of a convergent series. As a consequence

(14.3) $$x_h = u_h + [u_1, \ldots, u_n]_2, \; h = 1, 2, \ldots, n.$$

One may also write (14.3) in the vector form

(14.4) $$x = u + [u]_2 .$$

This relation represents a topological mapping of a region of $\mathfrak{B}_u$ into one of $\mathfrak{B}_x$ which sends origin into origin and is uniformly regular there (see IV, 5). As a consequence the stability properties of linear systems with constant coefficients (see IV,4) carry over bodily to the present case. This is in agreement with § 2.

(14.5) One may however go much further. Suppose merely that there is a solution of the first approximation of the form say

(14.6) $$u = (a_1 e^{\lambda_1 t}, \ldots, a_k e^{\lambda_k t}, 0, \ldots, 0)$$

where at first one assumes that the λ_k all have negative real parts. The same reasoning will bring out the existence of a solution of the form (14.4) where u is now (14.6) and it will be valid for $t \geq \tau$ and $||a||$ small enough.

Suppose now that the real parts of the $\lambda_1, \ldots, \lambda_k$ are positive. Upon making the change of variable $t = -t'$ we find that if

$$u = (a_1 e^{-\lambda_1 t'}, \ldots, a_k e^{-\lambda_k t'}, 0, \ldots, 0)$$

then

(14.7) $$x = u + [u]_2$$

is a solution valid say for $t' \geq -\tau$, $\tau > 0$ and $||a||$ small enough of the system

$$dx/dt' = -Px + [x]_2 .$$

Hence (14.7) will define a solution of the initial system valid for $t \lesssim \tau$ and $||a||$ sufficiently small.

(14.8) *Stability properties of analytical systems.* The Liapunov analysis through theorem (9.1) or with its extension (14.1) makes it evident that if all [only $k < n$] characteristic roots have negative real parts and satisfy (8.5), i.e. form a well behaved set, then all solutions form an analytical n dimensional [there is an analytical k dimensional] family of asymptotically stable solutions. Similarly with the stability properties (7.1), (7.10) save that now the solutions envisaged are all analytical.

CHAPTER VI

The Differential Equation

$$dx/dt = Px + q(x; t)$$

$$(P \text{ a constant matrix}; \ q(0, t) = 0)$$

(continued)

The general theme of the present chapter is the same as in the preceding. The topics dealt with are: (a) Poincaré's general method for the analytical case; (b) the very powerful stability theorems constituting the second method of Liapunov; (c) a general theorem (Dykhman) for the case when the leading matrix has zero characteristic roots, which rests upon a noteworthy theorem due to Persidskii; (d) Liapunov's stability theorem for the case of a single characteristic root zero.

References: The same as for (V) and in addition Dulac [1]; Dykhman [1]; Krassovskii [1], Malkin [1, 2, 3]; Persidskii [2].

§ 1. The Method of Poincaré

1. We return to a complex autonomous system with t real

$$dx/dt = X(x) \tag{1.1}$$

and assume that all functions are analytic wherever considered. If one finds a sufficiently large collection of manifolds $f(x) =$ const., which are loci of paths, their mutual intersections give rise to the paths. A n.a.s.c. for f to behave as stated is that $df(x(t))/dt = 0$, along any solution $x(t)$ of (1.1). This gives the explicit relation

$$\Sigma X_h \, \partial f/\partial x_h = 0 \tag{1.2}$$

and so one must find the solutions of this partial differential equation. There is a well known existence theorem (theorem of Cauchy-Kovalevsky) which answers the purpose in the regions where $X \neq 0$. However the problem under consideration is precisely to find what happens when $X = 0$. Let us follow then Poincaré's reasoning.

Suppose that $X(0) = 0$ and that X is holomorphic in $\Omega(A)$: $||x|| \leq A$. Write

$$X = Px + X^1, X^1 = [x]_2 \tag{1.3}$$

where P is a constant matrix whose characteristic roots $\lambda_1, \ldots, \lambda_n$ are all distinct, and satisfy (V, 8.5) and in place of (V, 8.4) the following condition, only apparently weaker (see 3.8):

(1.4) *Convexity condition: The smallest closed convex region Π in the complex plane containing all the points λ_h does not contain the origin.*

As before let the coordinate system be so chosen that P is in normal form. Thus $P = \text{diag}(\lambda_1, \ldots, \lambda_n)$ and so

$$X_h = \lambda_h x_h + X_h{}^1, \quad h = 1, 2, \ldots, n.$$

Instead of (1.2) take now the equation

$$\Sigma X_h \, \partial f/\partial x_h = \lambda_1 f. \tag{1.5}$$

We will endeavor to determine a function holomorphic at the origin and satisfying this equation. At all events one must have $f(0) = 0$. Upon substituting a power series for $f(x)$ one finds that in view of (1.4) the coefficients of a given degree p may be determined in terms of those of degree $< p$ and of the coefficient of x_1. The coefficients of $x_2, \ldots, x_n$ are all zero. There is thus determined a formal power series solution of (1.5) and we must prove its convergency.

2. Let us first derive an elementary property of the set $\{\lambda_h\}$. Since Π does not contain the origin it is at a positive distance σ from the origin. Since

$$\nu = \frac{1}{m} \Sigma m_j \lambda_j$$

is the affix of the centroid of the masses m_j located at the λ_j, $\nu \in \Pi$ and hence $|\nu| \geq \sigma$. If λ^* is the largest $|\lambda_h|$ we have then for $m > 1$:

$$(2.1)\quad \gamma = \left|\frac{-\lambda_h + \Sigma\, m_j \lambda_j}{m-1}\right| > \left|\frac{1}{m}\Sigma\, m_j \lambda_j\right| - \frac{\lambda^*}{m} \geq \sigma - \frac{\lambda^*}{m}.$$

Let m_0 be such that for $m > m_0$: $\lambda^*/m < \sigma/2$. Then for $m > m_0$ $\gamma > \sigma/2$. Since according to (V, 8.5) the numerators in (2.1) are never zero γ has a positive lower bound ε_1 for $1 < m \leq m_0$. Thus if ε is the least of ε_1 and $\sigma/2$, γ has the positive lower bound ε for every $m > 1$.

Returning to the function f, it has the form $f = -a\, x_1 + v\,(x)$, $v = [x]_2$, where a is arbitrary. Let us choose $a > 0$. Upon substituting in (1.5) there is obtained for v the equation

$$(2.2)\qquad \Sigma\,(\lambda_h x_h + X_h^1)\, \partial v/\partial x_h - a\, X_1^1 = \lambda_1 v\,.$$

Let M be an upper bound for all the X_h^1 in $\Omega\,(A)$ and set

$$u = \Sigma\, x_h,\ \varphi\,(u) = M\,\{(1 - u/A)^{-1} - 1 - u/A\}\,.$$

One may determine $V\,(x)$ from the relation

$$(2.3)\qquad \Sigma\,(\varepsilon\, x_h - \varphi\,(u))\, \partial V/\partial x_h - a\, \varphi\,(u) = \varepsilon\, V\,,$$

like one may determine $v\,(x)$ formally from (2.2), and it yields for $V\,(x)$ a formal power series whose coefficients are all positive, which begins with terms of degree at least two and is a majorante of the series $v\,(x)$.

Let us endeavor to satisfy (2.3) by a series $V\,(u)$ in powers of u. If this is possible $V\,(u)$ will satisfy the relation

$$(2.4)\qquad (\varepsilon\, u - n\, \varphi\,(u))\, dV/du - \varepsilon\, V = a\, \varphi\,(u)\,.$$

Notice that $\varphi\,(u) = u^2\, \psi\,(u)$, where $\psi\,(0) \neq 0$ and ψ is holomorphic at the origin. Setting now $V = uW$, we find for W an equation

$$(2.5)\qquad dW/du + \beta\,(u)\, W = \gamma\,(u)$$

where β, γ are holomorphic at the origin. If

$$\delta(u) = \int_0^u \beta(u)\,du$$

then by a well known process (2.5) has a solution

$$W = e^{-\delta(u)} \int_0^u e^{\delta(u)} \cdot \gamma(u)\,du .$$

Thus $W(u)$ is holomorphic at the origin and $W(0) = 0$. Hence $V = uW$ is a solution of (2.4) holomorphic at the origin and whose series in powers of u begins with terms of degree at least two. Since the formal series solution $V(x_1 + \ldots + x_n)$ has the same properties and is unique $V(u)$ is that series. This proves the convergence of the majorante, and hence of the formal power series solution, absolutely and uniformly in a suitable $\Omega(A)$.

3. From this point on we follow an argument due to Dulac [1]. To begin with it has been shown that there exists a function which satisfies (1.5) and is of the form $x_1 + [x]_2$. Similarly for every j there is a function, now written z_j such that

$$z_j = x_j + [x]_2 \tag{3.1}$$

and that

$$\Sigma X_h\, \partial z_j/\partial x_h = \lambda_j z_j . \tag{3.2}$$

Now since the Jacobian of the right hand sides in (3.1) is equal to 1 for $x = 0$, one may solve (3.1) for x as a power series in z. The identification of the first degree terms yields immediately for the solution

$$x = z + [z]_2 . \tag{3.3}$$

Observe now generally that, with z_j replaced by $\varphi(x)$, (3.2) is

$$dx/dt = X , \tag{3.4}$$

$$d\varphi/dt = \lambda_j \varphi . \tag{3.5}$$

Now under the change of variables (3.3), the system (3.4) goes into

$$dz/dt = Z(z) ,$$

but (3.5) remains the same. Replacing back φ by z_j, (3.5) becomes

$$(3.6) \qquad dz_j/dt = \lambda_j z_j .$$

Therefore $Z(z) = (\lambda_1 z_1, \ldots, \lambda_n z_n)$ and so the system (3.4) goes into (3.6) under our change of variables.

Now (3.6) is merely the first approximation to (3.4). Its general solution is $z = (a_1 e^{\lambda_1 t}, \ldots, a_n e^{\lambda_n t})$. Hence the general solution of (3.4) is

$$(3.7) \qquad x = (a_1 e^{\lambda_1 t}, \ldots, a_n e^{\lambda_n t}) + [a_1 e^{\lambda_1 t}, \ldots, a_n e^{\lambda_n t}]_2 .$$

Of course this solution is only valid within a range for which the transformation (3.3) is applicable, i.e. for $||(a_1 e^{\lambda_1 t}, \ldots, a_n e^{\lambda_n t})||$ small enough. If the λ_h all have negative real parts, the solution (3.7) of (3.4) will be valid for $||a||$ sufficiently small and $t \geq 0$. However if some of the λ_h have positive real parts the solution (3.7) will only be valid within a certain finite time interval: $t_1 \leq t \leq t_2$.

(3.8) *Remark.* The above result is to be compared with (V, 14.2) where the same result was obtained, following Liapunov, by means of series solutions of the differential equation.

The Liapunov restriction—the real parts of the characteristic roots are all negative—does not differ very much from Poincaré's convexity condition. For admitting the latter there is a line D in the complex plane such that all the λ_h are on one side of D. Now the change of variables $x \to xe^{i\omega t}$ replaces λ_h by $\mu_h = \lambda_h - i\omega$. The new system has thus the form

$$(3.9) \qquad dx_j/dt = \mu_j x_j + [x\, e^{-i\omega t}]_2 .$$

Since the μ_h are merely the λ_h rotated by the fixed angle $-\omega$, one may choose the latter so that D becomes vertical to the left of the origin and with the μ_h to its left. Thus the μ_j will have negative real parts and so (3.9) is amenable to the Liapunov method. The resulting series solutions differ from (3.7) but are valid for all time t.

§ 2. The Direct Stability Theorems of Liapunov

4. In his great Mémoire [1] Liapunov gave several theorems attacking directly the problem of stability. His method, inspired by Dirichlet's proof of Lagrange's theorem on the stability of equilibrium, is referred to by Russian authors as Liapunov's second method; his first method is to use the series solution of the differential system as exposed in (V, § 4).

Generally speaking Liapunov aims to set up something resembling a potential function which has an extremun at the given critical point and to which trajectories do or do not tend. Let us examine the question in more detail.

Take a real vector system

$$dx/dt = X(x; t) \tag{4.1}$$

where X is continuous and satisfies a Lipschitz condition in a set $\Omega(A, \tau)$ and $X(0; t) = 0$ for $t \geqq \tau$. Thus the origin is a critical point and its stability is to be discussed.

Let $V(x; t)$ be a scalar function such that $V(0; t) = 0$ whatever $t \geqq \tau$ and that V is of Class C^1 in $\Omega(A, \tau)$. Let Γ: $x(t)$ be a trajectory. Thus on Γ the function V becomes a function $V(t) = V(x(t); t)$ and on Γ its derivative is

$$dV/dt = V'(t) = \partial V/\partial t + \Sigma X_h \partial V/\partial x_h \tag{4.2}$$

where the x_h are to be replaced by the components $x_h(t)$ of $x(t)$. The whole argument of Liapunov rests upon the comparative signs of suitable functions V and their time derivatives V'. The idea is more or less that if one may choose V so that $V = \text{const.}$ represent tubes surrounding the line $x = 0$ such that all the Γ's cross through the tubes toward the line, the system is stable, while if they proceed in the other direction, the system is unstable.

5. The proper description of the functions V rests upon a certain number of definitions. In these definitions the scalar functions $V(x; t)$ and $W(x)$ under consideration will be defined and continuous, respectively, in $\Omega(A, \tau)$ and $\Omega(A)$ with

$V(0; t) = 0$ for $t \geqq \tau$ and $W(0) = 0$. We say that

$\left.\begin{matrix} W(x) \\ V(x; t) \end{matrix}\right\}$ is of *fixed positive* [*negative*] *sign in* $\left\{\begin{matrix} \Omega(A) \\ \Omega(A, \tau) \end{matrix}\right\}$

whenever it is $\geqq 0$ [$\leqq 0$] there;

$W(x)$ is *positive* [*negative*] *definite in* $\Omega(A)$ whenever it is > 0 in $\Omega(A)$ for $x \neq 0$;

$V(x; t)$ is *positive* [*negative*] *definite in* $\Omega(A, \tau)$ whenever it dominates [is dominated by] there [by] a positive [negative] definite function $W(x)$;

$\left.\begin{matrix} V(x; t) \\ W(x) \end{matrix}\right\}$ is a *Liapunov function over* $\left\{\begin{matrix} \Omega(A, \tau) \\ \Omega(A) \end{matrix}\right\}$ whenever, in that set,

it is positive definite, and with a derivative along the trajectories $\left\{\begin{matrix} dV/dt \\ dW/dt \end{matrix}\right\}$ which is continuous and of fixed negative sign.

Observe that the restriction to class C^1 imposed upon V is not essential but merely convenient. Of course when V is of class C^1 one may calculate V' directly from (4.2) and *without* any knowledge of the solutions. There are cases, however, when the weaker restriction on the Liapunov functions is decidedly convenient (see notably 24).

As an example, of the two functions over $\Omega(A, 2)$, $x_1^2 + x_2^2 - 2x_1x_2 \cos t$, $t(x_1^2 + x_2^2) - 2x_1x_2 \cos t$, the first is merely of fixed positive sign, but since the second dominates $x_1^2 + x_2^2$, it is positive definite.

6. We shall now consider Liapunov's theorems.

(6.1) STABILITY THEOREM. *If there exists a Liapunov function* $V(x; t)$ *over* $\Omega(A, \tau)$ *then the origin is stable.*

By hypothesis in $\Omega(A, \tau)$, $V(x; t)$ dominates a certain positive definite function $W(x)$. Let $0 < \varepsilon < A$ and on the compact set $\|x\| = \varepsilon$ let $W(x) \geqq \alpha > 0$. Take any $t_0 \geqq \tau$. Since $V(x; t)$ is continuous in x and $V(0; t_0) = 0$ there is a $0 < \eta(\varepsilon, t_0) \leqq \varepsilon$ such that $V(x_0; t_0) < \alpha$ for $\|x_0\| < \eta$. Let Γ be the trajectory of the solution $x(t)$ such that $x(t_0) = x_0$. Set also $V(x_0; t_0) = V_0$. Along Γ

$$V = V_0 + \int_{t_0}^{t} V' dt, \quad t \geqq t_0. \tag{6.2}$$

Since $V' \leqq 0$ we see that along Γ, $V \leqq V_0 < \alpha$. Hence Γ can never reach $||x|| = \varepsilon$ since on that set $V \geqq W \geqq \alpha$. This proves the theorem.

(6. 3) ASYMPTOTIC STABILITY THEOREM. *If there exists in $\Omega(A, \tau)$ a Liapunov function $V(x; t)$ dominated by a positive definite $W_1(x)$ and such that V' is negative definite then the origin is asymptotically stable.*

By hypothesis there are three positive definite functions $W(x)$, $W_1(x)$, $W_2(x)$ defined in $\Omega(A)$ such that for $(x; t) \in \Omega(A, \tau)$

$$W(x) \leqq V(x; t) \leqq W_1(x),\ W_2(x) \leqq -V'(x; t).$$

At all events the origin is stable. With all quantities as before, given any $0 < \xi < \eta$, one must show that the solution $x(t)$ is such that for t large enough $||x(t)|| < \xi$. Let λ be the lowest bound (positive) of $W(x)$ on the compact set $\xi \leqq ||x|| \leqq \varepsilon$. One may choose $0 < \zeta \leqq \xi$, such that $W_1 < \lambda$ in $||x|| < \zeta$, and hence $V(x; t) < \lambda$ on the same set and for all $t \geqq \tau$. Let μ, ν be lowest bounds (positive) for $W(x)$, $W_2(x)$ on $\zeta \leqq ||x|| \leqq \varepsilon$ and suppose that the trajectory Γ remains in that set for t large. Then on Γ from some point $(x_0; t_0)$ on we have from (6. 2) $V \leqq V_0 - \nu(t - t_0) < \mu$ for t above a certain value t_1. At that moment Γ is in $||x|| < \zeta$ Since $V < \lambda$ and decreases for $t \geqq t_1$, Γ cannot enter the set $\xi \leqq ||x|| \leqq \varepsilon$ and the theorem follows.

(6. 4) INSTABILITY THEOREM. *Let there exist a function $U(x; t)$ defined, bounded, and of class C^1 in $\Omega(A, \tau)$. Thus dU/dt is defined along the trajectories in $\Omega(A, \tau)$. Let $U > 0$ in a certain subregion Ω_1 of Ω whose portion B of the boundary contains the ray T: $x = 0$, $t \geqq \tau$, and let $U = 0$ on B. Suppose that:* (a) *whatever $t_0 \geqq \tau$ there exist points $(x_0, t_0) \in \Omega_1$ arbitrarily close to T;*

(b) *for every small $h > 0$ there is a $k(h) > 0$ such that $U \geqq h$ in Ω_1 implies $U' \geqq k(h)$ in the same set.*

Then the origin is unstable.

Given any $0 < \varepsilon < A$ and whatever $0 < \eta < \varepsilon$ and $t_0 \geqq \tau$ there is a point $(x_0; t_0)$, $||x_0|| < \eta$, such that $U(x_0; t_0) = h > 0$. Let Γ be a trajectory issuing from $(x_0; t_0)$. Since $U'(x_0; t_0) > 0$, U increases along Γ and so $U'(x; t) \geqq k(h)$ along the trajectory. Let λ be a (positive) upper bound of U in the closure of the set $\{||x|| < \varepsilon\} \times T$. Since U is bounded in $\Omega(A, \tau)$, λ is finite. Now along Γ by (6, 2) for U: $U \geqq h + k(t - t_0)$. Hence U, along Γ, will sometime exceed λ. Hence Γ will leave the set $U \geqq h$, $||x|| < \varepsilon$. Since this can only be through the sphere $||x|| = \varepsilon$, the origin is unstable.

(6.5) COROLLARY. *Under the same conditions for U but merely with*

$$U' = \mu U + U^*$$

where μ is a positive constant and $U^* \geqq 0$, the origin is again unstable.

For along Γ at once

$$U \geqq U_0 e^{\mu(t-t_0)}$$

from which instability follows.

(6.6) *Četaev's generalization of Liapunov's instability theorem.* Četaev observed that actually all that the theorem requires is not the whole of $\Omega(A, \tau)$ but merely a subregion Ω containing $\Omega_1 \cup B$. Otherwise the statement of the theorem is unchanged.

(6.7) *Autonomous case.* One may then state the corresponding theorems, but with functions $W(x)$ replacing $V(x; t)$ in the two stability theorems, and a function $U(x)$ in place of $U(x; t)$ in the instability theorems. Moreover, since one may replace everywhere t by $t + k$, one may take throughout $\tau = t_0 = 0$. There are several interesting consequences:

(a) *Stability theorem.* This time η will be a function $\eta(\varepsilon)$, independent of t_0: the stability is *uniform* (see IV, 2).

(b) *Asymptotic stability.* We have here $V(x; t) = W(x) = W_1(x)$ and $- V' = W_2(x)$, but otherwise there is no change.

(c) *Instability.* The conditions $U(x)$ bounded, and those related to h, $k(h)$ follow naturally from the compactness of $||x|| \leqq \varepsilon$.

(6.8) *Geometric interpretation.* It is most convenient to describe it for $n = 2$ and functions $W(x_1, x_2)$. Suppose that W is definite positive. Introduce a third coordinate y and consider the surface $F: y = W(x_1, x_2)$ over $\Omega(A)$, that is within the cylinder $x_1^2 + x_2^2 = A^2$. We think of it as an inverted cup over the region. In that region the surface is above the horizontal plane $y = 0$, except for touching it at the origin. The curves $W(x_1, x_2) = \varepsilon$ are the projections of the horizontal sections $H(\varepsilon)$ of F by the planes $y = \varepsilon$. The paths Γ in $\Omega(A)$ are imaged

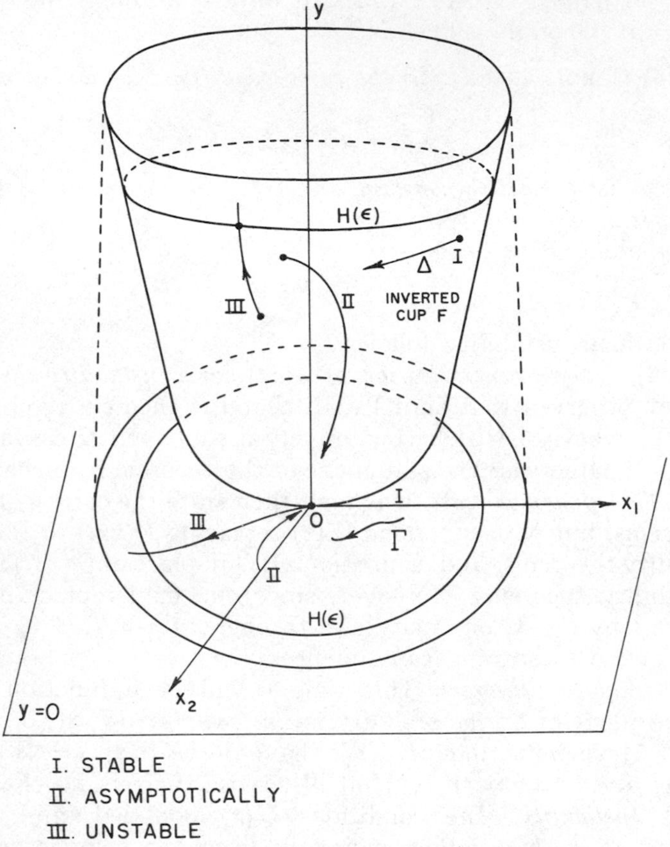

Fig. 1

into paths Δ on F. In any case the orthogonal projection between F and $\Omega(A)$ is actually topological.

Now let us examine how the sign of W' affects stability. To say that $W' \leq 0$ along a path is merely to affirm that on the surface F the path (I of figure 1) goes down sluggishly and need not reach the origin. This corresponds to mere stability. If W' is actually negative and even below a certain $-\alpha < 0$ then the path has a downward slope bounded away from zero and tends briskly to O. This is the case for path II and we have asymptotic stability. Finally, if along some path such as III: $W' > \beta > 0$ the reverse takes place: the path actually ascends at a rate bounded away from zero, and reaches $H(\varepsilon)$ no matter how low its starting point. Here we have instability.

One may observe that if W is of class C^1 and the surface F is tangent to the horizontal plane at the origin, then the curves $H(\varepsilon)$ for ε small enough comprise a set of ovals surrounding the origin.

The above representation is easily carried out for any dimension. For $V(x; t)$ something analogous may be done but the resulting inverted cup will touch $y = 0$ along the ray $t \geqq \tau$.

6.9 In the treatment of the theorems of Liapunov the coordinates have been (tacitly) assumed real. Suppose, however, that there are p conjugate complex pairs of coordinates and $n - 2p$ real coordinates as

$$x_1, \bar{x}_1, \ldots, x_p, \bar{x}_p, x_{2p+1}, \ldots, x_n .$$

Then, in the notations of (I, 13.10), V is a function $V(t; x_1, x_1^*, \ldots, x_{2p+1}, \ldots)$ which is *real* at the real points $x_1, \bar{x}_1, \ldots$. Thus

$$V(t; x_1, \bar{x}_1, \ldots, x_{2p+1}, \ldots) = V(t; \bar{x}_1, x_1, \ldots, x_{2p+1}, \ldots)$$

and we have

$$V' = \partial V/\partial t + \Sigma (X_h \, \partial V/\partial x_h + X_h^* \, \partial V/\partial \bar{x}_h) + \\ + \Sigma X_{2p+j} \, \partial V/\partial x_{2p+j} ,$$

where X_h^* has the meaning fully described in (I, 13.10).

(6.10) *Applications.* As a first application of the theorems of Liapunov let us prove a partly more general stability property than (V, 7.10). Of course the proof is startingly simpler and more direct.

(6.11) *Given the quasi-linear system*

$$dx/dt = Bx + q(x; t)$$

where, in $\Omega(A, \tau)$, q is continuous and satisfies a Lipschitz condition, $q(x; t) = \alpha(||x||)$, where $\alpha \to 0$ with $||x||$, and B is a constant matrix whose characteristic roots λ_h are all distinct. Then if the λ_h all have negative real parts the system is asymptotically stable; if they all have positive real parts the system is unstable.

Choose coordinates such that B is in normal form. If the λ_h are all real take

$$V = \Sigma x_h^2.$$

Then

$$V' = 2 \Sigma \lambda_h x_h^2 + o(||x||^2)$$

and so for x small V' has the sign of the λ_h. Since V is positive (7.2) is a consequence of (6.3), (6.4).

Suppose now that $\lambda_h, \bar{\lambda}_h, h = 1, \ldots, p$ are complex and λ_{2p+j}, $j = 1, 2, \ldots, n - 2p$ are real. Then choose

$$V = \sum x_h \bar{x}_h + \sum x^2_{2p+j}.$$

If $\lambda_h = \lambda_h' + i \lambda_h''$ we now find

$$V' = 2 \sum \lambda_h' x_h \bar{x}_h + 2 \sum \lambda_{2p+j} x^2_{2p+j} + o(||x||^2)$$

and the conclusion is the same.

(6.12) As a second application let us give a *second proof* of Lagrange's theorem (IV, 1.1) on the stability of equilibrium. Let us suppose that we have a system with n degrees of freedom and in the notation of (IV, 1) let $H = K + V$. Suppose also that the system depends on the usual variables $q_1, \ldots, q_n$ the positional variables, and $p_1, \ldots, p_n$ the kinetic variables. The potential energy V depends solely on q and we suppose that $V(0) = 0$ and that it is a minimum of V. Otherwise let V be holomorphic at the origin. Regarding K it is a positive definite quadratic form in the p_h with coefficients holomorphic in q. Thus H is a positive function in the sense of Liapunov. The equations of motion are

$$dq_s/dt = \partial H/\partial p_s, \quad dp_s/dt = - \partial H/\partial q_s.$$

If we take H as the V-function of (6) we find $dH/dt = 0$. Therefore the origin is stable. That is to say static equilibrium at $q = 0$

is stable. This is (IV, 1.1). The proof has been given under certain restrictive conditions which are however readily removed.

Liapunov has proved the counterpart of Lagrange's theorem for an isolated maximum of the potential energy. He restricted himself, however, to the analytical case. The complete statement is:

(6.13) THEOREM. *Let the Hamiltonian $H(p; q)$ be analytic at the origin O: $p = 0$, $q = 0$. If $q = 0$ is an isolated maximum of the potential energy, then this point is an unstable point of equilibrium.*

With the notations remaining the same, this time

$$K(p) = K_{2r}(p) + \ldots, V(q) = -V_2(q) + \ldots$$

where K_{2r}, V_2 are positive definite forms of degrees $2r$, 2 in their variables and ... represent, as usual, terms small relative to those written in a certain neighborhood Ω of O.

Choose now

$$U = \Sigma p_h q_h$$

Then

$$\frac{dU}{dt} = \Sigma p_h \frac{dq_h}{dt} + \frac{dp_h}{dt} q_h$$

$$= \Sigma p_k \frac{\partial K_{2r}}{\partial p_h} + \frac{\partial V_2}{\partial q_h} q_h + \ldots = 2(rK_2 + V_2) + \ldots$$

(by Euler's relation for forms). Thus in Ω: $dU/dt > 0$, except that it is zero at O. On the other hand in $p_h, q_h > 0$ (all h) we also have $U > 0$. Hence the equilibrium where V is a maximum, is unstable.

The simplest example is that of a vertical bar fixed at one end. If z is the distance from the centroid to the point of suspension, the potential energy is $V = -mgz$. It is a minimum where the bar hangs downward — stable equilibrium; a maximum when the bar stands upward — unstable equilibrium.

7. We have dealt (6.11) with the stability of a quasi-linear system in the case of distinct characteristic roots. We shall now attack the more general case of a *non-critical* matrix B, that is, one which has no zero or pure complex roots. The construction of suitable functions V is then definitely more arduous.

Consider again a system

$$\text{(7.1)} \qquad dx/dt = Bx + q(x; t)$$

where B is non-critical and as before in $\Omega(A, \tau)$: $q(x; t) = \alpha \, ||x||$, where $\alpha \to 0$ with $||x||$. The stability behavior of the first approximation

$$\text{(7.2)} \qquad dx/dt = Bx$$

is known, (see IV, 4): the origin is asymptotically stable if B is stable, and unstable otherwise. We propose to prove:

(7.3) THEOREM. *If the matrix B is non-critical the stability properties of* (7.1) *and of its first approximation* (7.2) *are the same.*

The method will consist of constructing simple functions V for (7.2) and showing that they also serve for (7.1).

Let us agree for the present that all vectors are column-vectors (one column matrices). If x is such a vector, x' stands for the row-vector (one-row matrix) with the same components. As a consequence, a quadratic form

$$P(x) = \sum f_{ij} x_i x_j,$$

with the matrix F symmetrical: $F = F'$ may be written simply $x'Fx$. In particular, the Euclidean distance squared is $x'x$ (inner vector product).

If our quadratic form is positive [negative] definite we shall denote the fact by $F > 0$ [$F < 0$].

If $f(x)$ is a scalar function $\partial f/\partial x = \text{grad} \, f$ will stand, for the present, for the row-vector with components $\partial f/\partial x_h$.

Let $P(x)$ be the same quadratic form as above. Then its derivative along the paths of (7.2) is

$$\text{(7.4)} \qquad dP(x)/dt = (\partial P/\partial x)Bx = Q(x)$$

where $Q(x)$ is likewise a quadratic form. We propose to prove the following property due more or less to Liapunov:

(7.5) *If the characteristic roots of B all have positive [negative] real parts and Q is positive [negative] definite, then* (7.4) *has a unique solution for $P(x)$ and it is a positive definite quadratic form.*

It is evident that this property for the "negative" case is an inverse of Liapunov's asymptotic stability theorem for the linear system (7.2), and that the "positive" part is a partial inverse of his instability theorem for (7.2) also.

Suppose that

$$P(x) = x'Fx,\ F' = F,\ Q(x) = x'Gx,\ G' = G.$$

Then

$$dP/dt = x'(B'F + FB)x.$$

Hence (7.4) yields

$$B'F + FB = G.$$

But then it is clear that (7.5) is an immediate consequence of (App. I, 8).

Suppose now that the characteristic roots λ do not have negative real parts, and that $p > 0$ have positive real parts. One may then choose coordinates $y_1, \ldots, y_p, z_1, \ldots, z_q$ such that $B = \text{diag}\,(B_1, B_2)$ where the characteristic roots of B_1 all have positive real parts and those of B_2 all have negative real parts (App. I, 11). Note that if $q = 0$, then $B_2 = 0$, but the modifications required in this case are obvious enough. At all events, (7.1) assumes now the form

$$dy/dt = B_1 y,\ dz/dt = B_2 z.$$

Take

$$Q_1(y) = \Sigma\, y_h^2,\ Q_2(z) = \Sigma\, z_k^2,\ Q(y; z) = Q_1 + Q_2.$$

According to (7.5) we can find positive definite quadratic forms $P_1(y)$, $P_2(z)$ such that

$$(\partial P_1/\partial y)\, B_1 y = Q_1(y),\ (\partial P_2/\partial z) B_2(z) = -Q_2(z).$$

Let then $P(y; z) = P_1(y) - P_2(z)$. Thus $P > 0$ in the region $P_1 > P_2$ of the $(y; z)$ space. This region exists since it contains the set $z = 0$. In the whole space however (origin excepted)

$$(d/dt) P(y; z) = Q_1(y) + Q_2(z) > 0.$$

Thus we have an inversion of the instability theorem (6.5).

We shall now apply the preceding results to the quasi-linear system (7.1).

If B is stable take $Q(x) = -\Sigma x_r^2$ and $P(x)$ as the unique positive definite quadratic form solution of (7.4). We have then in $\Omega(A, \tau)$

$$dP(x)/dt = Q(x) + \ldots,$$

along the solutions of (7.1) (the dots are as before). Hence in $\Omega(A, \tau)$: $P'(x) < 0$, $Q(x) > 0$, except at the origin where both vanish. Therefore, by (6.3) the origin is asymptotically stable for (7.1).

Suppose now that among the real parts of the characteristic roots λ_h some are positive. Under the same choice of coordinates as before (7.1) may be replaced by a system

$$\frac{dy}{dt} = B_1 y + q_1(y; z; t)$$

$$\frac{dz}{dt} = B_2 z + q_2(y; z; t)$$

where $q_i = \alpha_i(||y|| + ||z||)$ and $\alpha_i \to 0$ with $||y|| + ||z||$. Hence with $P(y; z)$, as before, we have again

$$dP/dt = Q_1(y) + Q_2(z) + \ldots,$$

so that in a certain region Φ: $P > 0$ and in $\Omega(A, \tau)$ itself (the ray T excepted) $dP/dt > 0$. Hence we have instability. This completes the proof of Theorem (7.3).

Concluding remarks about Liapunov's direct method. The few applications that have been discussed suffice to indicate the scope of the method. We emphasize the important fact that if one can obtain suitable functions V one may obtain stability information about the solutions directly from the equation itself, and without any knowledge of the solutions.

§ 3. Stability in Product Spaces

8. The title is somewhat misleading. What we have in mind is the discussion of the stability of certain systems of the form

$$(8.1) \quad \begin{cases} \text{(a)} \quad dy/dt = Y(y) + Y^*(y; z), \\ \text{(b)} \quad dz/dt = Z(z) + Z^*(y; z). \end{cases}$$

Here y and z are a p-vector and a q-vector; the functions Y, Y^*, Z, Z^* vanish at the origin which is thus a critical point for the system; Y^* and Z^* are small in some sense relative to Y and Z. The question arises as to what extent is the stability of the complete system governed by that of the partial systems

(8.2a) $\quad dy/dt = Y(y)\,, \qquad$ (8.2b) $\quad dz/dt = Z(z)\,.$

This problem has been dealt with at length by Liapunov and his successors in the Soviet Union. However, except for a general result due to Persidskii they usually confine their attention to $Z = Q\,z$, where Q is a constant or variable matrix such that (8.2b) is stable. Generally also they assume that $Y, \ldots,$ contain the time t. We feel, however, that even the autonomous case, upon which we mostly concentrate, will give a respectable notion of the scope of their work.

9. Let $\mathfrak{B}_x = \mathfrak{B}_y \times \mathfrak{B}_z$. The norms for $\mathfrak{B}_y$ and $\mathfrak{B}_z$ are naturally those induced by the norm $||x|| = \sup \{|x_h|\}$ for $\mathfrak{B}_x$, namely $||y|| = \sup \{|y_j|\}$ and $||z|| = \sup \{|z_k|\}$. The basic Ω regions of (V, 4) are $\Omega_y(A)$, $\Omega_y(A, \tau)$ for $\mathfrak{B}_y$ and $\Omega_z(A)$, $\Omega_z(A, \tau)$ for $\mathfrak{B}_z$. Thus $\Omega_z(A, \tau)$ is the set: $||z|| \leq A, t \geq \tau$.

We first treat a noteworthy result due to Persidskii [1]. He actually established it for vectors with countable components but for our purpose it will be sufficient to deal with finite dimensional vector spaces.

Consider a system in a p-vector y and a q-vector z

(9.1)
$$\begin{cases} \text{(a)} \ dy/dt = Y(y; z; t)\,, \\ \text{(b)} \ dz/dt = Z(y; z; t) \end{cases}$$

in a closed region $\Omega(A, \tau)$. It is assumed that the system is continuous and satisfies a Lipschitz condition and that $Y(0; 0; t)$, $Z(0; 0; t) = 0$ for $t \geq \tau$. Thus the origin is a critical point for the system.

Let now $\zeta(t)$ be a continuous q-vector. Take the associated system

(9.2)
$$dy/dt = Y(y; \zeta(t); t).$$

Let O_y, O_z denote the origins $y = 0$ and $z = 0$ for (9.1a) and (9.1b), and O the origin $y = 0$, $z = 0$ for the complete system

(9.1). We say that O_y is *quasi-stable* for (9.1a) whenever, given any $0 < \varepsilon < A$ and $t_0 \geqq \tau$, there exists $0 < \eta(\varepsilon, t_0) \leqq \varepsilon$ such that if $||\zeta(t_0)|| < \eta$ then any solution $y(t)$ of (9.2) with $||y(t_0)|| < \eta$ has the property that in any interval $t_0 \leqq t \leqq t_1$ in which $||\zeta(t)|| \leqq \varepsilon$ we have $||y(t)|| < \varepsilon$.

The origin O_y is said to be *quasi-unstable* for (9.1a) whenever given any ε, η such that $0 < \eta \leqq \varepsilon < A$ and any continuous $\zeta(t)$ such that $||\zeta(t_0)|| < \eta$ and $||\zeta(t)|| < \varepsilon$ for $t \geqq t_0$, then for some solution $y(t)$ of (9.2) such that $||y(t_0)|| < \eta$ we have $||y(t)|| = \varepsilon$ for some $t \geqq t_0$.

Of course, with the obvious changes, all the above "quasi" definitions apply to O_z and (9.1b).

(9.3) PERSIDSKII'S THEOREM. *If both O_y, O_z are quasi-stable for* (9.1a), (9.1b) *then O is stable for* (9.1.). *If O_y or O_z is quasi-unstable for the corresponding system, then O is unstable for* (9.1).

Stability. Given $\varepsilon > 0$ and $t_0 \geqq \tau$ there are corresponding η_1, η_2 of quasi-stability for (9.1a) and (9.1b). Let η be the least of η_1, η_2. Take a solution $[y(t), z(t)]$ of (9.1) such that $||y(t_0)||, ||z(t_0)|| < \eta$. Consider now (9.2) with $\zeta(t) = z(t)$ and let t_1 be the first value of t such that $||z(t)|| = \varepsilon$. Thus, $||z(t)|| < \varepsilon$ for $t_0 \leqq t < t_1$. Since O_y is quasi-stable for (9.1a), $||y(t)|| < \varepsilon$ for $0 \leqq t \leqq t_1$. Hence the first value t_2 of t for which $||y(t)|| = \varepsilon$ is $> t_1$. Hence, by the quasi-stability of O_z for (9.1b), and with $y(t)$ as the analogue of $\zeta(t)$, $||z(t_1)|| < \varepsilon$. The contradiction shows that $t_1 = +\infty$ and hence O is stable for (9.1).

Instability. Suppose that (9.1a) is quasi-unstable and that nevertheless O is stable for (9.1). Then with ε and t_0 as before, there is an $\eta(\varepsilon, t_0)$ such that any solution $[y(t), z(t)]$ of (9.1) with $||y(t_0)||, ||z(t_0)|| < \eta$ satisfies $||y(t)||, ||z(t)|| < \varepsilon$ for $t \geqq t_0$; this is however impossible since (9.1a) is assumed quasi-unstable.

10. Consider now a system

$$(10.1) \qquad \begin{cases} \text{(a)} \ \dfrac{dy}{dt} = Y(y; z) \\[2ex] \text{(b)} \ \dfrac{dz}{dt} = Qz + Z(y; z) \end{cases}$$

where the dimensions are as before, Q is a constant stable matrix, Y and Z are continuous; furthermore:

(10.2) I. $Y(0; 0) = 0$, $Z(0; 0) = 0$;

II. Y, Z satisfy Lipschitz conditions in $\Omega(A)$;

III. $||Z(y; z)|| = O\,(||y||^{\beta} + ||z||)$, $\beta > 1$ as $||y|| + ||z|| \to 0$.

(10.3) THEOREM. *Under the above assumptions the stability properties of* (10.1) *are the quasi-stability properties of* (10.1a) (Dychman [1]).

By Persidskii's theorem, (10.3) will follow if we can show that (10.1b) is quasi-stable. For the proof to follow we are indebted to J. K. Hale. It will be based upon the

(10.4) GENERALIZED GRONWALL CRITERIUM. *Let* $\varphi(t)$, $\psi(t)$ *be real, continuous, non-negative functions on* $[a, b]$, *and let* k *be a positive constant. If on* $[a, b]$

$$\varphi(t) \leqq \psi(t) + k\int_a^t \varphi(s)ds \tag{10.4a}$$

then

$$\varphi(t) \leqq \psi(t) + \int_a^t e^{k(t-s)}\psi(s)ds. \tag{10.4b}$$

Upon setting $R(t) = k\int_a^t \varphi(s)ds$ we have in succession:

$$\frac{dR}{du} - kR(u) \leqq k\psi(u),\quad e^{-k(u-a)}R(u)\Big]_a^t \leqq k\int_a^t e^{-k(u-a)}\psi(u)du;$$

$$R(t) \leqq k\int_a^t e^{k(t-s)}\psi(s)ds.$$

Since $\varphi \leqq \psi + R$, (10.4b) follows.

We shall also require the following two properties:

(10.5) *Given any* $\alpha > 0$ *there exists a* $\varrho > 0$ *and* $\leqq A/2$ *such that if* $||z||$, $||\eta|| < \varrho$, *then* $||Z(\eta, z)|| < K(\alpha\,||z|| + ||\eta||)$, K *and* $\alpha > 0$.

(10.6) *There exists positive numbers* B, γ *such that* $||e^{Qt}|| \leqq Be^{-\gamma t}$. [This is merely (V, 4.2)].

To simplify the calculations that follow, we shall assume that $t_0 = 0$, which is manifestly immaterial.

We have at once (III, 10.1):

$$z(t) = e^{Qt}z_0 + \int_0^t e^{Q(t-u)}Z[\eta(u),\ z(u)]du,\quad z_0 = z(0).$$

Therefore

$$||z(t)|| \leqq Be^{-\gamma t}||z_0|| + C\int_0^t e^{-\gamma(t-u)}||\eta(u)||du$$

$$+ Ca\int_0^t e^{-\gamma(t-u)}||z(u)||\,du,$$

where $C = \eta\, BK > 0$. This relation is well adapted to the use of the generalized Gronwall inequality (10.4b). We set then

$$||z(t)|| = e^{-\gamma t}\varphi(t),\ k = Ca,\ \psi(t) = B\,||z_0|| + C\int_0^t e^{\gamma u}||\eta(u)||du.$$

Hence (10.4b):

$$\varphi(t) \leqq \left\{B||z_0|| + C\int_0^t e^{\gamma u}||\eta(u)||du\right\}$$

$$+ Ca\int_0^t\left\{Bz_0 + C\int_0^s e^{\gamma u}||\eta(u)||\,du\right\}e^{Ca(t-s)}\,ds.$$

After some simplifications and an integration by parts one obtains

$$\varphi(t) \leqq B||z_0||e^{Cat} + C\int_0^t e^{Ca(t-s)+\gamma s}||\eta(s)||\,ds,$$

or, finally returning to $z(t)$:

(10.7) $$||z(t)|| \leqq B||z_0||e^{-(\gamma-Ca)t} + B\int_0^t e^{-(\gamma-Ca)(t-s)}\,||\eta(s)||\,ds.$$

Since we dispose of a let it be chosen $< \gamma/C$. It is then clear that if one maintains $\eta(s)$ and z_0 small enough one will be able to choose $||z(t)|| < \varrho$. Then (10.7) shows that O_z is quasi-stable for (10.1b) and our theorem follows.

§ 4. An Existence Theorem

11. Henceforth we confine our attention to analytical systems. Before attacking our main argument we must prove a preliminary result due to Malkin [1].

(11.1) THEOREM. *The notations being as before consider the system*

(11.1a) $$\partial y/\partial z \cdot \{Qz + L(y; z)\} = B(y; z)$$

where Q is a constant stable matrix and L, B are non-units such that

$$L = [y; z]_2, \quad B(y; 0) = [y]_2, \quad B(y; z) = [z]_r, \quad r \geqq 1.$$

Then the system has a unique solution $y(z)$ holomorphic at the origin, and $y(z) = [z]_r$.

Let us show first that there is a unique formal expansion satisfying (11.1a):

$$(11.2) \qquad y = \sum_{(m)} f^{(m)} z_1^{m_1} \dots z_q^{m_q}, \quad m_1 + \dots + m_q = m\,.$$

Here (m) represents the collection $m_1, \dots, m_q$ and the summation is for all (m) and $m = 0, 1, 2, \dots$. Let the collections (m) be ranged in lexicographic order. It is seen at once from (11.1a) that $f^{(m)} = 0$ for $m < r$. Thus if one obtains a solution (11.2) we will have $y(z) = [z]_r$. We assume from now on that $m \geqq r$.

Notice that the form of our system is unchanged by a transformation of coordinates in $\mathfrak{B}_z$. Taking advantage of this let the coordinates be so chosen that Q is in normal form. Let $\lambda_1, \dots, \lambda_q$ be its characteristic roots.

Write now (11.1a) explicitly as

$$(11.3) \qquad \sum_k \frac{\partial y_h}{\partial z_k} \{\sum_j q_{kj} z_j + L_k(y; z)\} = B_h(y; z)\,.$$

Suppose that all the $f^{(s)}$, where (s) precedes (m), have been determined. The identification of the coefficients of $z^{m_1} \dots z^{m_q}$ in (11.3) yields

$$(11.4) \qquad f_h^{(m)} \cdot \sum m_j \lambda_j = G_h^{(m)}(\dots, f^{(s)}, \dots)$$

where the right-hand side is a polynomial in the $f^{(s)}$ such that (s) precedes (m). A n.a.s.c. in order that one may solve successively for the $f_h^{(m)}$ is that the λ_j satisfy no relation in positive integral coefficients m_j of the form

$$(11.5) \qquad \sum m_j \lambda_j = 0\,.$$

This is certainly the case here since the λ_j all have negative real parts. We thus obtain a unique formal power series solution $y(z)$. Naturally this also yields a unique formal power series solution in terms of the initial z coordinates.

It may be shown that one may always assume $r > 1$. Indeed let $r = 1$ and let $B(0; z) = Cz + [z]_2$, where C is a constant matrix. The regular change of variables $y \to y + CQ^{-1}z$, $z \to z$, replaces (11.1) by a similar system with $r > 1$. This is assumed henceforth for (11.1) itself.

12. We must now prove that the formal power series represents a function holomorphic at the origin. To that end we fall back upon the standard device of the majorante.

Let us set

$$Y = \Sigma\, y_h,\ Z = \Sigma\, z_k,$$

$$F(Y, Z) = \frac{1}{1 - \frac{1}{A}(Y + Z)} - 1 - \frac{1}{A}(Y + Z),$$

$$G(Y, Z) = \frac{1}{1 - \frac{1}{A}(Y + Z)} - \left(1 + \frac{Y}{A}\right) - \left(\frac{Z}{A} + \cdots \frac{Z^{r-1}}{A^{r-1}}\right).$$

Let $\lambda = \inf \{-Re\, \lambda_j\}$. We recall that according to (App. I, 3.3), one may take as normal form for Q one in which the subsidiary diagonals $1, \ldots, 1$ may be replaced by $-\varepsilon, \ldots, -\varepsilon$ where $\varepsilon > 0$. As a consequence our system is seen to have the majorante, in which M is suitably large:

$$(12.1)^* \qquad \Sigma \frac{\partial y_h}{\partial z_k} \{\lambda z_k - \varepsilon\, z_{k-1} - MF(Y, Z)\} = MG(Y, Z),$$

where $z_1 = 0$. This system manifestly possesses a formal power series solution $y^*(z)$ whose coefficients are all positive. Let (12.1)* be modified to (12.1)** by making $z_{-1} = z_n$. The analogue Q^{**} of Q for (12.1)** has for equation of characteristic roots

$$(12.2) \qquad (\lambda - x)^n - \varepsilon^n = 0\,.$$

Thus the characteristic roots of Q^{**} are $\lambda - \varepsilon\,\omega^k$, $k = 0, 1, \ldots, n - 1$, $\omega = e^{2\pi i/n}$. Their real parts are $\lambda - \varepsilon \cos 2k\pi/n \geq \lambda - \varepsilon$. Hence if $0 < \varepsilon < \lambda$, the real parts all have the same sign. Consequently (12.1)** will have a unique formal power series solution $y^{**}(z)$. We shall prove:

(12.3) *The series $y^{**}(z)$ is holomorphic at the origin and its coefficients are all positive and at least as great as the corresponding coefficients of $y^*(z)$. Hence $y^{**}(z)$ is a majorante for $y(z)$. As a consequence $y(z)$ represents a solution of* (11.1) *of the desired type.*

Assume that it has already been proved that all the coefficients of $y^{**}(z)$ are positive. Let $\varphi^{(m)}$, $\psi^{(m)}$ be the corresponding coefficients of y^*, y^{**} and let the analogue of (11.4) for y^* be

$$(12.4) \qquad m\,\lambda\,\varphi_h{}^{(m)} = \Gamma_h{}^{(m)}(\ldots, \varphi_k{}^{(s)}, \ldots).$$

Here $\Gamma_h{}^{(m)}$ is a polynomial with all coefficients positive. For y^{**} there will result relations

$$(12.5) \qquad m\,\lambda\,\psi_h{}^{(m)} = \Gamma_h{}^{(m)}(\ldots, \psi_k{}^{(s)}, \ldots) + \Delta^{(m)}(\ldots, \psi_k{}^{(m)'}, \ldots)$$

where Δ is like Γ save that $(m)'$ may also be posterior to (m) in the ordering of $\{(m)\}$. Suppose that it has been proved that $\varphi_h{}^{(s)} \leq \psi_h{}^{(s)}$ for all (s) preceding (m). This follows at once for the first (m) from the comparison of (12.5) and (12.4). The same comparison yields then the same result for all (m) in succession. All that remains then is to show that (12.1)** has a holomorphic solution whose series expansion has only positive coefficients.

Let us endeavor to find a solution of (12.1)** in which $y_1 = \ldots = y_p = Y/p =$ a function of Z. For such a solution we will have

$$(12.6) \qquad \frac{1}{p}\frac{dY}{dZ}\{(\lambda - \varepsilon)\,Z - q\,MF\,(Y, Z)\} = MG(Y, Z).$$

The change of variables $Y = ZY_1$ reduces (12.6) to a form readily shown to be

$$(12.7) \qquad \frac{dY_1}{dZ} = \frac{-Y_1 + aZ + [Y_1, Z]_2}{Z(1 + [Y_1, Z]_1)}$$

where $a = 0$ if $r > 2$. With an auxiliary variable t, (12.7) is equivalent to a system.

$$(12.7\text{a}) \qquad \frac{dY_1}{dt} = -Y_1 + aZ + [Y_1, Z]_2; \qquad \frac{dZ}{dt} = Z(1 + [Y_1, Z]_1).$$

Anticipating on a later theme (IX, 4) independent of the present argument we have here a saddle point. It has two analytic solutions as curves through the origin, the one $Z = 0$, and the other with a different tangent, of the form $Y_1 + \beta Z + [Y_1, Z]_2 = 0$. This one gives a solution $Y_1 = [Z]_1$ or $[Z]_2$ to which corresponds a solution $Y = [Z]_2$ or $[Z]_3$ of (12.6). Since $\lambda - \varepsilon > 0$ upon substituting a power series

$$Y = \Sigma \eta_m Z^m \tag{12.8}$$

in (12.6) and identifying coefficients, we find that the η_m are uniquely determined and they are in fact zero for $m < r$, and positive for $m \geq r$. Thus we have a solution for our majorante (12.6) with all coefficients positive. Hence the formal solution $y(z)$ of (11.1a) is an actual solution. Its series is absolutely and uniformly convergent in a certain $\Omega(A)$ and theorem (11.1) follows.

(12.9) *Application. Consider the system*

$$\frac{\partial W}{\partial z} Qz = f(z) \tag{12.9a}$$

in the unknown scalar function $W(z)$ *and the known scalar function* $f(z)$. *Then*: (a) *if* $f(z)$ *is a form of degree* $m > 0$ *there is a unique form of degree* m *satisfying* (12.9a); (b) *if* $f(z)$ *is a polynomial and* $f(z) = f_r(z) + f_{r+1}(z) + \ldots + f_s(z)$ *where* $r > 0$ *and* f_h *is a form of degree* h, *there is a unique solution* $W(z)$ *which is a polynomial such that* $W = W_r + \ldots + W_s$ *where* W_h *is a form of degree* h. (Liapunov [1], p. 276).

Regarding (a) it is at once seen that the formal solution $W(z)$ has only terms of degree m and so it is an actual solution. Thus no majorante argument is needed. As for (b) W_h is merely the solution of (12.9a) corresponding to $f = f_h$.

§ 5. Stability in Product Spaces: Analytical Case

13. Take an analytical p-vector system

$$dy/dt = F_s(y) + F_{s+1}(y) + \ldots, \tag{13.1}$$

where the components of F_h (and similarly later) are forms of degree h in the $y_1, \ldots, y_p$, and the right-hand side is convergent in $\Omega(A)$.

(13.2) *Definition.* (Malkin [9], No. 91). The origin is said to be stable for **(13.1)** regardless of the terms of degree $> N$ whenever it is stable for

(13.2a) $$dy/dt = F_s(y) + \ldots F_N(y) + G(y; t)$$

in the following sense: whatever the continuous function $G(y; t)$ such that $||G|| < K||y||^{N+1}, K > 0$, in $\Omega(A)$, and corresponding to any $0 < \varepsilon \leqq A$, there is a $0 < \eta(\varepsilon, K)$, such that for any solution $y(t)$ of (13.2a) such that $||y(0)|| < \eta$, we have $||y(t)|| < \varepsilon$ for all $t \geqq 0$.

The origin is said to be unstable for (13.1) regardless of the terms of degree $> N$, whenever under the same circumstances as above and no matter how small $||y(0)||$ for some $y(t)$ sometimes $||y(t)|| = \varepsilon$.

Consider now an analytic system

(13.3) $$\begin{cases} \text{(a)}\ \dfrac{dy}{dt} = F(y) + Y(y; z) \\ \text{(b)}\ \dfrac{dz}{dt} = Qz + Z(y; z) \end{cases}$$

with the following properties all valid in some $\Omega(A)$;

I. $F(y) = F_s(y) + F_{s+1}(y) + \ldots + F_N(y), \quad s \geqq 2$;

II. *The stability properties of*

(13.4) $$dy/dt = F(y)$$

are independent of the choice of terms of degree $> N$ in the sense of definition (13.2);

III. *Q is a constant stable matrix;*

IV. $Y(y; z) = [y; z]_2, \ Y(y; 0) = [y]_{N+1}$;

V. $Z(0; z) = [z]_2; \ Z(y; 0) = [y]_N$.

We have then with Dychman [1]:

(13.5) THEOREM. *Under the preceding conditions the stability properties of the complete system* (13.3) *and of the reduced system* (13.4) [*in the sense of* (13.2)] *are the same.*

The proof will essentially consist in applying a regular transformation of coordinates suppressing in $Y(y; z)$ the terms in y whose degree $\leqq N$ and then applying Dychman's criterium (10.3).

There are (trivially) no terms of degree —1 in y in $Y(y; z)$. Let us suppose that there are no terms of degree $< m \leqq N$ in y in $Y(y; z)$ and consider a regular transformation

(13.5) $$z \to z, \ \bar{y} = y - \sum_{(m)} u^{(m)}(z)\, y_1^{m_1} \ldots y_p^{m_p}, \ \sum m_j = m,$$

where (m) is any combination of the summands m_j and the $u^m(z)$ are non-units, holomorphic in $\Omega_z(A)$. Such a transformation has an inverse

$$z \to z, y = \bar{y} + v(\bar{y}; z), \tag{13.6}$$

where $v(\bar{y}; 0) = 0$, $v(\bar{y}; z) = [\bar{y}]_1$, and $v(0; z) = 0$. By substituting the expansion of v in (13.5) it is found that $v = [\bar{y}]_m$.

We find now from (13.5) that

$$d\bar{y}/dt = F(y) + Y(y; z) \tag{13.7}$$

$$- \sum_{(m), j} m_j u^{(m)} y_1^{m_1} \ldots y_j^{m_j - 1} \ldots y_p^{m_p} (Y_j(y) + Y_j(y; z))$$

$$- \sum_{(m)} y_1^{m_1} \ldots y_p^{m_p} \partial u^{(m)}/\partial z \cdot (Qz + Z(y; z)).$$

14. We may write

$$Y(y; z) = Y_0(z) + Y_1(y; z) + \ldots, \tag{14.1}$$

where the components of Y_h are forms of degree h in those of y with coefficients non-units in z for $h \leq N$. Let the powers $y_1^{m_1} \ldots y_p^{m_p}$ be ranged in lexicographic order and let $m = m_1 + \ldots + m_p$, $(m) = \{m_1, \ldots, m_p\}$. Suppose that by successive application of regular transformations (13.6) we have succeeded in reducing (13.1) to the same form with the same F and Q, but with $Y(y; z)$ lacking all terms corresponding to an (s) preceding (m). Upon applying a regular transformation

$$z \to z, \; y \to y + y_1^{m_1} \ldots y_p^{m_p} v(z), \tag{14.2}$$

we find that in addition the term corresponding to (m) will disappear provided that v satisfies a relation

$$\frac{\partial v}{\partial z} [Q(z) + Z(0; z)] = B(v; z) \tag{14.3}$$

where B behaves like the analogous term of (11.1). Hence, (14.3) has a unique, non-unit solution $v(z)$ and we have the desired transformation (14.2).

By repetition of the process just described we will finally

reduce the system (13.1) to the same form, with F and Q unchanged but with $Y(y; z) = [y]_{N+1}$.

Under the circumstances, in view of the convergency of the series $Y(y; z)$ in $\Omega(A)$, we shall have for $||y|| < A$ and any $||z(t)|| \leqq A$: $||Y(y; z)|| < K\, y^{N+1}$ for some positive K.

It is now a consequence of property II, that the quasi-stability properties of (13.1) are the stability properties of (14.3) in the sense of (13.2). Our theorem follows then from Dychman's criterium (10.3).

§ 6. System with a Single Characteristic Root Zero and the Rest with Negative Real Parts

15. Let us apply the preceding considerations to the system investigated by Liapunov ([1], p. 301) where there is a single characteristic root zero and the rest have negative real parts. Such a system may be reduced to the form

$$(15.1) \qquad \begin{cases} \text{(a)} \ dy/dt = \Phi(y) + Y(y; z), \\ \text{(b)} \ dz/dt = Qz + Z(y; z). \end{cases}$$

This time y is just a scalar variable and Q is as before. The functions Φ, Y, Z are non-units holomorphic in $\Omega(A)$ and such that

$$\Phi = [y]_N,\ N \geq 2;\ Y(y; z) = [y; z]_2 = [y]_{N+1};\ Z(0; z) = [z]_2 .$$

We will set also $Z(y; 0) = [y]_T$. It may be observed that if Y does not have the form shown at the outset then it may be reduced to that form without affecting the first term of Φ, (14).

Since Q is non-singular one may solve uniquely

$$Q\zeta + Z(y; \zeta) = 0$$

for $\zeta(y)$ holomorphic at the origin and such that $\zeta(0) = 0$.

As a consequence the change of variables

$$y \rightarrow y,\ z = z^* + \zeta(y)$$

is regular at the origin. In terms of the new variables (15.1a) is replaced by

(15.2) $dy/dt = \Phi(y) + Y(y; z^* + \zeta) = \Phi^*(y) + Y^*(y; z^*)$,

which is of the same type as before save that now $\Phi^* = [y]_{N^*}$ where $N^* \geq 2$, but possibly $N^* \neq N$.

The relation (15.1b) now becomes

(15.3) $dz^*/dt = Qz^* + W(y; z^*) - d\zeta/dy \cdot (\Phi^*(y) + Y^*(y; z^*))$

where

$$W = W_1(y) z^* + W_2(y) z^{*2} + \ldots,$$

the coefficients $W_h(y)$ being non-units. Thus (15.3) assumes the form

(15.4) $$dz^*/dt = Qz^* + Z^*(y; z^*),$$

where Z^* is a non-unit and $Z^*(0; z^*) = [z^*]_2$, $Z^*(y; 0) = [y]_{T^*}$, $T^* > N^*$ (requiring perhaps repetition of the operation). In other words our transformation preserves the form of (15.1) but may change N and the new system will have a $T \geqq N$. Let us assume then that (15.1) already satisfies this condition.

An exceptional case is when identically $\Phi^*(y) = \Phi(y) + Y(y; \zeta(y)) = 0$, $Z^*(y; 0) = 0$. When this happens the system (15.4) has the solution $z^* = 0$, $y = c$, where c is an arbitrary constant. In other words the y axis is a *line* of critical points. We shall assume that we are not in presence of this case. Thus effectively $N \geqq 2$.

16. Let $\Phi(y) = gy^N + \ldots$ then

(16.1) *The stability properties of*

$$dy/dt = \Phi(y)$$

are independent of the terms of degree $> N$.

That is, if

(16.2) $$\frac{dy}{dt} = gy^N[1 + h(y; t)y]$$

where h is continuous and $|h(y; t)| < K$, for y suitably small, then the stability or instability of the origin for (16.2) is independent of h. In fact let $K = 1$ and $|y| < \frac{1}{2}$. Then the sign of dy/dt in (16.2) is fixed on each side of $y = 0$ and independent of h, that is, it is the same as for

$$\frac{dy}{dt} = gy^N. \tag{16.3}$$

Since dy/dt is zero only for $y = 0$ in $|y| < \frac{1}{2}$, y can never stop its motion to or away from the origin. Thus (16.3) and (16.2) behave alike.

We have now three possibilities:

(a) *N even*. Then in both systems for t large y can only take the sign of $-g$ and then $\rightarrow 0$ as $t \rightarrow +\infty$. Hence both systems are conditionally asymptotically stable.

(b) *N odd and g positive*. Then in either system for t large y cannot be small, hence both are unstable.

(c) *N odd and g negative*. Then in either system y may be small of both signs for t large and $\rightarrow 0$ as $t \rightarrow +\infty$. Hence both systems are asymptotically stable.

This proves (16.3).

Since the system (14.1) may be written

$$dy/dt = F(y) + Y(y; z)$$
$$dz/dt = Qz + Z(y; z)$$

which is of the type of (13.1), the application of (13.4) yields the following result due to Liapunov:

(16.4) THEOREM. *Let the system* (15.1) *be reducible to the same form with*

$$F(y) = gy^N + \ldots, \; Z(y; 0) = [y]_T, \; T > N.$$

Then: (a) *If N is odd and g negative* [*positive*] *the system is asymptotically stable* [*is unstable*]; *if N is even the system is unstable.*

17. Liapunov obtained his theorem as an application of his second method. It is interesting to show in some detail how he went about it.

If we assume that the reductions of (14) and (15) have already been carried out the system will be in the form:

$$\text{(17.1)} \quad \begin{cases} \text{(a)} \; dy/dt = gy^N + Y(y; z), \\ \text{(b)} \; dz/dt = Qz + Z(y; z), \end{cases}$$

where g is a constant and

$$Y(y;z) = [y;z]_2 = [y]_{N+1};\ N \geq 2;$$
$$Z(y;z) = [y;z]_2;\ Z(y;0) = [y]_T,\ T \geq N.$$

Consider first the system

$$\partial W/\partial z \cdot Qz = \Sigma z_k^2. \tag{17.2}$$

According to (11.1) it has a unique solution $W(z)$ which is a quadratic form. Consider now the system:

$$dz/dt = Qz. \tag{17.3}$$

In relation to that system

$$W' = dW/dt = \partial W/\partial z \cdot Qz = \Sigma z_k^2,$$

and so W' is definite positive. Observe incidentally that since $W \to 0$ uniformly with z and (17.3) is stable at the origin, W cannot take the sign of V' (6.4) and hence W is definite negative.

Suppose now that N is *even* and set $V = y + gW$. Then in relation to (17.1):

$$V' = gy^N + Y(y;z) + g\,\partial W/\partial z\,\{Qz + Z(y;z)\}$$
$$= g(y^N + \Sigma z_k^2) + \ldots,$$

where the terms omitted are small in comparison with the parenthesis. Hence V' is definite near the origin. On the other hand V takes the sign of y near the origin, hence sometimes the sign of V' and so by (6.4) the system is unstable.

Suppose now that N is *odd* and set $V = 1/2\,y^2 + gW$. This time

$$V' = y\{gy^N + Y(y;z)\} + g\,\partial W/\partial z\,\{Qz + Z(y;z)\}$$
$$= g(y^{N+1} + \Sigma z_k^2) + \ldots.$$

Near the origin V has variable sign if $g > 0$, while V' has the sign of g. Hence (17.1) is unstable when $g > 0$ (6.4) and asymptotically stable when $g < 0$ (6.3).

Needless to say these results agree with (16.4).

§ 7. The converse of Liapunov's theorems.

18. Is the stability situation of the origin completely characterized by the existence of suitable functions V, W or U? The problem has been dealt with by many authors and largely solved, in the affirmative, by Krassovskii [1]. There are also notable contributions by Persidskii [1] (first to deal with the question), Malkin [9], Massera [2, 3], Kurzweil [1], and others. Especially ample references are given by Krassovskii [1] and Antosiewicz [2].

The complete discussion of this converse problem is quite complicated and difficult. We shall do no more here than give a taste of the situation and deal with the converse for stability and for asymptotic stability.

19. Take our usual system

$$dx/dt = X(x; t), \tag{19.1}$$

with $X(0; t) = 0$ for $t \geqq \tau$ in a region $\Omega(A, \tau)$ where we assume, however, that X is of class C^1.

Preliminary observation: It is evident that the proofs of the two Liapunov theorems on stability and asymptotic stability are still valid under much weaker assumptions on the functions $W(x)$. In particular the following type is quite sufficient. Let $r(x) = ||x||$ stand for the Euclidean modulus and choose as $W(x)$, a step function $F(r)$ defined thus: Divide the interval $[0, A]$ into a sequence of subintervals by means of a decreasing sequence $A_0 = A, A_1, A_2, \ldots$, where A_{i+1} is the midpoint of $[0, A_i]$. Thus $|A_{i+1}A_i| \to 0$ with $1/i$. As a matter of fact, any decreasing sequence $\to 0$ with the preceding property would do. On $[A_{i+1}A_i]$ define $F(r) = \varrho_i > 0$, where $\varrho_{i+1} < \varrho_i$ and $\varrho_i \to 0$ with $1/i$.

It is clear that the adequacy of the new functions $W(x)$ is much more easily verified than that of the original functions. Notice furthermore, that as a consequence one has the following characterization (due to Antosiewicz [2]) of the function $V(x; t)$ of Liapunov's theorem: V is of class C^1 in $\Omega(A, \tau)$; it is zero for $x = 0$ and $t \geqq \tau$; for every $0 < \varepsilon < A$ there is a $\varrho(e) > 0$ such that $V \geqq \varrho(\varepsilon)$ in $||x|| \in [\varepsilon, A]$; $V' \leqq 0$ in $\Omega(A, \tau)$. It is such a function that is henceforth called a Liapunov function.

20. We shall now take up the proof given by Persidskii [1]

of the inverse of Liapunov's stability theorem (6.1). We follow more or less the detailed formulation of Antosiewicz [2] p. 146.

It will be convenient to write $F(t; x_0; t_0)$ for the solution of (19.1) with trajectory Γ passing through x_0 at time t_0. We also assume that the domain of existence is of the form $I \times \Omega(A)$: $t \geqq 0$, $||x|| < A$. That is, it is assumed that the solutions extend indefinitely from any $(x_0; t_0)$ down to $t = 0$ and up to infinity.

(20.1) THEOREM. *Let $X(x; t)$ be of class C^1 in $I \times \Omega(A)$. If the origin is stable there exists a Liapunov function $V(x; t)$ in some $I \times \Omega(A')$, $A' \leqq A$.*

It is convenient to write for the present x^*, x; t for x, x_0; t_0 and s for the "variable" time. This amounts to taking the basic equation as

$$dx^*/ds = X(x^*; s). \tag{20.2}$$

The solution

$$x^*(s) = F(s; x; t) \tag{20.3}$$

is characterized by

$$x = F(t; x; t). \tag{20.4}$$

Let ε, $\eta(\varepsilon, t_0)$ be the constants of the definition of stability. The function $f(x; t) = F(0; x; t)$ is defined in a region $\Omega(A')$ where $A' \leqq A$ and is such that $||f(x; t)|| < \eta(A, 0)$ for $||x|| < A'$. This is the A' of the theorem.

Since $x^*(s)$ does not change as $(x; t)$ decribes the trajectory Γ defined by F,

$$\frac{\partial F}{\partial t} + \left(\frac{\partial F_i}{\partial x_j}\right) X(x; t) = 0. \tag{20.5}$$

Define now as applied to (19.1):

$$V(x: t) = \left[1 + \int_t^{+\infty} h(t)\, dt\right] \cdot ||f(x; t)||^2,$$

where $h(t)$ is positive on I, decreasing, and such that $\int^{+\infty} h(t)\, dt$ converges. Evidently V is of class C^1 in $I \times \Omega(A')$ and $V > 0$ on the set except that it is zero for $x = 0$ and $t \geqq 0$.

Also by (20.2) in $I \times \Omega(A')$:

$$V' = -h(t)||f(x; t)||^2 \leqq 0.$$

Take now any $0 < \varepsilon$ small enough. Since the origin is stable there is an $\eta(\varepsilon) \geqq 0$ such that if $||f(x; t)|| < \eta$ then $||x|| < \varepsilon$ for $t \,\varepsilon\, I$. Hence, $||x|| \geqq \varepsilon$ implies $||f(x; t)|| \geqq \eta$ and $V \geqq l\eta^2 > 0$. Thus V is a positive definite function and the theorem is proved.

21. We pass now to the converse of Liapunov's asymptotic stability theorem (6.3). We shall give Massera's proof [2] (the first) for the autonomous system

(21.1) $$dx/dt = X(x), \; X(0) = 0, \; x \in \Omega(A)$$

where X is of class C^1.

Observe that a Liapunov function in $\Omega(A)$ continues to be characterized as before by these properties: $W(0) = 0$; $W(x) > 0$ in $\Omega(A) - 0$; $W' \leqq 0$ in $\Omega(A)$.

(21.2) THEOREM. *If the origin is asymptotically stable for* (21.1) *there exists in a suitable* $\Omega(A')$, $A' \leqq A$, *a Liapunov function* $W(x)$ *such that* $W' < 0$ *along the paths of* (23.1).

22. Certain preliminary results are required. Let $S(\varrho)$ denote the sphere $||x|| < \varrho$. The region of operation is thus $S(A)$.

Let ε, $\eta(\varepsilon)$ be the constants of the definition of stability (IV, 2). At the cost of taking a smaller η we may assume that stability means that any path Γ emanating from the closure $\bar{S}(\eta)$ remains in $S(\varepsilon)$. In particular, if we take $\varepsilon = A$ then $\eta(\varepsilon) = \eta(A) = A'$ of the theorem.

Notice that a path may be defined by a solution $F(t; x_0; 0) = f(t; x_0)$. Let U_t denote the transformation $x_0 \to f(t; x_0)$, $x_0 \in \bar{\Omega}(A')$.

That $f(t; x_0) \to 0$ as $t \to +\infty$ means that given $0 < \xi < \eta$ and $x_0 \in \bar{S}(\eta)$ there is a time $T_0 = T(x_0)$ such that $U_t x_0 \,\varepsilon\, S_\xi$ for $t \geqq T_0$.

(22.1) LEMMA. *There is a time* T *such that* $U_t S(\eta) \subset ((S(\xi)$ *for all* $t \geqq T$.

This is an obvious uniformity property as to $x_0 \in S(\eta)$.

Let $\eta_1 = \eta(\xi)$. Evidently there is a time T'_0 such that $U_t x_0 \subset (S(\eta_1)$ for $t \geqq T'_0$. Since $x_1(x_0) = U_{T'_0} x_0$ is a continuous function of x_0, the latter has a neighborhood N_0 in $\bar{S}(\eta)$ such that $U_{T'_0} N_0 \subset S\ (\eta_1)$. It follows that $U_t N_0 \subset S\ (\xi)$ for $t \geqq T'_0$ (stability property). Since $\bar{S}\ (\eta)$ is compact it has a finite covering $\{N_1, \ldots, N\}_s$ where N_i is like N_0 with T'_i instead of T'_0. Hence $T = \sup \{T'_i\}$ behaves as required by the lemma.

23. (23.1) LEMMA. (Massera) *Let $\varphi\ (t)$ be a function defined, positive, continuous on $(0, +\infty)$, $\rightarrow 0$ as $t \rightarrow +\infty$. There exists a function $G\ (\varrho)$ defined, positive, continuous, increasing on the same interval, such that the integral*

$$\int^{+\infty} G\ [\varphi^*\ (t)]dt \tag{23.2}$$

is convergent, and this uniformly for all $0 \leqq \varphi^*(t) \leqq \varphi(t)$.

Let $0 < t_1 < t_2 < \ldots < t_n < \ldots, t_{n+1} \geqq 1 + t_n$, be a divergent sequence so chosen that for $t \geqq t_n$ also $\varphi\ (t) \leqq 1/(n+1)$. Define $\varrho\ (t)$ such that $\varrho\ (t_n) = 1/n$ and that $\varrho\ (t)$ is linear on $[t_n, t_{n+1}]$. On $(0, 1)$ set $\varrho\ (t) = (t_1/t)^p$ where the positive integer p is chosen so high that for $t = t_1$ the slope of the curve $\varrho = \varrho(t)$ in the (t, ϱ) plane is larger to the left than to the right. As a consequence $\varphi\ (t) < \varrho\ (t)$ on $(t_1, +\infty)$ and $\varrho\ (t)$ has a unique inverse $t\ (\varrho)$ on $(0, +\infty)$.

Define now

$$G\ (\varrho) = \int_0^{\varrho} \varepsilon^{-t(\varrho)} d\varrho.$$

Clearly, $G\ (\varrho) > 0$ for $\varrho > 0$ and $G\ (0) = 0$. For $t > t_1$ we have $\varphi^*(t) < \varphi\ (t) < \varrho\ (t)$ and since $t\ (\varrho)$ is a decreasing function, $t\ [\varphi^*(t)] \geqq t[\varrho\ (t)] = t$. Regarding the convergence of (23.2) it is the same as that of

$$< \int_{t_1}^{+\infty} dt \int_{+\infty}^{t} \varrho'(t)\ e^{-t}\ dt.$$

$$J = \int_{t_1}^{+\infty} dt \int_0^{\varphi^*(t)} e^{-t(\varrho)}\ d\varrho$$

Since $\varrho'(t)$ is bounded and has only finite discontinuities at the t_n and a slope < 1 in absolute value, the last integral exists. Our inequality proves then that J, or the integral (23.2), is uniformly convergent relative to $\varphi^*(t)$.

24. *Proof of* (21.2). Introduce again the new equation

$$dx^*/ds = X(x^*) \tag{24.1}$$

and let first A be finite. Define

$$W(x) = \int_0^{+\infty} G\,[||F(s; x; 0)||^2]\,ds. \tag{24.2}$$

Evidently $W(x)$ is positive definite and of class C^1 in $\Omega(A')$. Now in view of $F(s; x; t) = F(s - t; x; 0)$, we have

$$W(x) = \int_t^{+\infty} G\,[||F(s; x; t)||^2]\,ds = \lim_{h\to+\infty} \int_t^h G\,[||F(s;x;t)||^2]\,ds,$$

and hence, along the paths in $\Omega(A')$:

$$\frac{dW(x)}{dt} = \lim_{h\to+\infty} \left\{ \int_t^h \frac{d}{dt} G\,[||F(s; x; t)||^2]\,ds - G\,[||F(t; x; t)||^2] \right\}$$
$$= -G\,(||x||^2) < 0$$

for $x \neq 0$ and 0 for $x = 0$. This proves the theorem when A is finite.

Suppose now A infinite and let $0 < A_1 < A_2 < \ldots < A_n < \ldots$, be a divergent sequence. For each A_i there is an $A'_i = \eta(A_i)$ with $W(x)$, given by (24.2) defined in $\Omega(A'_i)$ where $\Omega(A'_i) \subset \Omega(A'_{i+1})$, and behaving as required. Hence, $W(x)$ behaves in appropriate manner in $\Omega(A')$ where A' is the finite or infinite limit of the nondecreasing sequence $\{A'_i\}$. This completes the proof of the theorem.

CHAPTER VII

The Differential Equation

$$dx/dt = P(t)\,z + q(x; t),$$

$$(P(t) \text{ a variable matrix; } q(0; t) = 0)$$

This time we are discussing systems such as those of the preceding two chapters but with the all important difference that the matrix $P(t)$ is not constant any more. Contrary to what happens when P is constant the stability properties of the first approximation are insufficient to govern those of the system itself. This was indeed brought out by Perron who showed by an example ([3], p. 705, see problem 11) that the first approximation could be stable and yet the original system be unstable. A good part of our discussion will therefore be devoted to the determination of sufficient conditions to be imposed upon the matrix $P(t)$ to guarantee the stability of the given system.

We first prove a theorem of Perron which states that one may apply a linear transformation of variables not affecting the stability properties and reducing $P(t)$ to a triangular matrix. Several equivalent sufficient conditions are then developed for asymptotic stability of the first approximation. One of these via Liapunov's second method yields a satisfactory asymptotic stability condition for the system under discussion. In the last section we discuss briefly the Liapunov numbers and their relation to stability.

References: Liapunov [1]; Malkin [1], Chs. 2, 3; Perron [2], [3]; Persidskii [1].

§ 1. Perron's Reduction Theorem

1. Consider the linear vector system of dimension n

$$(1.1) \qquad dx/dt = P(t)\, x$$

where for $t \geq \tau$ the matrix $P(t)$ is continuous and bounded. If we make a change of variables $x \to U^{-1}(t)\, x$, where in the appropriate range $U(t)$ is bounded and continuous together with its derivative and its inverse exists and is also bounded, then (1.1) is replaced by

$$(1.2) \qquad dx/dt = U^{-1}(PU - dU/dt)\, x$$

where the coefficient matrix is of the same nature as before.

(1.3) THEOREM. *There exists for $t \geq \tau$ a real, continuously differentiable, orthogonal matrix $U(t)$ such that $y = U(t)\, x$ satifies*

$$(1.3a) \qquad \frac{dy}{dt} = Q(t)\, y, \quad Q(t) = U'(t) \left[P(t)\, U(t) - \frac{dU}{dt} \right]$$

where the matrix $Q(t)$ is triangular.

(1.4) REMARK: *If the column-vectors of $U(t)$ are denoted by $u_i(t)$, $1 \leq i \leq n$, and the elements of $Q(t)$ by $q_{ij}(t)$, $1 \leq i, j \leq n$, then*

$$q_{ij}(t) = \begin{cases} [1/(1 + \delta_{ij})]\, u_i'(t) \cdot [P(t) + P'(t)]\, u_j(t), & i \leq j, \\ 0, & i > j. \end{cases}$$

Hence, if $P(t)$ is bounded for $t \geq \tau$, so is $Q(t)$.

Let $x_1(t), \ldots, x_n(t)$ be n linearly independent solutions of (1.1). Set

$$v_1(t) = x_1(t), \; u_1(t) = v_1(t)/\|v_1(t)\|$$

and define successively, for $k = 2, 3, \ldots, n$,

$$v_k(t) = x_k(t) - \sum_{i=1}^{k-l} (x_k'(t) \cdot u_i(t))\, u_i(t), \; u_k(t) = v_k(t)/\|v_k(t)\|.$$

Because of the linear independence of the set of solutions $x_1(t), \ldots, x_k(t)$, $k = 1, 2, \ldots, n$, we have $\|v_k(t)\| \neq 0$ for $t \geq \tau$ and thus the vectors $u_k(t)$, $1 \leq k \leq n$, are defined for $t \geq \tau$. Obviously, they are continuously differentiable for $t \geq \tau$, and we easily verify

$$u_i'(t) \cdot u_j(t) = \delta_{ij}, \quad 1 \leq i, j \leq n.$$

We also notice that the vectors $u_1(t), \ldots, u_k(t)$, $k = 1, 2, \ldots, n$, depend solely upon the vectors $x_1(t), \ldots, x_k(t)$. Hence, in particular, if $U(t) = (u_1(t), \ldots, u_n(t))$ and $X(t) = (x_1(t), \ldots, x_n(t))$, then

$$U(t) = X(t)\, S(t) \tag{1.5}$$

where $S(t)$ is triangular. Clearly, $S(t)$ is continuously differentiable and non-singular for $t \geq \tau$.

Differentiating (1.5), we find

$$\frac{dU}{dt} = \frac{dX}{dt} \cdot S + X \frac{dS}{dt} = PU + US^{-1} \frac{dS}{dt} \tag{1.6}$$

and hence

$$Q = U' \left[PU - \frac{dU}{dt} \right] = - S^{-1} \frac{dS}{dt}.$$

Since S is triangular, so is S^{-1}. Therefore Q is triangular, and the proof is completed.

As to the Remark, note that since $U'U = E$ implies $U' \dfrac{dU}{dt} = - \dfrac{dU'}{dt} U$, we also have

$$Q = \left[U'P + \frac{dU'}{dt} \right] U$$

and, using (1.6),

$$Q = \left[U'P + U'P' + \left(S^{-1} \frac{dS}{dt}\right)' U'\right] U = U'(P' + P)\, U - Q'.$$

Thus

$$Q + Q' = U'(P + P')\, U,$$

which yields (1.4) because Q is triangular.

Perron's theorem means in substance that *regarding stability of the origin, asymptotic or otherwise, one may replace the system* (1.1) *by a similar system in which $P(t)$ is a triangular matrix.* We shall therefore assume throughout the sequel that $P(t)$ is triangular.

Observe that by making the change of coordinates $x_k \to x_{n-k}$ one may replace a triangular matrix P with zeros below the main diagonal by one with zeros above that diagonal and vice versa.

§ 2. Various Stability Criteria

4. Let there be given a system

$$dx/dt = X(x; t), \; X(0; t) = 0 \tag{4.1}$$

valid in a region $\Omega(A, \tau)$. We will say that $V(x; t)$ is a *strong Liapunov function* of the system (4.1) if it satisfies the conditions for asymptotic stability at the origin of Liapunov's second theorem (VI, 6.3). That is to say V, V' are both definite but of opposite signs and $|V|$ is dominated by a $W(x)$. Unless otherwise stated we will generally assume that V is positive definite.

Let us now take a system (1.1) where the matrix $P(t) = (p_{ij}(t))$ is continuous, bounded and triangular. The following functions will play an important role in the sequel:

$$G_s(t, \tau) = \exp \int_\tau^t p_{ss}(t')\, dt', \; t \geqq \tau, \; s = 1, 2, \ldots, n. \tag{4.2}$$

A noteworthy set of solutions of (1.1) is the system $y_{jk}(t, t_0)$, $t_0 \geqq \tau$, defined by the initial conditions

$$y_{jk}(t_0, t_0) = \delta_{jk}. \tag{4.3}$$

One finds by direct calculation the values:

$$y_{11} = G_1(t, t_0),$$

$$\cdots\cdots\cdots\cdots$$

$$y_{s1} = G_s(t, t_0) \int_{t_0}^t \sum_{i=1}^{s-1} p_{si}\, y_{i1} \cdot G_s^{-1}(t', t_0)\, dt',$$

$$\cdots\cdots\cdots\cdots$$

$$y_{12} = 0,$$

$$y_{22} = G_2(t, t_0),$$

$$\cdots\cdots\cdots\cdots$$

$$y_{s2} = G_s(t, t_0) \int_{t_0}^t \sum_{i=1}^{s-1} p_{si}\, y_{i2} \cdot G_s^{-1}(t', t_0)\, dt',$$

$$\cdots\cdots\cdots\cdots$$

$$y_{1n} = \ldots = y_{n-1,\, n} = 0,$$

$$\cdots\cdots\cdots\cdots$$

$$y_{nn} = G_n(t, t_0).$$

5. The system (1.1) possesses solutions $x(t, t_0)$, $t_0 \geq \tau$, such that $||x(t_0, t_0)|| = 1$. Thus the solutions $(y_{1s}, \ldots, y_{ns})$ have that property. Notice in fact that if $||x(t_0, t_0)|| = 1$, then the solution $x^*(t, t_0) = \varrho\, x(t, t_0)$, where ϱ is any positive number, is such that $||x^*(t_0, t_0)|| = \varrho$. To impose $||x(t_0, t_0)|| = 1$ is thus merely a way of normalizing the solution $x(t, t_0)$.

We shall now prove the following property due to Persidskii [1]:

(5.1) *If the system* (1.1) *possesses a strong Liapunov function and* $x(t, t_0)$ *is a solution such that* $||x(t_0, t_0)|| = 1$, $t_0 \geqq \tau$, *then*

(a) $||x(t, t_0)||$ *is bounded as* $t \to +\infty$;

(b) *there exist positive constants* $M < 1$, *but otherwise arbitrary, and* $T(M)$ *depending solely on* M *such that* $||(x(t_0 + t, t_0)|| \leqq M$ *for* $t \geqq T$.

We will refer to (a), (b) as *Persidskii's conditions.*

Since the system is stable (a) follows from (IV, 3.3).

Let now $V(x; t)$ be a strong Liapunov function of (4.1) and $W(x)$, $W_1(x)$ the two associated positive definite functions such that $V \geqq W$, $V' \leqq -W_1$ for $t \geqq \tau$. Choose any positive $M < 1$. Since $V(x; t) \to 0$ uniformly in t, for ϱ small $l = \sup V(x; t)$ when $||x|| = \varrho$ and all $t \geq \tau$, is finite. Let also $l' = \inf W_1(x)$ for $M\varrho \leq ||x|| \leq A$, $0 < \varrho < A$, and set $T = l/l'$. Then if $x(t)$ is such that $x(t_0) = \xi$, $||\xi|| = \varrho$, we have with $t \geq t_o$:

$$\begin{aligned} W(x) &\leq V(\xi; t_0) + \int_{t_0}^{t} V'(x(t'); t')\, dt' \\ &\leq l - (t - t_0)\, l' . \end{aligned}$$

It follows that at time $t_0 + T$: $||x(t_0 + T)|| < M\varrho$, since otherwise

$$W(x(t_0 + T)) \leq l - T\, l' = 0 ,$$

which is ruled out since W is definite positive.

Since the system is linear we conclude that if $||x(t_0, t_0)|| = 1$ then $||x(t_0 + T, t_0)|| < M$. Thus condition (b) holds likewise.

6. Besides G_s introduce the function for $t \geqq t_0$:

$$H_s(t, t_0) = G_s(t, t_0) \int_{t_0}^{t} G_s^{-1}(t', t_0)\, dt'.$$

A first and elementary result is:

(6.1) *If G_s and H_s are both bounded as $t \to +\infty$ then $G_s \to 0$.* (Perron [3]).

Suppose $t \geq \tau$ and that for all $t \geq \tau$:

(6.1a) $$G_s(t, \tau) < R, \ H_s(t, \tau) < S,$$

where R and S are positive constants. Then

$$S \geq G_s(t, \tau) \int_\tau^t \frac{dt'}{R} \geq \frac{G_s(t, \tau)(t - \tau)}{R}.$$

Hence $G_s(t, \tau) \leq \dfrac{RS}{t - \tau} \to 0$ with $1/t$ as asserted.

We will refer to "G_s and H_s are bounded for all s" as the *Perron conditions*. We shall prove with Malkin [1] the following two properties:

(6.2) *The Persidskii conditions imply the Perron conditions.*

(6.3) *Introduce the functions*

$$K_s^{(m)}(t, \tau) = G_s^{-m}(t, \tau) \int_t^{+\infty} G_s^m(t', \tau)\, dt',$$

where m is any positive integer > 1. Then if the Perron conditions hold there exist positive constants α, β such that for all $t \geq \tau$:

(6.3a) $$\alpha \leq K_s^{(m)}(t, \tau) \leq \beta.$$

Let us write everywhere $p, G, \ldots$, for $p_{ss}, G_s, \ldots$.

Proof of (6.2). It follows from the expressions of the y_{sj} in (4) that as a consequence of the second Persidski condition $G(t_0 + T, t_0) < M$ for all $t_0 \geq \tau$. Hence

$$G(t_0 + mT, t_0 + (m-1)T) < M.$$

As a consequence

$$G(t_0 + mT, t_0) = \prod_{r=0}^{m} G_s(t_0 + (r+1)T, t_0 + rT) < M^m.$$

Consequently

$$G(t_0 + mT, \tau) < M^m G(t_0, \tau),$$

for $\tau \leqq t_o \leqq \tau + T$. Hence if $\gamma = \sup G(t_0, \tau)$ for t_o in the segment just stated:

$$G(t, \tau) < \gamma M^m \tag{6.4}$$

when t is between $\tau + mT$ and $\tau + (m+1)T$. Since $M < 1$, $G(t, \tau) \to 0$ as $t \to +\infty$.

The argument for $H(t, \tau)$ is more difficult. First observe that

$$G(t, \tau) G^{-1}(t', \tau) = \exp\{\int_\tau^t p(u)\,du - \int_\tau^{t'} p(u)\,du\}$$

$$= \exp \int_{t'}^t p(u)\,du .$$

Hence

$$H = \int_\tau^t G(t, t')\,dt' .$$

From this follows that for $\tau + (m-1)T \leqq t \leqq \tau + mT$:

$$H \leqq \int_\tau^{\tau+mT} G(t, t')\,dt' = \int_\tau^{\tau+T} + \int_{\tau+T}^{\tau+2T} + \cdots + \int_{\tau+(m-1)T}^{\tau+mT}$$

Set now

$$\varphi_k = G((m-1)T + \beta, (k-1)T + \alpha),$$
$$\alpha = t - (k-1)T, \beta = t - (m-1)T .$$

Then

$$\varphi_k = G^{-1}((k-1)T + \alpha, (k-1)T) \cdot$$
$$G((m-1)T + \beta, (m-1)T) \cdot G((m-1)T, (k-1)T) .$$

Let C be an upper bound for all the p_{ss}. Remembering the upper bound (6.4) for $G(t, t')$ (t' takes the place of τ) we have

$$\varphi_k \leqq e^{2CT} G((m-1)T, (k-1)T) \leqq DM^{m-k} .$$

Hence as $t \to +\infty$

$$H < DT \Sigma M^{m-k} \leqq DT/(1-M) .$$

This completes the proof of (6.2).

Proof of (6.3). In view of the continuity of the functions G_s, H_s for $t \geqq \tau$, and of their boundedness as $t \to +\infty$, we have (6.1a) and so as in (6.1):

$$G(t, \tau) \leqq RS/(t - \tau)\,.$$

Therefore since $m > 1$ the integral in (6.3) converges and hence $K^{(m)}(t, \tau)$ is a continuous function of t for all $t \geqq \tau$. Applying now the mean value theorem to the expression that follows, which can be calculated and found to be $-1/m$, we have:

$$G^{-m}(t, \tau) \int_t^\infty p(t')\, G^m(t', \tau)\, dt' = -1/m = p(\theta)\, K^{(m)}(t, \tau)\,,$$
$$\tau \leqq \theta < +\infty\,.$$

Since $p(\theta)$ is bounded, $K^{(m)}(t, \tau)$, which is positive, is bounded away from zero, or $K^{(m)} \geqq a > 0$.

Consider now the expression

$$H^m(t, \tau)\, K^m(t, \tau) = \frac{\displaystyle\int_t^\infty G^m(t', \tau)\, dt'}{\left[\dfrac{1}{\displaystyle\int_\tau^t G^{-1}(t', \tau)\, dt'}\right]^m}\,.$$

As $t \to +\infty$ it takes the form 0/0. By the rule of l' Hospital its value as $t \to +\infty$ is the same as for the quotient of the derivatives or of the expression

$$\frac{-G^m(t, \tau)}{\dfrac{-mG^{-1}(t, \tau)}{\left[\displaystyle\int_\tau^t G^{-1}(t', \tau)\, dt'\right]^{m+1}}} = \frac{H^{m+1}(t, \tau)}{m}\,.$$

Hence as $t \to +\infty$

$$\frac{K^{(m)}(t, \tau)}{H(t, \tau)} \to \frac{1}{m}\,.$$

Since $H(t, \tau)$ is bounded so is $K^{(m)}(t, \tau)$ and this completes the proof of (6.3).

7. The results just obtained will now be applied to the proof of the following proposition also due to Malkin [1]:

(7.1) *If the Perron conditions hold then the system* (1.1) *possesses a strong Liapunov function which is a quadratic form in the* x_i.

Observe that the existence of such a Liapunov function is not affected by a linear homogeneous transformation of coordinates (with constant coefficients). Consider a transformation

$$x_1 = \sigma^{n-1} y_1,\ x_2 = \sigma^{n-2} y_2,\ \dots,\ x_n = y_n\,.$$

The system for y will be

$$\begin{aligned} dy_1/dt &= p_{11}\, y_1 \\ dy_2/dt &= \sigma\, p_{21}\, y_1 + p_{22}\, y_2 \\ &\dots\dots\dots\dots\dots\dots \\ dy_n/dt &= \sigma^{n-1} p_{n1}\, y_1 + \dots + p_{nn}\, y_n\,. \end{aligned}$$

By taking σ small enough we may dispose of the situation so that the coefficients below the main diagonal are in absolute value arbitrarily small for $t \geqq \tau$.

Let us suppose then that in the initial system (1.1) the p_{jk}, $j \neq k$, are suitably small. The Perron conditions still hold and so we can define

$$V(x;t) = \Sigma\, K_s^{(2)}(t,\tau)\, x_s^2\,,$$

and we note that since there are positive constants α, β such that

$$\alpha \leqq K_s^{(2)}(t,\tau) \leqq \beta,\ s = 1, 2, \dots, n$$

we will have

$$W(x) = \beta\, \Sigma x_s^2 \geqq V(x;t) \geqq \alpha\, \Sigma x_s^2 = W_1(x)\,.$$

Hence V is positive definite, and $\to 0$ with x uniformly in t. Moreover

$$\begin{aligned} V' &= \Sigma\,(-2p_{ss}(t)\, K_s^{(2)}(t,\tau) - 1)\, x_s^2 \\ &\quad + \Sigma\, K_s^{(2)} \cdot 2\, x_s\, \Sigma\, p_{sr}\, x_r \\ &= -\Sigma\, x_s^2 + \sum_{r<s} q_{sr}\, x_s\, x_r\,, \end{aligned}$$

where the q_{sr} may be assumed arbitrarily small. Let them be chosen such that for $\Sigma x_s^2 = 1$ the maximum of $\Sigma\, q_{sr}\, x_s\, x_r$ is $< 1/2$. Under these conditions $V' < -1/2\, \Sigma\, x_s^2$ and so V' is definite negative. Thus $V(x; t)$ is a Liapunov function such as asserted.

8. The preceding results have a number of noteworthy consequences. First there is the following sequence of properties for the system (1.1):

(8.1) *Existence of a strong Liapunov function* → *Persidskii conditions* → *Perron conditions* → *Existence of a quadratic Liapunov function.*

Now the extreme properties are independent of the choice of variables under transformations such as occur in the proof of Perron's "triangular" matrix theorem. Therefore:

(8.2) *Given a system*

$$dx/dt = P(t)\, x, \tag{8.2a}$$

where the matrix $P(t)$ is merely continuous and bounded for $t \geqq \tau$ (but not necessarily triangular), if the system possesses a strong Liapunov function, it also possesses a quadratic Liapunov function.

From (8.1) coupled with theorem (VI, 6.3) of Liapunov we find:

(8.3) *For a system* (1.1) *with triangular matrix $P(t)$ the Persidskii conditions, the Perron conditions and the existence of a strong Liapunov function are all equivalent. Any one of the three implies that the system is asymptotically stable.*

We are now in position to prove:

(8.4) Stability Theorem of Perron. *Consider a general system*

$$dx/dt = P(t)\, x + q(x; t)\,, \tag{8.4a}$$

where $P(t)$ is continuous and bounded for $t \geqq \tau$, but not necessarily triangular, where in $\Omega(A, \tau)$ the function $q(x; t)$ is continuous, $q(0; t) = 0$, and

$$\|q\| = o(\|x\|)\,. \tag{8.4b}$$

If the first approximation

$$dx/dt = P(t)\, x \tag{8.4c}$$

has a strong Liapunov function then the given system (8.4a) *is asymptotically stable for all q behaving as described.*

We may assume that the Liapunov function $V(x; t)$ of the first approximation is a quadratic in the x_i. Then, taken along the solutions of (8.4a)

$$V' = \Sigma\,(q_s + \Sigma\, p_{sj} \cdot x_j)\,\partial V/\partial x_s + \partial V/\partial t\,.$$

We have shown in fact in (7) that there is a positive definite quadratic $W(x)$ such that

$$V' - \Sigma\, q_s\, \partial V/\partial x_s < - W(x)\,.$$

In view of (8.4b) we will have for x small say $V' < - 1/2\, W$. Hence V is a strong Liapunov function for (8.4a) also and the theorem follows.

(8.5) *Remark.* We emphasize once more Perron's observation that the mere asymptotic stability of the first approximation is not sufficient *a priori* to guarantee the stability of the system itself. A stronger condition is required: existence of a strong Liapunov function for the first approximation, or equivalently, fulfillment of the conditions of Perron or Persidskii.

§ 3. The Liapunov Numbers. Application to Stability

9. In his fundamental mémoire, to which we have often referred, Liapunov introduced ([1], p. 223) certain numbers analogous in a sense to the characteristic roots, and which he no doubt designed as a substitute for the characteristic roots. And indeed when characteristic roots do exist their negatives are the Liapunov numbers. It may be said that much later Perron [1] unaware of Liapunov's work defined the same numbers and obtained analogous results to those of Liapunov. We shall discuss very sketchily the Liapunov numbers and indicate—without proofs—their application to stability.

10. The Liapunov number of a function $\varphi(t)$ of the real variable t is a number which measures as it were the "exponential" growth of $\varphi(t)$ as $t \to + \infty$.

Let $\{\lambda'\}$ be the set of all numbers such that $e^{\lambda' t} \cdot \varphi(t) \to 0$ as

$t \to +\infty$. If λ_1' is in the set and $\lambda_2' < \lambda_1'$ then λ_2' is likewise in the set. Hence $\{\lambda'\}$ is an interval infinite to the left. Let λ_0 be its right end-point. Similarly the set $\{\lambda''\}$ of all numbers such that $e^{\lambda'' t} \cdot |\varphi(t)| \to +\infty$ with t, is an interval infinite to the right. Let λ_1 be its left end-point. If both λ_0, λ_1 are finite:

(10.1) $$\lambda_0 = \lambda_1$$

If the contrary holds then there is an interval $\lambda_0 < \mu < \lambda_1$. Let α be in the interval. Then $e^{\alpha t} \cdot |\varphi(t)|$ is bounded as $t \to +\infty$. Hence if $\beta < \alpha$ is in the interval then $e^{\beta t} \cdot \varphi(t) \to 0$ as $t \to +\infty$. It follows that β is a λ'. Hence $\beta \leqq \lambda_0$ contrary to assumption. Hence $\lambda_0 = \lambda_1$. The common value $\lambda_0 = \lambda_1$ is the *Liapunov number* $\lambda(\varphi)$ of φ. If one of them is infinite it is by definition $\lambda(\varphi)$.

If $\{\varphi_1, \ldots, \varphi_r\}$ is a finite set of functions of t then the least of the numbers $\lambda(\varphi_h)$ is defined as the *Liapunov number* of the set. If f is a vector then the Liapunov number of f, written again $\lambda(f)$, is the Liapunov number of the set of components of f. In particular, if we have an n-dimensional system

(10.2) $$dx/dt = P(t)\, x$$

where $P(t)$ is continuous, then the Liapunov number $\lambda(x)$ of a vector solution $x(t)$ is well defined.

(10.3) *If the matrix $P(t)$ is continuous and bounded for t large then the Liapunov number $\lambda(x)$ of every solution $x(t)$ of* (10.2) *is finite.*

Let μ be a real number and set $z = e^{\mu t} \cdot x$. Thus z satisfies

$$dz/dt = (P + \mu E)\, z\,,$$

or explicitly

$$dz_h/dt = p_{h1} z_1 + \ldots + (p_{hh} + \mu) z_h + \ldots + p_{hn} z_n\,.$$

From this follows with $S = \Sigma z^2_h$ that

$$1/2\, dS/dt = \Sigma\, (p_{hh} + \mu)\, z_h^2 + \sum_{h \neq k} (p_{hk} + p_{kh})\, z_h z_k\,.$$

Let $-\varrho$ be the minimum, for t large enough, of

$$\Sigma\, p_{hk}\, z_h z_k \text{ for } S = 1\,,$$

and let $\mu > \varrho + 1/2\, N$. It is readily shown that ϱ is finite. We have then $dS/dt > NS$, and hence $S > Ce^{Nt}$, $C > 0$, for t large enough. Similarly $S < Ce^{-Nt}$ for t large enough. Hence for some finite μ there are functions $|z_h(t)| \to +\infty$, and for some (other) finite μ there are functions $z_h(t) \to 0$. Hence the Liapunov number of some x_h is finite and it is $-\infty$ for none. This proves (10.3).

11. Let there be given a system (10.2) with $P(t)$ continuous and bounded for $t \geqq \tau$. A system of n linearly independent solutions is said to be *normal* if the sum S of its Liapunov numbers is equal to its maximum value.

Let $\mu = \lambda\,(\exp\,(-\int_\tau^t \Sigma p_{ss}(t')\,dt'))$. At all events $S \leqq -\mu$. We can only have $S = -\mu$ if

$$\lambda\,(\exp \int_\tau^t \Sigma\, p_{ss}(t')\,dt') + \lambda\,(\exp - \int_\tau^t \Sigma\, p_{ss}(t')\,dt') = 0\,.$$

If $S = -\mu$ the system is said to be *regular*. We have:

(11.1) *If the matrix $P(t)$ of* (10.2) *is triangular then n.a.s.c. for regularity are*

$$\lambda\,(G_s(t, \tau)) + \lambda\,(G_s^{-1}(t, \tau)) = 0,\; s = 1, 2, \ldots, n\,.$$

Liapunov considers now a *regular* analytic system (10.2). If $\lambda_1, \ldots, \lambda_n$ are the characteristic numbers of a set of n linearly independent solutions of the first approximation, he derives an expansion theorem analogous to (V, 9.1) and deduces from it stability relations similar to those of (V, 14.2).

This suffices to give a slight idea of the scope of Liapunov's results.

CHAPTER VIII

Periodic Systems and Their Stability

In the present chapter we discuss general periodic systems and related stability problems. They are closely related to the stability theory developed in the preceding chapters.

References: Liapunov [1]; Malkin [2]; Poincaré [1, 2, 3].

§ 1. Linear Homogeneous Systems with Periodic Coefficients

1. Let there be given a general real system in an n-vector x,

$$dx/dt = X(x; t) \tag{*}$$

where X has appropriate continuity and differentiability properties. Suppose moreover that X is periodic with period ω in t and let there be known a solution $\xi(t)$ likewise periodic and of period ω. Upon substituting $x = \xi + y$ above, one obtains the variation equation of ξ and it has the form

$$dy/dt = P(t)\, y$$

where P is periodic with period ω in t. Thus linear systems with periodic coefficients arise naturally as variation systems of periodic solutions of general periodic systems.

Let us take then a system now written in terms of x:

$$dx/dt = P(t)\, x \tag{1.1}$$

where the matrix $P(t)$ is continuous, periodic and has period $\omega > 0$. Side by side with (1.1) we also consider the matrix equation

$$(1.2) \qquad dX/dt = P(t)\, X\,.$$

These systems have already been discussed in (III, § 5). We recall the following properties proved there:

I. If $X(t)$ is a non-singular solution of (1.2) then $X(t+\omega) = X(t)\, C$, where C is constant, non-singular and its characteristic roots $\mu_1, \ldots, \mu_n$ are the characteristic exponents of (1.2).

II. There is a matrix B such that $C = e^{\omega B}$. The matrix $B = 1/\omega \log C$ is not unique (see Appendix I, § 4). If $C = \operatorname{diag}(C(\mu_1), \ldots, C(\mu_r))$ then $B \sim \operatorname{diag}(C(\lambda_1), \ldots, C(\lambda_r))$ where $\lambda_h = 1/\omega \log \mu_h + 2\, k_h \pi i/\omega$. (The notation $C(\lambda)$ utilized here is the same as in Appendix I and stands for the blocks of the normal form.)

III. The matrix $Z(t) = e^{Bt} \cdot X^{-1}$ has the period ω. It is continuous, non-singular and its terms and those of $Z^{-1}(t)$ are uniformly bounded.

IV. The transformation of variables

$$(1.3) \qquad y = Z(t)\, x$$

reduces (1.1) to the linear system with constant coefficients

$$(1.4) \qquad dy/dt = By\,.$$

V. The system (1.1) has a base $\{x^h\}$, $x^h = \psi^h(t)\, e^{\lambda_h t}$, where the components of the vector ψ^h are polynomials in t with coefficients periodic in t. These coefficients are continuous and uniformly bounded in t.

Let us bring out also an interesting relation due to Liapunov ([1], p. 394) between the periodic matrix $P(t)$ and the matrix B. From (III, 4.3) and the periodicity of X we infer the relation

$$|X(t+\omega)| = |X(t)|\, e^{\int_0^\omega \operatorname{trace} P(t)\, dt}\,.$$

Hence by (I, 10.11)

$$|X^{-1}(t)\, X(t+\omega)| = |e^{\omega B}| = e^{\omega \operatorname{trace} B} = e^{\int_0^\omega \operatorname{trace} P(t)\, dt}\,.$$

Since the last two exponents are real they are equal or

$$\text{trace } B = \frac{1}{\omega}\int_{0}^{\omega} \text{trace } P(t)\, dt\,. \tag{1.5}$$

This is the relation which we had in view. It yields the following two relations for the characteristic roots λ_h and the characteristic exponents μ_h:

$$\Sigma\, \lambda_h = \frac{1}{\omega}\int_{0}^{\omega} \text{trace } P(t)\, dt\,, \tag{1.6}$$

$$\Pi\, \mu_h = \exp \int_{0}^{\omega} \text{trace } P(t)\, dt\,. \tag{1.7}$$

Notice finally that if one starts with the system (*) and if $\xi(t)$ is a periodic solution of the system then the variation equation is (1.1) with $P(t) = \partial X(\xi(t))/\partial\xi$. Thus (1.5) and (1.6) become

$$\text{trace } B = \Sigma\, \lambda_h = \frac{1}{\omega}\int_{0}^{\omega} \text{trace } \partial X/\partial\xi \cdot dt\,. \tag{1.8}$$

Recall that

$$\text{trace } \partial X/\partial x = \Sigma\, \partial X_h/\partial x_h\,.$$

This last sum is commonly known as the *divergence* of the vector X, written div X, with varied physical interpretations notably in fluid mechanics. If γ is the closed path corresponding to $\xi(t)$, (1.8) may be given the form

$$\text{trace } B = \Sigma\, \lambda_h = \frac{1}{\omega}\int_{\gamma} \text{div } X \cdot dt\,. \tag{1.9}$$

2. Upon examining property V in more detail a noteworthy conclusion may be drawn. A non-singular solution of the matrix equation corresponding to (1.4) is represented by e^{Bt}. By a suitable choice of coordinates we may assume that B is in normal form with terms corresponding to the characteristic number zero first. Let this characteristic number be of multiplicity s. Now the associated x^h are the columns of the matrix

$$X = Z^{-1}(t)\, e^{Bt}\,.$$

Thus the first of these columns is pure periodic and the last $n - s$ are periodic-exponential multiplied by polynomials. It is readily shown that no linear combination of the last $n - s$ column vectors can have the period ω. On the other hand at least one linear combination of the first s column vectors and may be more have that period. If we remember that $\lambda_h = 0$ when and only when $\mu_h = 1$, we may state:

(2.1) *If the system* (1.1) *has s characteristic exponents equal to unity then it has a maximal d-dimensional linear family of periodic solutions with period* ω, *where* $1 \leq d \leq s$.

One may mention here also another interesting result regarding the characteristic exponents due to Liapunov. Let $C(\mu_h)$ designate as before the blocks of the normal form of the matrix C. Thus

$$C \sim \operatorname{diag}(C(\mu_1), \ldots, C(\mu_r)). \tag{2.2}$$

Now Liapunov' result ([1], p. 398) reads:

(2.3) *If D is the analogue of C for the equation adjoint to* (1.1) *then*

$$D \sim \operatorname{diag}(C(\mu_1^{-1}), \ldots, C(\mu_r^{-1})).$$

The adjoint matrix equation to (1.2) is

$$dY/dt = -YP(t) \tag{2.4}$$

and D is defined by $Y(t + \omega) = DY(t)$ for a non-singular solution $Y(t)$ of (2.4). Let this solution be chosen, as one may, so that $Y(t)X(t) = E$. As a consequence

$$Y(t + \omega) X(t + \omega) = E = DY(t) X(t) C = DC.$$

Hence $D = C^{-1}$ and so (2.3) follows from Appendix I, (4.1).

The *stability properties* of (1.1) at the origin are readily obtained. For the transformation (1.3) is linear and uniformly regular at the origin and so (1.1) and (1.4) behave alike as regards stability at the origin. Moreover "$Re\,\lambda_h < 0$ or > 0" and "$|\mu_h| < 1$ or > 1" are equivalent properties. Hence we deduce from (V, 14.8):

(2.5) STABILITY THEOREM FOR LINEAR PERIODIC SYSTEMS. *If the characteristic exponents are all* $< 1 [> 1]$ *in absolute value the system*

(1.1) *is asymptotically stable [is unstable] at the origin. If k of the exponents are* $< 1\,[> 1]$ *in absolute value there is a linear k-dimensional family of asymptotically stable [unstable] solutions at the origin.*

§ 2. Analytic Systems with Periodic Coefficients

3. We turn now our attention to real systems in an n-vector x, of type (notations of (I, 14.5))

$$dx/dt = P(t)\,x + [x;t]_2 \tag{3.1}$$

where [...] is holomorphic in x for all t in a fixed $\Omega(A)$. Furthermore the terms of $P(t)$ and the coefficients of the series [...] are continuous and of period ω in t. Thus the question of the stability of the origin arises again and there are now two possibilities.

I. *The matrix P is constant.* The system falls then under the theory already treated (see V, 14.8).

II. *The matrix P is periodic but not constant.* Then the change of variables (1.3) will reduce the system (3.1) to one of the same form but with P replaced by B. This system falls under the type studied under (V, § 4) and (VI, § 1).

Observe now that property (V, 8.5) for the λ_h is equivalent to the following for the characteristic exponents μ_h:

(3.2) *There exists no combination*

$$\mu_j = \Pi\,\mu_h^{m_h},\ m_h \geqq 0,\ \Sigma m_h > 1\,.$$

It may be observed that since the λ_j are only determined mod $2\pi i/\omega$, condition (V, 8.5) for them reads:

(3.3) *The* λ_j *satisfy no relation*

$$\lambda_j = \Sigma m_h\,\lambda_h,\ \text{mod}\ 2\pi i/\omega;\ m_h \geqq 0,\ \Sigma m_h > 1\,.$$

We have now from (V, 9.1, 14.8):

(3.4) EXISTENCE THEOREM. *If the characteristic exponents are all* < 1 *in absolute value and satisfy* (3.2) *then theorem* (V, 9.1) *holds for the system* (3.1) *provided that the* $X^{(m)}(t;a)$ *are now polynomials in t with coefficients periodic and of period* ω *in t. The* λ_h *occurring in the series are the characteristic roots of the matrix B.*

(3.5) STABILITY THEOREM. *If the characteristic exponents are all* < 1 *in absolute value the system* (3.1) *is asymptotically stable at the origin. If* k *of them are* < 1 *in absolute value there is a family of solutions depending analytically on* k *parameters which is asymptotically stable at the origin.*

§ 3. Stability of Periodic Solutions

4. The preceding results have an immediate application to the stability of a periodic solution $\xi(t)$ of a system

$$dx/dt = X(x; t), \tag{4.1}$$

where X is holomorphic in x in a certain region $\mathfrak{R}$ of $\mathfrak{B}_x$ for all t and is continuous and periodic with period ω in $\mathfrak{R}$ and $\xi(t) \in \mathfrak{R}$. Suppose in addition that if $x = \xi + y$, then for y small enough and t arbitrary $X(\xi + y; t)$ may be expanded in power series of y. Upon substituting in (4.1) there results an equation in y of the form

$$dy/dt = P(t)\,y + [y; t]_2 . \tag{4.2}$$

The matrix

$$P(t) = \left(\frac{\partial X}{\partial x}\right)_{x=\xi(t)} \tag{4.3}$$

is merely the Jacobian matrix of the X_j as to the x_k taken on the known closed trajectory. If $P(t) \neq 0$ and is non-constant we are under Case II. There will be a set of characteristic exponents $\{\mu_j\}$ and we may state:

(4.4) *If the Jacobian matrix* (4.3) *is not constant and* $\{\mu_j\}$ *are its characteristic exponents, then theorem* (3.5) *holds with orbital stability of the trajectory* $\xi(t)$ *instead of stability.*

§ 4. Stability of the Closed Paths of Autonomous Systems. The Method of Sections of Poincaré

5. It so happens that autonomous systems escape the result just obtained. Let indeed

$$dx/dt = X(x) \tag{5.1}$$

be such a system, and let it behave otherwise like (4.1) and in particular let it possess a solution $\xi(t)$ of period ω. We form again (4.2) but unfortunately:

(5.2) *At least one of the characteristic exponents μ_j is now unity.* (Poincaré)

We notice first that

$$dy/dt = P(t)\, y \tag{5.3}$$

is the variation equation of the periodic solution $\xi(t)$. Since $\xi(t+\tau)$ is also a solution of (5.1), $\partial\xi(t+\tau)/\partial\tau$ is likewise a solution of (5.3). Since this solution is periodic of period ω, (5.2) is a consequence of (2.1).

Thus in the all important autonomous case one cannot apply the full stability part of the general stability theorem and one must look elsewhere to complement the argument. This is where Poincaré's method of sections enters the problem. It may be said that this method was extensively utilized by G. D. Birkhoff ([1,] II, 268).

Speaking first in very general terms suppose that the paths of an autonomous system, say (5.1), intersect repeatedly a differentiable manifold M^{n-1}, and do so "without contact," i.e. without ever being tangent to M^{n-1}. This means really that the vector $X(x)$ is never a differential vector of M^{n-1}. Let a path γ intersect M^{n-1} consecutively in points Q, Q'. Then $Q \to Q'$ defines a transformation S of M^{n-1} into itself. The fixed points of S correspond to closed paths. Moreover if F is such a point and γ the corresponding closed path it follows readily from the definitions in (IV, § 6) that the stability behavior of the transformation S in the neighborhood of F characterizes the orbital stability properties of γ. We will observe the application of this principle in a moment. The manifold M^{n-1} is called a *manifold of section* of the system, and the method just exposed is Poincaré's method of sections.

6. Returning now to our problem let γ be the closed path of $\xi(t)$ and let Q be any point of γ. Take Q as origin and coordinates such that the x_n axis is the tangent to γ at Q. Thus in the new coordinates $X(Q) = (0, \ldots, 0, X_n(Q))$. We may further choose units such that $X_n(Q) = 1$. Thus $X(Q)$ is merely the unit vector

along the x_n axis. To simplify matters we will also suppose that the period is 2π and that Q is $\xi(0)$ so that $\xi(0) = \xi(2\pi) = 0$.

Let now the present disposition be identified with the disposition of (II, 16) and so that the transverse cell E^{n-1} there considered is in the hyperplane H: $x_n = 0$. One may now take as the parameters u_h the coordinates $x_1, \ldots, x_{n-1}$ in H. However it will be actually more convenient to preserve for those coordinates the designations u_h.

According to (II, 16) the path γ' passing through the point u (u small) will meet H next at a point u' where

$$u' = Du + [u]_2 \tag{6.1}$$

where D is a constant matrix of order $n - 1$.

Now the general solution, near $\xi(t)$, starting at the point u in H at time $t = 0$ is analytic in u and so it has the form (notations of (I, 14.5))

$$x(t; u) = \xi(t)u + n(t) + \varphi(t; u), \quad \varphi = [u]_2 \tag{6.2}$$

where η is an $n \times n - 1$ matrix.

If $y = (y_1, \ldots, y_n)$ is an n-vector or A an $n \times m$ matrix, let y^* denote the $(n - 1)$-vector $(y_1, \ldots, y_{n-1})$ and A^* the $(n - 1) \times m$ matrix of the first $n - 1$ rows of A. Let also A_n denote the m-vector of the last row of A. With these notations we obtain from (6.2):

$$x^*(0; u) = u = \xi^*(0) + \eta^*(0)u + \varphi^*(0; u).$$

Hence $\eta^*(0) = E_{n-1}$. Observing now the relations

$$\xi_n(0) = \xi_n(2\pi) = 0; \quad \left(\frac{d\xi_n}{dt}\right)_0 = \left(\frac{d\xi_n}{dt}\right)_{2\pi} = X_n(0) = \alpha \pm 0$$

and if $2\pi + \varepsilon$ is the next time that γ' crosses H, we have

$$x_n(2\pi + \varepsilon; u) = 0 = \xi_n(2\pi + \varepsilon) + \eta_n(2\pi + \varepsilon)u$$
$$+ \varphi_n(2\pi + \varepsilon, u) = \alpha\,\varepsilon + \eta_n(2\pi)u + \psi(\varepsilon; u), \quad \psi = [\varepsilon; u]_2.$$

As a consequence, $\varepsilon(u)$ is holomorphic at $u = 0$ and $\varepsilon(0) = 0$. Observing also the relations

$$\xi^*(0) = \xi^*(2\pi) = 0; \left(\frac{d\xi^*}{dt}\right)_0 = \left(\frac{d\xi^*}{dt}\right)_{2\pi} = X^*(0) = 0,$$

we find again from (6.2):

$$x^*(2\pi + \varepsilon; u) = u' = \xi^* (2\pi + \varepsilon) + \eta^*(2\pi + \varepsilon) u$$
$$+ \varphi^*(2\pi + \varepsilon; u) = \eta^*(2\pi) u + [u]_2 = Du + [u]_2,$$

and therefore $\eta^*(2\pi) = D$.

Recall now that according to (III, **3**. 8) and the remark leading to (5.2), the variation equation of (5.1) has the fundamental matrix solution

$$C(t) = \begin{pmatrix} \eta^*(t), & a^{-1} \left[\dfrac{\partial \xi^*(t+\tau)}{\partial \tau}\right]_{\tau=0} \\ \eta_n(t), & a^{-1} \left[\dfrac{\partial \xi_n^*(t+\tau)}{\partial \tau}\right]_{\tau=0} \end{pmatrix}.$$

Since $C(0) = E_n$, the characteristic roots of $C(2\pi) = C$ are the characteristic exponents. Now

$$C = \begin{pmatrix} D, & 0 \\ \eta_n(2\pi), & 1 \end{pmatrix}.$$

Hence the $n - 1$ characteristic exponents, other than unity corresponding to the solution $\xi(t)$, are the characteristic roots of the matrix D. That is:

(6.3) *The characteristic exponents of the autonomous system* (5.1) *other than one equal to unity are merely the characteristic roots of the constant matrix D of the transformation* (6.1).

7. By applying if need be a linear transformation to the variables u_h (a linear transformation in the hyperplane Π) we may reduce D to normal form. Let this be done and let us suppose that the characteristic exponents are all less than unity in absolute value. Let S denote (6.1) and let its linear approximation S_1 be

(7.1) $$u' = Du.$$

Take one of the constituent blocks of D say

$$D_1 = \begin{pmatrix} \mu & 0 & \cdots & \\ 1 & \mu & & \\ \cdots & \cdots & \cdots & \\ \cdots & \cdots & 1 & \mu \end{pmatrix}$$

and if k is its order let $R_k = D_1 - \mu E_k$. We recall that $R_k{}^j$ is merely R_k with its diagonal of units lowered by j steps. Hence for s sufficiently high

$$D_1{}^s = (\mu E_k + R_k)^s = \begin{pmatrix} \mu^s, & 0, \ldots, & 0 \\ s\mu^{s-1}, & \mu^s, & \\ \cdots & \cdots & \cdots \\ \cdots & \cdots & \cdots \\ \binom{s}{k-1}\mu^{s-k+1}, & \binom{s}{k-2}\mu^{s-k+2}, \ldots, & \mu^s \end{pmatrix}.$$

From this we conclude that if μ is the largest $|\mu_j|$, so that $\mu < 1$, then the terms of D^s are in absolute value $< \beta_s = a\, s^n \mu^s$ where a is independent of s. Also $\log \beta_s = \log a + n \log s - s \log (1/\mu) \to -\infty$ with $-s$. Hence $\beta_s \to 0$ as $s \to +\infty$. It follows that if $||u|| < \varrho$ then $||S_1{}^s u|| < n\, \beta_s \varrho$.

Now referring to (6.1): $||[u]_2||/||u|| \to 0$ with u. Hence one may take ϱ so small that when $||u|| < \varrho$ then $||[u]_2|| < ||u||/10$. One may then choose s so large that $n\beta_s < 1/2$. Under the circumstances if U is the spheroid $\mathfrak{S}(0; \varrho)$ of the hyperplane H then $S^s U \subset U$. Therefore the path γ is orbitally stable. Moreover diam $S^s U \to 0$ as $s \to +\infty$ and hence the orbital stability is asymptotic.

Suppose that only $k < n - 1$ of the $|\mu_j| < 1$. One may assume that they are the first k. Then the same reasoning confined to paths initiated on $u_{k+1} = \ldots = u_{n-1} = 0$ close to the point Q shows that the family of such paths which depends upon $u_1, \ldots, u_k$ is orbitally asymptotically stable. If on the contrary $|\mu_j| > 1$ for $j \leq k$ then the same family is negatively orbitally asymptotically stable and therefore positively unstable.

To sum up then we have:

(7.2) *The stability theorem* (3.5) *is applicable to the orbital stability properties of a closed path* γ *of an autonomous system provided that the characteristic exponents* μ_j *in the statement are the* $n - 1$ *other than unity.*

§ 5. Systems of Periodic Solutions

8. We now turn our attention to another problem. Consider an analytical system

$$(8.1) \qquad dx/dt = X(x; v; t)$$

of the same type as discussed in (II, 11.1) except that in addition X has the period ω in t. Let us suppose that for $v = 0$ there is a known solution $\xi(t)$ of period ω. One asks whether this periodic solution is member of an analytic family of solutions with the same period ω. The method of attack goes back to Poincaré ([4], vol. I, Ch. IV). We first apply his theorem (II, 11.1). The solution which starts at time $t = 0$ at $\xi(0) + \eta$ is analytic in η and v for $0 \leqq t \leqq \tau, \tau \geqq \omega, ||\eta|| < a, ||v|| < a$. Let this solution be

$$x(t; \eta; v) = \xi(t) + y(t; \eta; v), \ y(t; 0; 0) = 0.$$

The analytic function y satisfies

$$(8.2) \qquad \frac{dy}{dt} = f(t)\, y + g(t)\, v + \psi(y; v; t), \ \psi = [y; v]_2,$$

$$f(t) = \left.\frac{\partial X(x, v, t)}{\partial x}\right]_{(\xi; 0; t)}, \ g(t) = \left.\frac{\partial X(x; v; t)}{\partial v}\right]_{(\xi; 0; t)}$$

Thus f and g have the period ω. Moreover,

$$y(0; \eta; v) = \eta, \ y(\omega; \eta; v) = \eta + \varphi(\eta; v).$$

Since $y(\omega; 0; 0) = 0$ we have $\varphi(0; 0) = 0$. The variation equation of (8.2) is

$$(8.3) \qquad \frac{du}{dt} = f(t)\, u$$

and we notice that it is periodic of period ω. From Poincaré's theorem for (8.2) one infers an expansion

$$(8.4) \qquad y(t; \eta; v) = \alpha(t) \cdot \eta + \beta(t) \cdot v + \ldots,$$

where α, β are matrices. Substituting in (8.2) one finds

$$\frac{d\alpha}{dt} \cdot \eta + \frac{d\beta}{dt} \cdot v + \ldots = f(t)\, (\alpha \cdot \eta + \beta \cdot v + \ldots) + g(t) \cdot v +$$

and hence by comparison of terms

$$\frac{d\alpha}{dt} = f(t) \cdot \alpha.$$

Thus α is a "matrix-solution" of (8.3). Also,

$$y\,(0;\eta;v) = \eta = \alpha(0)\cdot\eta + \beta\,(0)\cdot v + \ldots,$$

$$y\,(\omega;\eta;v) = \eta + \varphi\,(\eta;\;v) = \alpha\,(\omega)\cdot\eta + \beta\,(\omega)\cdot v + \ldots$$

Hence,

$$\alpha\,(0) = E_n,\ \alpha\,(\omega) = E_n + J,\ J = \partial\varphi\,(0;\,0)/\partial\eta.$$

A *n.a.s.c.* for y to have the period ω is

$$\varphi\,(\eta;v) = 0. \tag{8.5}$$

If the determinant $|J| \neq 0$, (8.5) will have a unique (real) solution $\eta(v)$ holomorphic at $v = 0$ such that $\eta\,(0) = 0$. To this will correspond in (8.1) a unique family of periodic solutions $\xi\,(t) + y\,[t;\eta\,(v);v]$ depending upon the vector parameter v and reducing to $\xi\,(t)$ for $v = 0$.

Let μ now be a characteristic exponent of (8.3) and let $\alpha(t)\bar{\eta}$ ($\bar{\eta}$ a constant vector) be an associated solution $\neq 0$ of (8.3). Thus,

$$\alpha\,(\omega)\,\bar{\eta} = \mu\,\alpha(0)\,\bar{\eta} = \mu\,\bar{\eta} = (E_n + J)\,\bar{\eta},$$

and hence,

$$[(1-\mu)\,E_n + J]\,\bar{\eta} = 0.$$

Since $\bar{\eta} \neq 0$ this implies that the determinant

$$|(1-\mu)\,E_n + J| = 0. \tag{8.6}$$

Let $\mu_1, \ldots, \mu_n$ be the solution of (8.6). The existence of a periodic solution of (8.3) is equivalent to having some $\mu_h = 1$, and hence to $|J| = \Pi\,(1-\mu_n) = 0$. Comparing with the above this yields:

(8.7) *If the variation equation has no solutions of period ω then the system* (8.1) *possesses a solution of period ω depending analytically upon the vector parameter v and tending to $\xi(t)$ as $v \to 0$.*

9. The preceding considerations may be extended to an autonomous analytical system

$$dx/dt = X\,(x;\,v)\,. \tag{9.1}$$

Let the system possess a solution $\xi(t)$ of period ω corresponding to a closed path γ. Everything will depend upon the choice of a special system of coordinates for a small neighborhood of γ. Since γ is analytic it has an arc length s. Let s be counted from a certain point A of γ and positively in the direction of increasing time. Let also the scale for s be so chosen that the total length of γ is unity. The variable s will be one of our coordinates. We propose to prove the following property in whose statement $n = \dim \mathfrak{V}_x$ and γ is in fact any analytical Jordan curve in $\mathfrak{V}_x$:

$(9.2)_n$ *There can be assigned, uniquely, to points in a suitably small neighborhood* $U(\gamma)$ *of* γ, *coordinates* $y_1, \ldots, y_{n-1}$, s *such that*: (a) *for a given* s *the coordinates* $y_1, \ldots, y_{n-1}$ *are suitable coordinates in a subspace* $\mathfrak{V}^{n-1}$ *transverse to* γ *at the point* $P(s)$ *and with origin at* P; (b) *the transformation* $x \rightarrow (y; s)$ *is regular analytic at* P.

The dimension $n = 2$ is the lowest of interest so we first prove $(9.2)_2$, i.e. when γ is a plane curve.

Since γ is analytical the radius of curvature R of γ has a positive lower bound 2ϱ. For otherwise γ would contain a point Q with $R(Q) = 0$ and Q would be a cusp, i.e. a point where γ ceases to be analytical. It follows that there is an $\alpha > 0$ such that if $P', P'' \in \gamma$ and $|s(P') - s(P'')| < \alpha$ then the normals at P', P'' meet at a point whose distance from P' or P'' exceeds ϱ. Hence if M is in the ϱ neighborhood of γ the shortest normal PM from M to γ, with foot at P, is unique. We take $s(M) = s(P)$ and $y_1 = \pm PM$ according as the normal points inward or outward at γ, and the coordinates y_1, s manifestly satisfy $(9.2)_2$.

The case $n = 3$ is still special and we must consider it separately. The directions of the tangents to γ form a system of dimension one and so one may choose the direction of the x_3 axis—the vertical axis—not among them. Hence γ has at most a finite number of vertical chords. For otherwise it would have a continuous family and therefore a vertical tangent. Moreover by a familiar argument each vertical chord meets γ in a finite number of points. Thus γ has at most a finite number of points $Q_1, \ldots, Q_p$ through each of which there passes a vertical chord.

Let now $A(a_1, a_2, a_3) \in \gamma$ and let δ be an open arc of γ containing A, with at most one of the Q_h on $\bar{\delta}$, and admitting a parametric representation by convergent power series

$$x_i = a_i + b_i \sigma + c_i \sigma^2 + \dots . \tag{9.3}$$

The tangent at A is

$$x_i = a_i + b_i \sigma ,$$

and, since it is not vertical, b_1 and b_2 are not both zero. Hence the first two equations of (9.3), which represent the vertical projection δ' of δ on $x_3 = 0$, define δ' as an analytical arc of that plane. Since δ possesses no vertical chord the vertical projection maps δ topologically on δ'.

Since γ is compact it has a finite covering by open arcs $\delta_1, \dots, \delta_s$ such as δ. Let δ_h' be the projection of δ_h. One may treat δ_h' like γ in the plane case provided that one only deals with points M' near δ_h' such that the foot P' of the shortest normal is interior to δ_h' and at a certain positive distance from the end-points of δ_h'. Assign now to M near enough δ_h the value of s at $P \in \delta_h$ projected on P', the coordinate y_1 equal to $\pm$ the shortest normal from M' to δ_h, and $y_2 = x_3(M) - x_3(P)$. It is clear that for y_1, y_2, s property $(9.2)_3$ holds.

Let now $n > 3$ and suppose that $(9.2)_{n-1}$ holds. The directions of the tangents and chords of γ make up a system of dimension two. Since $n > 3$ one may choose the vertical direction outside this system. As a consequence the projection γ' of γ on $x_n = 0$ will be such that the operation of projection is a topological mapping $\gamma \to \gamma'$. Let now M be a point very near γ and M' its projection. By hypothesis coordinates $y_1, \dots, y_{n-2}, s$ may be assigned to points such as M', where we may in fact choose for s the arc length on γ itself. There is then a point P' of γ' such that $s(M') = s(P')$. The point P' is the projection of a unique point P of γ and we define $s(M) = s(P') = s(P)$, then $y_{n-1}(M) = x_n(M) - x_n(P)$. This gives a uniquely defined assignment of coordinates $y_1, \dots, y_{n-1}, s$ to a suitably small neighborhood U of γ in $\mathfrak{B}_x^n$ for which $(9.2)_n$ holds. Thus (9.2) is proved.

One may transfer directly the system (9.1) to the new coordinates and one obtains the analytical system

$$\begin{aligned} dy/dt &= Y_0(y; v; s) , \\ ds/dt &= S(y; v; s) , \end{aligned} \tag{9.4}$$

where Y_o is a vector, S a scalar and both have period 1 in s. Moreover $Y_o(0; 0; s) = 0$ and $S(0; 0; s)$ is positive and bounded away from zero. Hence upon dividing Y_o by S one obtains a relation

$$dy/ds = Y(y; v; s) \tag{9.5}$$

where Y behaves like Y_o. This system is exactly of the same type as (8.1) save that t is replaced by s, the period is unity and the related $\xi(s) = 0$. The analogue of the variation equation (8.3) is here

$$\frac{du}{ds} = \left.\frac{\partial Y(y; 0; s)}{\partial y}\right|_{y=o} \cdot u \tag{9.6}$$

and so (8.7) applies. Thus:

(9.7) *If the variation equation* (9.6) *has no periodic solution of period unity then the system* (9.1) *has a closed path* γ_v *depending analytically upon* v *and such that* $\gamma_v \to \gamma$ *as* $v \to 0$. *Moreover the solution* $\xi(t; v)$ *corresponding to* γ_v *is such that* $\xi(t; 0) = \xi(t)$.

This is in substance the generalization of (8.7) which we had in view.

Since the coordinates y make up a coordinate system for a space transverse to γ at a point P with origin at P, they may serve as the coordinate system u of (6). This carries as a consequence that the characteristic exponents of (9.6) are all the $n - 1$ of (9.1) for $v = 0$ other than the special exponent unity corresponding to $\xi(t)$. Thus:

(9.8) *If the system* (9.1) *for* $v = 0$ *has only one characteristic exponent unity then* (9.1) *possesses an analytic family of periodic solutions* $\xi(t; v)$ *such that* $\xi(t; 0) = \xi(t)$.

(9.9) *A general remark on autonomous systems.* Let us consider generally the system (9.1) with known solution $\xi(t)$ whose period we now denote by ω_0. Let $\alpha(v) = 2\pi/\omega(v)$, $\omega(0) = \omega_0$, and make the change of variables $t = \theta/\alpha$, then add as $(n + 1)$-st coordinate $\alpha(v)$. As a consequence (9.1) is replaced by the new system in the unknown vector $(x; \alpha)$:

$$dx/d\theta = X(x; v)/\alpha, \; d\alpha/d\theta = 0 \, .$$

For $v = 0$ it has the periodic solution $(x(\theta; 0); \alpha(0))$ of period 2π and one looks for periodic solutions $(x(\theta; v); \alpha(v))$ of the same period 2π. The fact that this variable periodic solution $(x(\theta; v); \alpha(v))$ has the fixed period 2π makes it possible to apply the non-autonomous procedures to the autonomous case (see notably the developments of § 8).

§ 6. Quasi-Linear Systems and Their Periodic Solutions

10. One understands by quasi-linear system one of the form

$$dx/dt = Px + \mu q(x; \mu; t) \tag{10.1}$$

where P is our familiar constant matrix, μ is a small parameter and q is a power series in x and μ. The series is convergent for x and μ small and its coefficients are continuous and with period ω in t. Furthermore the characteristic roots of P all have negative real parts and satisfy condition (3.3). We propose to examine directly the existence of periodic solutions for μ small which $\to 0$ with μ.

The first approximation has the matrix solution e^{Pt} and the vector solution $x^0 = e^{Pt} \cdot a$, with initial vector $x^0(0) = a$. Let us adopt the same initial value for the general solution of (10.1). By Poincaré's expansion theorem (II, 11.1) this general solution is given by a series

$$x(t) = x^0(a; t) + \mu x^1(a; t) + \dots,$$

with the x^h analytic in a. Upon substituting in (10.1) and identifying powers of μ we obtain a system

$$\begin{cases} x^{o\prime} - Px^o = 0 \\ x^{h\prime} - Px^h = f_h(a; x^0, \dots, x^{h-1}; t), \; h > 0 \end{cases} \tag{10.2}$$

where f_h is analytic in a and its components are polynomials in those of the x^j, $j < h$. Together with the initial conditions $x^o(0) = a$, $x^h(0) = 0$ for $h > 0$, this system may be solved by recurrence. Of course $x^0 = e^{Pt} \cdot a$.

The condition for the existence of a solution of period ω is

$$\Phi(a; \mu) = x^0(a; \omega) - x^0(a; 0) + \mu\{x^1(a; \omega) - x^1(a; 0)\} + \dots = 0.$$

Now

$$x^0(a;\omega) - x^0(a;0) = (e^{P\omega} - E)\,a\,.$$

Since P has no pure complex characteristic roots the determinant $D = |e^{P\omega} - E| \neq 0$.

Now the Jacobian

$$\left|\frac{\partial \Phi(a;\mu)}{\partial a}\right|_{\substack{a=0\\ \mu=0}} = D \neq 0\,.$$

Hence $\Phi(a;\mu) = 0$ has a unique solution $a(\mu)$, holomorphic in μ near $\mu = 0$ and such that $a(0) = 0$. This proves the existence of the periodic family of trajectories.

Notice that as $\mu \to 0$ the closed trajectory γ_μ tends to the origin. For γ_0 is a periodic solution of the first approximation and under our assumptions on the λ_j, $e^{Pt}\cdot a$ has the period ω if and only if $a = 0$, i.e. if γ_0 is the origin.

11. Let us consider the stability of the periodic trajectory γ_μ, always for μ small. The corresponding solution has the form $\mu\xi(t;\mu)$. Substituting $x = \mu\xi + y$ in (10.1) one obtains the variation equation of γ_μ:

$$(11.1)\qquad dy/dt = (P + \mu\,Q(\mu;t))\,y\,.$$

Applying (II, 11.1) to the corresponding matrix equation, its solution assumes the form

$$Y(t) = Y_0(t) + \mu\,Y_1(t) + \ldots\,.$$

By substitution in the matrix form of (11.1) and identifying powers of μ we find that Y_0 is merely a solution of

$$dY_0/dt = PY_0\,,$$

or $Y_0 = e^{Pt}\cdot K$, K constant, non-singular. Hence

$$\begin{aligned} Y(t+\omega) &= e^{P(t+\omega)}\cdot K + \mu\,Y_1(t+\omega) + \ldots \\ &= (e^{Pt}\cdot K + \mu\,Y_1(t) + \ldots)\,C(\mu)\,. \end{aligned}$$

For μ small the parenthesis at the right represents a matrix with an inverse. Applying the inverse to both sides we find

$$C(\mu) = e^{P\omega} + \mu G + \dots .$$

Hence for μ small the characteristic exponents of the periodic solution $\mu\xi$ of (10.1) differ very little from the numbers $e^{\omega\lambda_j}$. Thus under our assumptions these characteristic exponents are all < 1 in absolute value. Hence γ_μ is asymptotically stable.

To sum up then we have proved:

(11.2) *A quasi-linear system* (10.1) *with the properties described possesses a unique family of closed trajectories: one and only one* γ_μ *for each small value* μ, *such that* γ_μ *depends analytically on* μ *and tends to the origin as* $\mu \to 0$. *The trajectory* γ_μ *is asymptotically stable for* μ *sufficiently small.*

§ 7. A Class of Periodic Solutions Studied by Liapunov

12. Liapunov made a detailed study for stability purposes of a system

$$dx/dt = Px + [x]_2 \tag{12.1}$$

with two pure complex characteristic roots $\pm i\lambda$ and the rest with negative real parts. It is in this connection that he discovered the periodic motions under consideration.

One may choose coordinates $y, y^*, z_1, \dots, z_q$ $(q = n - 2)$ such that (12.1) assumes the form

$$\begin{cases} \text{(a)}\ dy/dt = \lambda iy + Y(y, y^*; z), \\ \text{(b)}\ dy^*/dt = -\lambda iy^* + \overline{Y}(y^*, y; z), \\ \text{(c)}\ dz/dt = Qz + [y, y^*; z]_2 . \end{cases} \tag{12.2}$$

Here Y is a power series beginning with terms of degree at least two, $\overline{Y}$ is the series with conjugate coefficients, and Q is a real constant matrix whose characteristic roots are all those of P other than $\pm \lambda i$. The real solutions of (12.2) are obtained by taking $y^* = \bar{y}$.

Let us suppose now with Liapunov ([1] p. 361; see also Malkin [2] p. 152), that the system (12.2) has an exact integral,

(12.3) $$H(y, y^*; z) = \text{const.},$$

where H is analytic in the indicated variables. The existence of functions such as H is very natural for instance in dynamical systems: energy integral, Hamilton's function, etc. When there is a function H then along any path

$$dH/dt = (\lambda iy + Y(y, y^*; z))\, \partial H/\partial y + (-\lambda iy^* + \overline{Y}(y^*, y; z)) \times \partial H/\partial y^* + \Sigma\, (q_{hk} z_k + [y, y^*; z]_2{}^h)\, \partial H/\partial z_h = 0\,.$$

Let explicitly

$$H = ay + a^* y^* + byy^* + c\,(y^2 + y^{*2}) + y \cdot Cz + y^* \cdot \overline{C}z + W(z) + S(y, y^*; z)$$

where a, a^*, b, c are constants, C is a constant row vector, W is a quadratic form and $S = [y, y^*; z]_3$. Substituting above we find by identification to zero, and making $b = 1$, that

(12.4) $$H = yy^* + W(z) + S(y, y^*; z) = \mu^2\,,$$

where μ is an arbitrary constant.

Confining now our attention to real systems let $y^* = \overline{y}$, $y = re^{i\varphi}$. Set also $z = ru$, where u is a q-vector. Thus (12.2) is replaced by a real system

(12.5) $$\begin{cases} \text{(a)}\;\; dr/dt = r^2 R(r; u; \varphi)\,, \\ \text{(b)}\;\; d\varphi/dt = \lambda + r\,\Phi(r; u; \varphi)\,, \\ \text{(c)}\;\; du/dt = Qu + r\,\Psi(r; u; \varphi)\,, \end{cases}$$

where R, Φ, Ψ are real and holomorphicin r, u for r, u small, and as power series in r, u their coefficients are polynomials in $\sin\varphi$, $\cos\varphi$. As for (12.4) it yields

(12.6) $$r^2[1 + W(u) + rF(r; u; \varphi)] = \mu^2$$

where F is of the same type as R.

It is to be noted that as regards closed trajectories near the origin and their stability the system (12.5) may take the place of the initial system (12.1).

13. By differentiating (12.6) one finds that (12.5a) is a consequence of (12.5bc). Thus our system reduces to (12.5bc) together with (12.6). One may also eliminate dt from (12.5bc) and thus reduce the system to (12.6) together with

$$du/d\varphi = Qu + rS\,(r;\,u;\,\varphi) \tag{13.1}$$

where S is of the same type as R and Q behaves as before. We will have the same closed paths as (12.5) near the origin, i.e. for μ small and also the same stability properties. The closed paths will now correspond to periodic solutions $u\,(\varphi)$ with period 2π.

Upon extracting the square root in (12.6) and noting that only $r \geqq 0$ matters we find

$$r\,(1 + [u]_2 + r\,[r;\,u;\,\varphi]_1) = \mu \geqq 0\,, \tag{13.2}$$

where $[\ldots]_1$ refers to a series in r and u. The implicit function theorem applied to (13.2) yields for r when μ is small:

$$r = \mu + \mu^2\,G\,(\mu;\,u;\,\varphi)\,. \tag{13.3}$$

Substituting now in (13.1) we obtain finally

$$du/d\varphi = Qu + \mu\Omega\,(\mu;\,u;\,\varphi) \tag{13.4}$$

where Ω is like R save that r is replaced by μ. Our initial system is thus replaced by (13.4) together with (13.3).

Now the system (13.4) is really like (10.1) but with t replaced by φ and 2π as the period. Therefore by (11.2):

(13.5) *The system* (12.1) *under the assumptions made possesses a unique family of closed paths* γ_μ, *depending analytically on* μ *and such that* γ_μ *tends to the origin as* $\mu \to 0$ *from above. The family disappears as* μ *decreases through zero. The closed path* γ_μ *is asymptotically stable for* μ *sufficiently small.*

§ 8. Complete Families of Periodic Solutions

14. It is not too much to say that Poincaré, and other authors as well, have limited their treatment of various problems to

particularly simple situations: when a certain Jacobian does not vanish and the like. This has been notably true in connection with the determination of the families of periodic solutions of systems with a vector parameter (see § 5) containing a given periodic solution $\xi(t)$. We propose to return to this problem when the parameter is a scalar variable μ and give a treatment covering all cases. Our treatment is merely a mild adaptation of Kronecker's classical method of elimination. (See Lefschetz [5].)

The search for the periodic solutions $\xi(t; v)$ of a system (8.1) which as $v \to 0$ tend to the given periodic solution $\xi(t)$ of the system corresponding to $v = 0$ has been reduced (loc. cit.) to the solution of an analytic system

$$\varphi(\eta; v) = 0$$

where η is the unknown initial value of the solution. Let the vectors η, φ be denoted by y, Y. Taking then $v = \mu$ the system to be solved assumes the general form

$$Y_{1h}(y_1, \ldots, y_n, \mu) = 0,\ h = 1, 2, \ldots, n_1. \tag{14.1}$$

In point of fact here $n_1 = n$, but this is immaterial. The Y_{1h} are real holomorphic functions of y and μ near $y = 0$, $\mu = 0$. We shall first determine all the families of solutions of (14.1), whether real or complex, and then determine the real families.

Now if the Y_{1h} are all divisible by some power μ^j, $j > 0$, the general solution of (8.1) with $v = \mu = 0$, is periodic. Since we are interested in the families depending on μ, we divide every Y_{1h} by the highest possible power of μ and continue to call the quotient Y_{1h}. Thus $Y_{1h}(y; 0) \neq 0$. By a real linear transformation of the variables y_j one may dispose of the situation so that in Y_{1h} y_1 appears indeed to the lowest power $y_1^{m_h}$, where $m_h > 0$ is the lowest degree of any term in Y_{1h}. The Weierstrass preparation theorem (I, 14.1) yields then

$$Y_{1h} = Y_{1h}^1 E(y; \mu)$$

where Y_{1h}^1 is a real special polynomial in y_1. It is clear that we may replace (14.1) by the system

$$Y_{1h}^1 = 0,\ h = 1, 2, \ldots, n_1.$$

To simplify matters we may suppose then that in (14.1) the Y_{1h} are already special polynomials in y_1.

The Y_{1h} may have a common factor $D_1(y_1, \ldots, y_n, \mu)$. As is readily shown it will be a real special polynomial in y_1. Moreover the quotient $Y_{1h}/D_1 = Y^*_{1h}$ will also be, up to a unit factor, a special polynomial in y_1. Then $D_1 = 0$ represents for each μ an (n-1)-dimensional family of periodic solutions.

Consider now the system

$$(14.2) \qquad Y_{1h}^* = 0, \; h = 1, 2, \ldots, n_1 .$$

Following Kronecker we introduce two linear combinations with arbitrary parameters u, v

$$U = \Sigma\, u_h\, Y_{1h}^*, \quad V = \Sigma\, v_h\, Y_{1h}^* ,$$

and form the resultant $R(U, V)$ as to y_1. We will have

$$R = \Sigma\, W_k(u, v)\, Y_{2k}(y_2, \ldots, y_n, \mu) ,$$

where the W_k are monomials in the u_h and v_j. A n.a.s.c. in order that the system (14.2) possess a solution in y_1 is that

$$(14.3) \qquad Y_{2k}(y_2, \ldots, \mu) = 0, \; k = 1, 2, \ldots, n_2.$$

This system is entirely analogous to (14.1) but with one variable less. If the Y_{2k} are all divisible by μ then the $Y_{1h}^*(y_1, \ldots, y_n, 0)$ have a common factor $\Delta(y_1, \ldots, y_n)$ and $\Delta = 0$ represents a special (n-1)-dimensional family of periodic solutions of the system corresponding to $\mu = 0$. Let us suppose this out of the way and the possible μ factors already suppressed in the Y_{2k}.

One may now make a linear change to new variables $y_2, \ldots, y_n$, relative to which the Y_{2k}, to within unit factors, are regular in y_2 (its highest coefficient unity), take an irreducible factor $D_2(y_2, \ldots, \mu)$ of the H. C. F. of the Y_{2k} as to y_2, etc. If $D_2 = 0$ the Y_{1h}^* will have one or more irreducible common factors. If $Y_1(y_1, \ldots, y_n, \mu)$ is one of them we will have an (n-2)-dimensional system of periodic solutions:

$$Y_1(y_1, \ldots, y_n, \mu) = 0, \; D_2(y_2, \ldots, \mu) = 0$$

and so on.

The general scheme, in suitable notations, may be described as follows: There may exist for each $k \leq n$ and each μ sufficiently small a finite number of $(n\text{-}k)$-dimensional families of periodic solutions each represented by a system

$$(14.4)\qquad \begin{cases} Y_1(y_1, \ldots, y_n, \mu) = 0 \\ Y_2(y_2, \ldots, y_n, \mu) = 0 \\ \ldots\ldots\ldots\ldots\ldots\ldots \\ Y_k(y_k, \ldots, y_n, \mu) = 0 \end{cases}$$

where Y_h is a polynomial in y_h with non-unit coefficients and in particular Y_k is special in y_k. Furthermore Y_k is irreducible as a polynomial in y_k (it is not the product of two polynomials of its type), and Y_h, $h < k$, is irreducible in the same sense.

One may even go further. Let d_h be the degree of Y_h in y_h and let $d = \Pi d_h$. Thus given $y_{k+1}, \ldots, y_n$ arbitrary and complex, the system $(14.4)_k$ has d distinct solutions. Choose now real constants $c_1, \ldots, c_k$ such that the d values of $c_1 y_1 + \ldots + c_k y_k$ are distinct. Upon making the change of variables

$$y_h \to y_h,\ h < k,\ c_1 y_1 + \ldots + c_k y_k \to y_k\,,$$

we will have in place of $(14.4)_k$ a system such that all but the last equation in $(14.4)_k$ are of the first degree in the lowest indexed variable. The system $(14.4)_k$ assumes then the form

$$(14.5)_k\qquad \begin{cases} Y_k(y_k, \ldots, y_n, \mu) = 0 \\ A_0(y_{k+1}, \ldots, y_n, \mu)\, y_{k-h} - A_h(y_k, \ldots, y_n, \mu) = 0, \\ \qquad h = 1, 2, \ldots, k-1\,, \end{cases}$$

where Y_k is a polynomial in y_k and the A_j are non-units and real. Furthermore by a real linear transformation of the variables $y_k, \ldots y_n$ together with an application of the Weierstrass theorem one may dispose of the situation so that Y_k is a special polynomial in y_k.

We have thus obtained a complete, and even constructive, description of the real or complex families of periodic solutions which may arise and which contain the given solution $\xi(t)$ as a member corresponding to $\mu = 0$.

15. We pass now to the problem of reality of the solutions. We must find then the real solutions of the system $(14.5)_k$ in a small neighborhood of the origin. Let us first distinguish three types of solutions or points:

I. The points where the Jacobian matrix J of the left-hand sides is of maximum rank k at the same time as $A_0 \neq 0$. These are the *ordinary points*. If M is their set and $P \in M$ then there is a neighborhood U of P in M which is a complex analytical cell of real dimension $2(n - k + 1)$. This follows at once from the fact that about P one may express the coordinates and μ as power series in $n - k + 1$ of them.

II. The points where J is of rank $< k$. These are the *singular points* of M and we denote their set by S.

III. The points where $A_0 = 0$ or *exceptional points* and we denote their set by E.

Each of the three types may yield real points. If $Y_k = 0$ has a real solution in y_k for $y_{k+1}, \ldots, y_n, \mu$ arbitrary real and small, then the system (14.1) represents a continuous family of real periodic solutions, of dimension $n - k$ for each small real μ. This will certainly occur if Y_k is of odd degree in y.

Regarding the singular points let $Z_j(y; \mu)$, $j = 1, 2, \ldots, r$ be the minors of order $n - k + 1$ of J. The set S is then defined by the system

$$(15.1) \qquad Y_k = 0,\ A_0 y_{k-h} - A_h = 0,\ Z_j = 0\,.$$

This may be subjected to the same treatment as (14.1). It will yield a finite number of families of complex dimension $< n - k + 1$ whose real points are to be found. The treatment is the same as for the real ordinary points but with k replaced by $k + 1$.

For the exceptional points the argument is the same save that (15.1) is replaced by

$$(15.2) \qquad Y_k = 0,\ A_0 = A_1 = \ldots = A_{k-1} = 0\,.$$

It is clear from the preceding argument that the complete determination of all real periodic solutions may be accomplished in a finite number of steps.

16. As a mild application let us determine the families of periodic solutions of period 2π of the system

$$\text{(16.1)} \qquad \begin{cases} dx_1/dt = -x_2 + \mu g_1 (x_1, x_2, \sin t, \cos t, \mu) , \\ dx_2/dt = x_1 + \mu g_2 (x_1, x_2, \sin t, \cos t, \mu) \end{cases}$$

where the g_j are polynomials in the indicated variables. If we set $x = x_1 + i\, x_2$, $g = g_1 + i\, g_2$ then (16.1) assumes the form

$$dx/dt - ix = \mu g\, (x, \bar{x}, e^{it}, e^{-it}, \mu)$$

with g still a polynomial. To simplify matters we shall assume that g does not contain $\bar{x}$ so that the system to be treated is

$$\text{(16.2)} \qquad dx/dt - ix = \mu g\, (x, e^{it}, e^{-it}, \mu)$$

with g a polynomial in the indicated variables. This system has recently been discussed by Friedrichs [2].

We are then looking for solutions $x\,(t, \mu)$ of period 2π of (16.2) which tend to a solution of the first approximation

$$dx/dt - ix = 0$$

as $\mu \to 0$. Let $x\,(0, \mu) = \xi$, whatever μ. We will have

$$x = A_0\,(t) + \mu\, A_1\,(t) + \dots .$$

By substituting in (16.2) and identifying powers of μ we obtain the recursive system

$$\text{(16.3)} \qquad \begin{cases} A_0'\,(t) - i\, A_0\,(t) = 0 \\ A_1'\,(t) - i\, A_1\,(t) = g\,(A_0, e^{it}, e^{-it}, 0) \\ \dots\dots\dots\dots\dots\dots\dots\dots\dots \end{cases}$$

with initial conditions

$$A_0\,(0) = \xi; \; A_h\,(0) = 0, \; h > 0 .$$

Hence first

$$A_0\,(t) = \xi\, e^{it} .$$

The periodicity condition yields then $x\,(2\pi) = x\,(0)$, or

$$\text{(16.4)} \qquad A_1\,(2\pi) + \mu\, A_2\,(2\pi) + \dots = 0 .$$

From (16.3) we obtain

$$A_1(t) = \int_0^t e^{i(t-t')} g(\xi e^{it'}, e^{it'}, e^{-it'}, 0)\, dt',$$

and hence

$$A_1(2\pi) = \int_0^{2\pi} e^{-it} g(\xi e^{it}, e^{it}, e^{-it}, 0)\, dt = F(\xi),$$

where F is a polynomial in ξ. Thus (16.4) becomes

$$F(\xi) + \mu F_1(\xi) + \dots = 0 \tag{16.5}$$

where the F_h are readily shown to be polynomials. If they are all identically zero then the general solution for small μ is periodic. Let this not be the case and let $F_j(\xi)$ be the first $F_h \neq 0$. Writing now for simplicity $F, F_1, \dots,$ for $F_j, F_{j+1}, \dots,$ we will still have (16.5). With this convention if $\xi(\mu)$ is a solution and $\xi(0) = \xi_0$, then ξ_0 must be a root of the equation

$$F(\xi) = 0. \tag{16.6}$$

Let ξ_0 be a root of order p. We have then from the Weierstrass theorem

$$\begin{aligned} &F(\xi) + \mu F_1(\xi) + \dots \\ &= \{(\xi - \xi_0)^p + f_1(\mu)(\xi - \xi_0)^{p-1} + \dots + f_p(\mu)\} E(\xi - \xi_o, \mu), \end{aligned}$$

where the $f_h(\mu)$ are non-units. Hence the solution of (16.5) for $\xi(\mu)$ such that $\xi(0) = \xi_0$ reduces to that of

$$(\xi - \xi_o)^p + f_1(\mu)(\xi - \xi_0)^{p-1} + \dots = 0. \tag{16.7}$$

The required solutions may be obtained in a systematic manner by the Puiseux process (see for instance Picard [1] vol. II). In the present case there will be s so-called circular systems each consisting of q conjugate sets

$$\xi - \xi_o = \mu^{r/q} E(\mu^{1/q}). \tag{16.8}$$

The values $\xi(\mu)$ defined by (16.8) correspond to a single periodic family $x(\xi(\mu), t)$ such that $x(\xi(\mu), 0) = \xi_0$ and $\Sigma q = p$. Thus we have obtained a complete solution of our problem.

CHAPTER IX

Two Dimensional Systems. Simple Critical Points. The Index. Behavior at Infinity

Hereafter we confine our attention to two dimensional systems and equations of the second order. Abandoning the vector notation we shall consider more particularly in the present chapter autonomous systems

$$dx/dt = X(x, y), \quad dy/dt = Y(x, y) .$$

The principal goal is to obtain the *phase-portrait*, or the global qualitative description of the totality of the paths of the systems in the x, y plane or *phase-plane*. All that one can do in general is to provide a good deal of information about the paths and it is only in some cases that one may really obtain the full phase-portrait. The index, the closed paths, the behavior at infinity all contribute important, if not final, material to the general objective.

Our more detailed program is as follows: First there is given the "local" phase-portrait near a critical point when the first degree terms in the expansions of X, Y are present in its vicinity. Some details regarding the index of a closed curve and of a point are then given, the topological fine points being relegated to Appendix II.

To determine the behavior at infinity when X and Y are polynomials one must extend the concept of index to the sphere and the projective plane, a task carried out at length in Appendix II. Regarding infinity the given system is imbedded as it were into a broader system valid for the projective plane. As a consequence of this the infinite region may be treated like the rest of the plane. This is explicitly brought out by an example at the end of the chapter.

References: Bendixson [1]; Coddington, Levinson [1]; Hurewicz [1]; Poincaré [2].

§ 1. Generalities

1. Consider a real analytical system

$$(1.1) \qquad dx/dt = X(x, y), \; dy/dt = Y(x, y)$$

valid in a region Ω and possessing only a finite number of critical points there. If O is such a point one may choose it as origin and in its neighborhood

$$(1.2) \quad X = ax + by + X_2(x, y), \; dy/dt = cx + dy + Y_2(x, y)$$

where X_2, $Y_2 = [x, y]_2$. The characteristic equation of (1.2) is

$$(1.3) \qquad (a - \lambda)(d - \lambda) - bc = 0 .$$

We shall consider more particularly the case when the characteristic roots $\lambda_1, \lambda_2 \neq 0$, i.e. when $ad - bc \neq 0$. The corresponding critical points are known as *elementary*.

As the topological character of the critical point is the same, except in one case, as for the first approximation, which is linear homogeneous, we shall first discuss such systems.

(1.4) *Path-rectangle.* Referring to (II, 16.4) if γ is a true

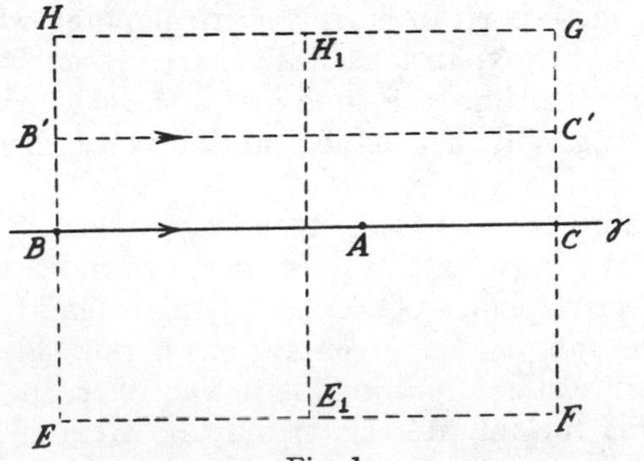

Fig. 1

path (not a critical point) and A is a point of γ, a certain neighborhood of A may be mapped topologically on a rectangle so as to produce the configuration of fig. 1: (for convenience the paths and their arcs are identified with their topological images) where BC, $B'C'$ are arcs of paths and EH, FG are subarcs of preassigned arcs transverse to γ at B, C. Moreover every path passing sufficiently near A contains an arc such as $B'C'$. The dotted rectangle $EFGH$ is a *path rectangle* and the arc BC of γ is its axis.

(1.5) *Closed paths.* According to (II, 15.1) the path γ is closed whenever it is a Jordan curve containing no critical point or equivalently whenever its solution $(x(t), y(t))$ is periodic.

§ 2. Critical Points of Linear Homogeneous Systems

2. Consider the real system

$$dx/dt = ax + by,\ dy/dt = cx + dy,\ ad - bc \neq 0\,. \tag{2.1}$$

The characteristic roots λ_1, λ_2 are still the solutions of (1.3) and the reduced normal form of the coefficient matrix $\begin{pmatrix} ab \\ cd \end{pmatrix}$ will completely determine the behavior of the paths, and more particularly their behavior in the neighborhood of the critical point at the origin. If the characteristic roots are real then a real transformation of coordinates will reduce the system to one of the same form but with coefficient matrix of one of the two types

$$A: \begin{pmatrix} \lambda_1 & 0 \\ 0 & \lambda_2 \end{pmatrix}; \quad B: \begin{pmatrix} \lambda & 0 \\ 1 & \lambda \end{pmatrix},$$

while if the roots are complex a complex transformation will reduce the matrix to the form

$$C: \begin{pmatrix} \lambda & 0 \\ 0 & \bar{\lambda} \end{pmatrix},$$

with a certain subcase. All told there are five cases which we shall now discuss separately.

(2.2) *First case. Real roots of same sign, matrix of type A.* The reduced form is

$$du/dt = \lambda_1 u, \; dv/dt = \lambda_2 v \,. \tag{2.3}$$

Suppose first λ_1, λ_2 both *negative*: $\lambda_i = -\mu_i$, $\mu_i > 0$. Then the general solution is

$$u = \alpha \, e^{-\mu_1 t}, \;\; v = \beta \, e^{-\mu_2 t}$$

where α, β are arbitrary constants. Evidently the path γ tends to the origin as $t \to +\infty$. It reduces to the u axis when $\beta = 0$, to the v axis when $\alpha = 0$. Supposing $\alpha\beta \neq 0$ and $\mu_1 < \mu_2$ the ratio $v/u \to 0$ as $t \to +\infty$. Hence γ is tangent to the u axis at the origin. If $\alpha = 0$ [$\beta = 0$] γ is the v axis [the u axis]. The form of the characteristics is that shown in fig. 2. The arrows indicate the direction of motion on the paths. The critical point thus arising is called a *stable node*.

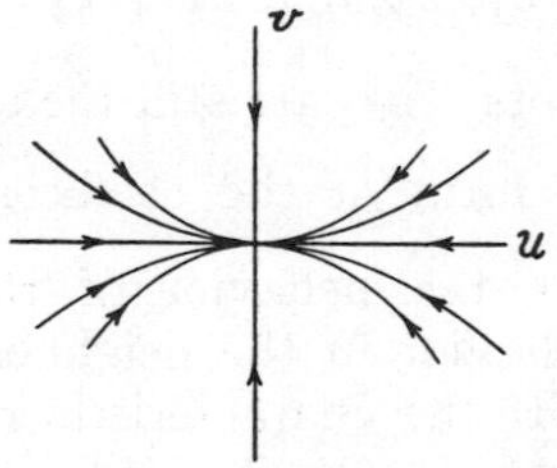

Fig. 2

If λ_1, λ_2 are real and positive, the preceding behavior corresponds to $t \to -\infty$. Hence assuming again $\lambda_1 < \lambda_2$ we have the situation of fig. 3 and the critical point known as *unstable node*.

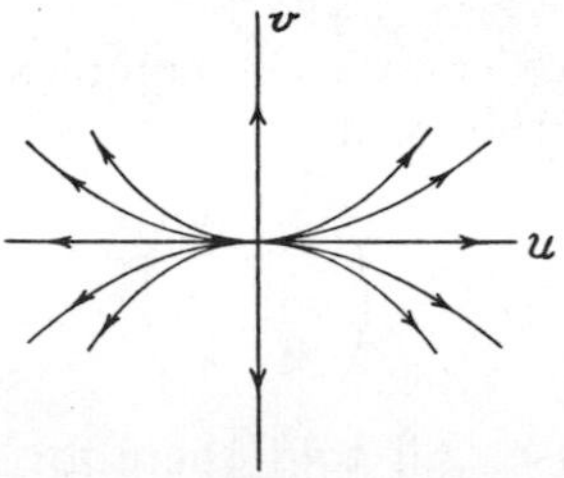

Fig. 3

When $\mu_1 > \mu_2$ the role of the two axes is reversed but the general aspect of the paths is the same. When $\lambda_1 = \lambda_2$, hence $\mu_1 = \mu_2$, all the paths are straight lines through the origin.

(2.4) *Second case, matrix of type B.* There is only one characteristic root λ and it is of course real. The reduced form is

$$du/dt = \lambda u, \quad dv/dt = u + \lambda v \,. \tag{2.5}$$

Suppose first $-\lambda = \mu > 0$. The general solution is

$$u = \alpha\, e^{-\mu t}, \quad v = (\alpha\, t + \beta)\, e^{-\mu t} \,.$$

For $\alpha = 0$ it represents the positive v axis if $\beta > 0$, the negative v axis if $\beta < 0$. Whatever α, β both $u, v \to 0$ as $t \to +\infty$. Hence the path γ tends to the origin as $t \to +\infty$. Assume now $\alpha \neq 0$. Since $v/u \to +\infty$ when $t \to +\infty$, γ is tangent to the v axis at the origin. It also crosses the u axis at $t = -\alpha/\beta$. The coordinate v has an extremum when $dv/dt = 0$ or $t = (\alpha - \beta\mu)/\alpha\mu$. Hence the paths behave as indicated in fig. 4. When $\lambda > 0$ the situation is that of fig. 5, the arrows being merely reversed.

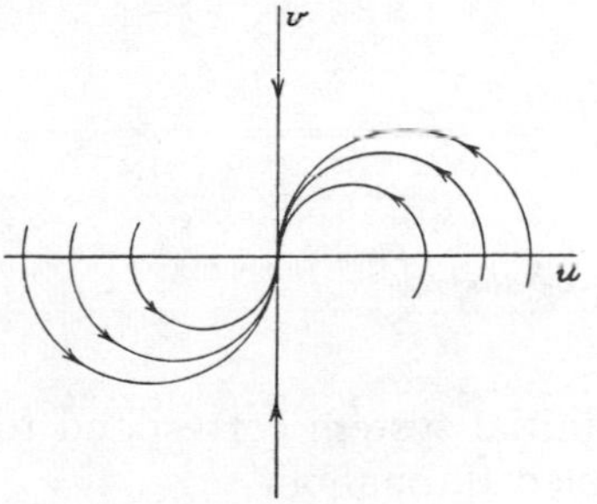

Fig. 4

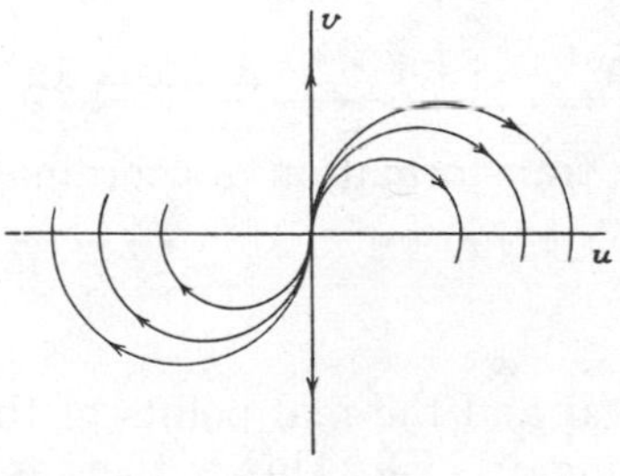

Fig. 5

The critical point is still called a stable or unstable node. Thus the earmark of the node is a set of paths tending to the origin when $t \to +\infty$ for the stable node or $t \to -\infty$ for the unstable node.

(2.6) *Third case. Roots real and of opposite sign.* The matrix is then necessarily of type A. The reduced form is still (2.3). Assuming $\lambda_1 = -\lambda$, $\lambda_2 = \mu$, λ and $\mu > 0$, the paths are

$$\gamma\colon u = \alpha e^{-\lambda t}, \quad v = \beta\, c^{\mu t} \,. \tag{2.7}$$

The semi-axes u, v are still the paths corresponding to $\beta = 0$, $\alpha = 0$. If $\alpha\beta \neq 0$ then $u \to 0$, $v \to +\infty$, as $t \to +\infty$. Hence the paths have the general form of fig. 6. The origin is then called a *saddle point*.

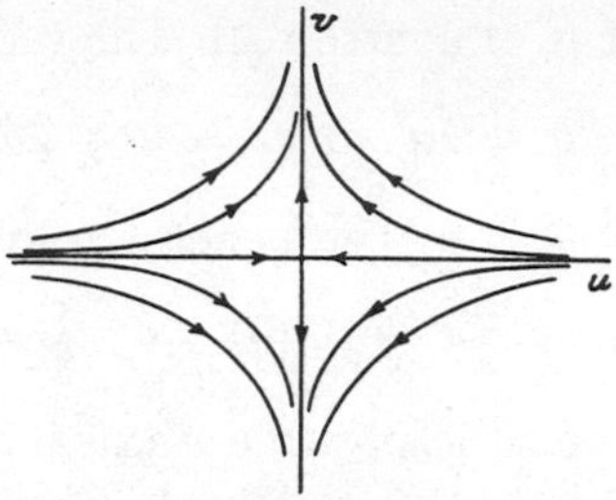

Fig. 6

When the signs of λ_1, λ_2 are reversed the role of the axes is reversed, or equivalently all the arrows in fig. 6 are reversed. However the essential aspect of the critical point remains the same.

(2.8) *Fourth case: Complex roots with non-zero real part.* The reduction is to the form

$$du/dt = \lambda u, \ d\bar{u}/dt = \bar{\lambda}\bar{u} \,. \tag{2.9}$$

The transformation of coordinates

$$x \to \frac{u + \bar{u}}{2}, \ y \to \frac{u - \bar{u}}{2i}$$

is real and the real points of the initial system correspond to $\bar{u}$ conjugate of u. This is then assumed throughout.

Let us suppose first $\lambda = -\mu + i\omega$, where ω and μ are positive. The general solution of (2.9) is

$$u = \gamma\, e^{\lambda t} = a\, e^{(-\mu + i\omega)\, t + i\beta}, \ a \text{ and } \beta \text{ real.}$$

Setting $u = re^{i\theta}$ we have then

$$r = a\, e^{-\mu t}, \ \theta = \beta + \omega\, t$$

which represents a logarithmic spiral. Thus the aspect of the paths is that of fig. 7. The critical point is then called a *stable focus*. When $\lambda = \mu + i\omega$, $\mu > 0$, the situation is the same with arrows reversed (fig. 8) and we have the *unstable focus*.

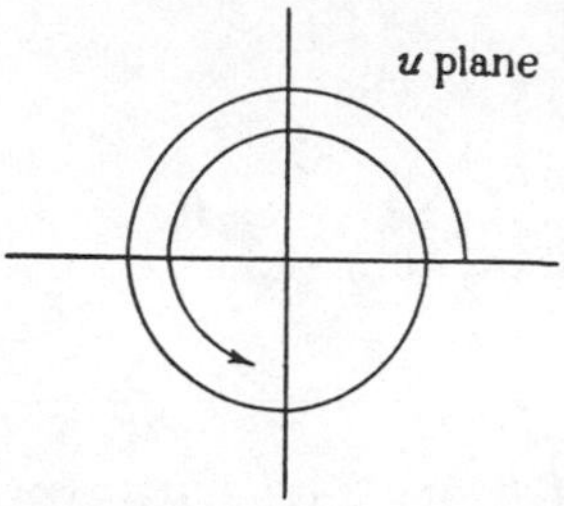

Fig. 7

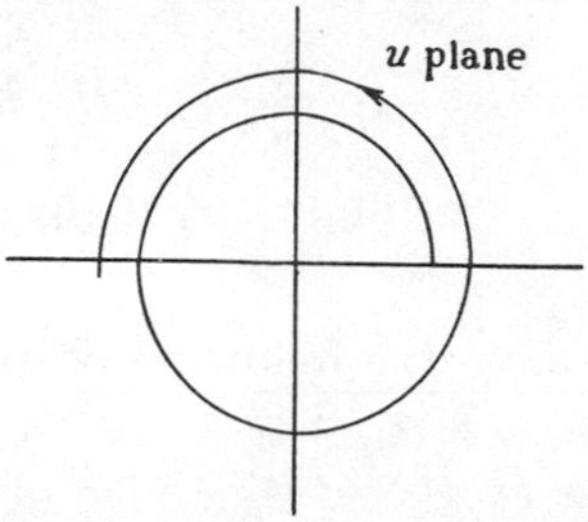

Fig. 8

(2.10) *Fifth case: Pure complex characteristic roots*. The situation is the same save that $\mu = 0$. Hence the paths are given by $r = a$, $\theta = \beta + \omega t$. In other words in the (u, v) plane they are circles with the critical point as center and all described with the same angular velocity ω. The critical point is then known as a *center* (fig. 9).

It should be kept in mind that the terms logarithmic spirals, circles, applied to the paths in the (u, v) plane are only partly appropriate. For even if the original x, y coordinates had been chosen rectangular, the transformations utilized were not orthogonal but merely linear homogeneous. Thus the circles in the (u, v) plane of fig. 9 would merely correspond in the original (x, y) plane to a family of concentric similar ellipses.

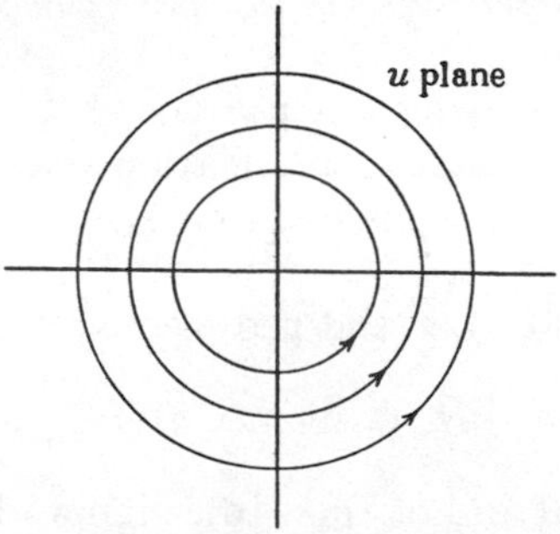

Fig. 9

§ 3. Elementary Critical Points in the General Case

3. Before taking up our main theme let us say a few words regarding a transformation S regular at the origin:

$$x = f(u, v),\ y = g(u, v),$$

$$f(0, 0) = g(0, 0) = 0,\quad \left|\frac{\partial (f, g)}{\partial (u, v)}\right|_{(0,\,0)} \neq 0 .$$

We note the following properties:

(3.1) *There exist neighborhoods M, N of the origins in the planes Π, Π^* of x, y and u, v, which are 2-cells mapped topologically into one another by S.*

(3.2) *S is regular on N and S^{-1} on M.*

(3.3) *S tranforms* (1.1) *into a new system and in such a manner that paths and* (*isolated*) *critical points go into the same for the new system. Moreover S does not change the matrix in normal form to which $\begin{pmatrix} a & b \\ c & d \end{pmatrix}$ may be reduced.*

In substance then we may freely use transformations such as S without affecting the behavior of the paths about the origin.

We shall assume now that we have the general system (1.1) with $ad - bc \neq 0$. A linear transformation of variables will reduce it to a system whose first approximation is of one of the forms considered in (2). The same terminology: node, focus, etc., is used as before. The basic result is:

(3.4) THEOREM. *The behavior of the paths in the neighborhood of a critical point is the same as for the first approximation except that when the characteristic roots are pure complex there may arise a center or a focus.*

The nature of the coefficient matrix of (2.1) and of the roots λ_1, λ_2 gives rise to the same classification into five cases as before and we must again examine each case separately.

4. (4.1) *First case.* Let the reduced system be written

$$(4.2)\qquad dx/dt = \lambda_1 x + p_2(x, y),\ dy/dt = \lambda_2 y + q_2(x, y)$$

where λ_1, λ_2 are real and of the same sign and generally $p_h(x, y)$, ..., are series $[x, y]_h$. The first approximation is (2.3). According to the complements (V, **14.1**, **14.2**) to Liapunov's expansion theorem we notice this: the first approximation has the special solutions

(4.3a) $\qquad u = e^{\lambda_1 t},\ v = 0;$ (4.3b) $u = 0,\ v = e^{\lambda_2 t}.$

Hence, if $\lambda_1, \lambda_2 < 0$, there correspond for t sufficiently large, and respectively to (4.3a, 4.3b), the solutions

(4.4a) $\qquad x = u + [u]_2,\ y = [u]_2,\ u = e^{\lambda_2 t},$

(4.4b) $\qquad x = [v]_2,\ y = v + [v]_2,\ v = e^{\lambda_2 t}.$

Now (4.4a) represents an analytical arc through the origin which is an integral of (4.2) and is tangent to the x axis at the origin since $dy/dx = 0$ for $u = 0$. Hence the arc has an equation

$$y + \varphi\,(x, y) = 0,\ \varphi = [x, y]_2.$$

Similarly (4.4b) yields an arc solution

$$x + \psi\,(x, y) = 0,\ \psi = [x, y]_2.$$

Now the regular transformation

$$X = x + \psi\,(x, y),\ Y = y + \varphi\,(x, y)$$

reduces (4.2) to the form

(4.5) $$\begin{cases} dX/dt = \lambda_1 X + F_1(X, Y),\ F_1 = [X, Y]_2, \\ dY/dt = \lambda_2 Y + G_1(X, Y),\ G_1 = [X, Y]_2. \end{cases}$$

Since (4.5) has for integrals $X = 0$ and $Y = 0$, necessarily $F_1(0, Y) = 0$, $G_1(X, 0) = 0$. Hence $F_1 = \lambda_1 X F$, $G_1 = \lambda_2 Y G$, where F, G are $[X, Y]_1$. Thus (4.5) has the actual form

(4.6) $$\frac{dX}{dt} = \lambda_1 X\,E_1(X, Y),\ \frac{dY}{dt} = \lambda_2\,Y\,E_2(X, Y),\ E_i(0, 0) = 1.$$

Since $\lambda_1 < 0$, X decreases in the first and fourth quadrants and increases in the second and third. Since $\lambda_2 < 0$, Y decreases in the first two quadrants and increases in the last two. The semiaxes are paths which tend to the origin. Hence all the paths tend to the origin and we have a stable node.

If $\lambda_1, \lambda_2 > 0$ the result is the same, save that in (4.4) one assumes $-t$ large. One arrives again at (4.6), the paths all move away from the origin and one has an unstable node.

(4.7) *Second case.* The basic system is now

$$(4.8) \quad dx/dt = \lambda x + p_2(x, y), \; dy/dt = x + \lambda y + q_2(x, y)$$

and the first approximation is (2.5). Suppose first $-\lambda = \mu > 0$ and let $u = ae^{-\mu t}$, $v = (at + b)\, e^{-\mu t}$ be the general solution of (2.5). Then according to Theorem (V, 9.1) the general solution of (4.4) is given in the neighborhood of the origin by expressions

$$x = \sum_{m=1}^{+\infty} P_m(a, b; t)\, e^{-m\mu t}, \quad y = \sum_{m=1}^{+\infty} Q_m(a, b; t)\, e^{-m\mu t},$$

where P_m, Q_m are polynomials in t whose coefficients are forms of degree m in a, b. The first terms of the x, y series are respectively equal to u, v. Hence we may write

$$x = ae^{-\mu t} + \sum_{m>1} P_m\, e^{-m\mu t}, \quad y = (at + b)\, e^{-\mu t} + \sum_{m>1} Q_m\, e^{-m\mu t}.$$

This shows that as $t \to +\infty$ (x, y) tends to the origin and

$$\lim y/x = \lim t + b/a = +\infty.$$

From this follows readily that the behavior of the paths is the same as for the first approximation and is represented near the origin by fig. 4 when $\lambda < 0$.

If $\lambda > 0$ we replace again t by $t' = -t$, and as in the preceding case, show that the behavior of the paths is the same as in fig. 4 with arrows reversed, i.e., the same as in fig. 5. Thus we have the same stable or unstable node as for the first approximation.

(4.9) *Third case.* The basic system is still (4.2) with $\lambda_1 = -\lambda$ $\lambda_2 = \mu$, λ and $\mu > 0$. The treatment is the same as for the first case save that $\lambda_1 = -\lambda$, $\lambda_2 = \mu$ in (4.6). Thus we have

$$(4.10) \quad dX/dt = -\lambda X E_1(X, Y), \; dY/dt = \mu Y E_2(X, Y), \quad E_i(0, 0) = 1.$$

Thus for (4.10) the axes are paths, and X decreases and Y increases in the first quadrant. Now near any segment of the axes near the origin (origin excepted) the system of paths forms a path rectangle. Hence in the first quadrant, near the origin, the general behavior can only be in accordance with fig. 6. Similarly for the other three quadrants. Hence the complete behavior in the X, Y plane, and hence also in the x, y plane is that of fig. 6, i.e., we have a saddle point as in the first approximation.

5. *Fourth and fifth cases.* These last two cases may be conveniently taken up together. The fundamental system is now

$$dx/dt = \lambda x + p_2(x, \bar{x}), \quad d\bar{x}/dt = \bar{\lambda}\bar{x} + \bar{p}_2(\bar{x}, x) \tag{5.1}$$

and the first approximation is (2.9).

The most convenient method for dealing with the paths is to pass to polar coordinates r, θ. Write explicitly $\lambda = \mu + i\omega$. Then (5.1) yields

$$e^{i\theta}\, dr/dt + ire^{i\theta}\, d\theta/dt = (\mu + i\omega)\, re^{i\theta} + p_2(re^{i\theta}, re^{-i\theta}) .$$

Upon dividing by $e^{i\theta}$ and equating real and complex parts we obtain relations

$$dr/dt = r(\mu + \alpha_1(\theta)\, r + \alpha_2(\theta)\, r^2 + \ldots) , \tag{5.2}$$

$$d\theta/dt = \omega + \beta_1(\theta)\, r + \beta_2(\theta)\, r^2 + \ldots \tag{5.3}$$

where $\alpha_n(\theta)$, $\beta_n(\theta)$ are polynomials in $\cos\theta$, $\sin\theta$.

The series converge for r sufficiently small and any θ. Since $\omega \neq 0$, we obtain by division a relation

$$dr/d\theta = r(\mu/\omega + \gamma_1(\theta)\, r + \ldots) \tag{5.4}$$

where the coefficients and convergence are as before. Since the system is analytic the solution, taking the value ϱ for $\theta = \theta_0$, may be represented in the form (see II, 10.1)

$$r(\theta, \theta_0, \varrho) = c_1(\theta, \theta_0)\, \varrho + c_2(\theta, \theta_0)\, \varrho^2 + \ldots$$

the series being valid for an arbitrary θ range and ϱ small. The term independent of ϱ is missing since $\varrho = 0$ must yield the solution $r = 0$. Since the series may be assumed valid for $\theta_0 = 0$, the solutions near the origin may be assumed to have their initial value ϱ for $\theta_0 = 0$. Hence we may write our solutions $r(\theta, \varrho)$ and choose for them a representation

(5.5) $$r(\theta, \varrho) = c_1(\theta)\,\varrho + c_2(\theta)\,\varrho^2 + \cdots.$$

The c_n are determined by substituting in (5.4) and identifying equal powers of ϱ. We thus obtain a system

(5.6) $$\begin{cases} dc_1/d\theta = \mu/\omega\, c_1 \\ dc_2/d\theta = \mu/\omega\, c_2 + \gamma_1(\theta)\, c_1{}^2 . \\ \cdots\cdots\cdots\cdots\cdots\cdots \end{cases}$$

Since $r(0, \varrho) = \varrho$ identically we must have

(5.7) $$c_1(0) = 1, \quad c_n(0) = 0 \text{ for } n > 1 .$$

The differential equations (5.6) together with the initial conditions (5.7) enable one to determine the $c_n(\theta)$ one at a time. In particular

(5.8) $$c_1(\theta) = e^{\mu\theta/\omega} .$$

In order that $r(\theta, \varrho)$ be periodic of period 2π, or which is the same in order that its path be an oval surrounding the origin we must have $r(2\pi, \varrho) = \varrho$ or

(5.9) $$\varrho = \Sigma c_n(2\pi)\,\varrho^n = \varphi(\varrho) .$$

There are now two possibilities:

(5.10) *All the paths for ϱ sufficiently small are ovals surrounding the origin, i.e. the origin is a center.* Then (5.9) is satisfied identically. Since $\varphi(\varrho) - \varrho$ is holomorphic at $\varrho = 0$, all its coefficients vanish. Hence

$$c_1(2\pi) = 1, \quad c_n(2\pi) = 0 \text{ for } n > 1 .$$

The first relation yields $e^{2\pi\mu/\omega} = 1$, hence $\mu = 0$ and the rest yield the condition that every $c_n(\theta)$ has the period 2π. Thus the center can only arise when the characteristic roots are pure complex and the $c_n(\theta)$ are all periodic. Conversely when these two conditions hold $\varphi(\varrho) - \varrho = 0$, the paths near the origin are ovals surrounding it and the origin is a center.

Notice that when the origin is a center the time period T for the description of the path $r(\theta, \varrho)$ calculated by means of (5.3) is

$$T = \int_0^{2\pi} \frac{d\theta}{\omega + \beta_1(\theta)\, r(\theta, \varrho) + \dots}$$

and will generally depend upon ϱ. Thus it is not necessarily constant, contrary to what happens in the linear case. Expressed also in another way the point $r(t), \theta(t)$ of a given path γ describes it with an instantaneous or even an average angular velocity which are not generally independent of γ.

(5.11) *The coefficients* $c_n(\theta)$ *do not all have the period* 2π. The closed paths, if any exist in the range considered, will correspond to the solutions in ϱ of (5.9). Under our hypothesis the coefficients of $\varphi(\varrho) - \varrho$ are not all zero. Hence $\varrho = 0$ is an isolated root of (5.9). This means that a $\sigma > 0$ may be selected such that (5.9) has no roots in the interval $0 < \varrho < \sigma$. Let ϱ be confined henceforth to this interval.

Starting at the point $P(0, \varrho)$ of a path, as θ varies by $2\pi, 4\pi, \dots$, there will be reached points $P_1(0, \varrho_1), P_2(0, \varrho_2), \dots$. Suppose $\varrho_1 < \varrho$. Then necessarily $\varrho > \varrho_1 > \varrho_2 \dots$. Thus $\{\varrho_n\}$ has a limit η, $0 \leqq \eta < \sigma$. Let us suppose $\eta > 0$. Since $\varrho_{n+1} = \varphi(\varrho_n)$, the sequence $\{(\varphi(\varrho_n) - \varrho_n)\} \to 0$. Since $\varphi(\varrho) - \varrho$ is continuous at $\varrho = \eta$, we have $\varphi(\eta) - \eta = 0$ or (5.9) has a root in the interval $0 < \varrho < \sigma$. Since this contradicts the assumption on σ we must have $\eta = 0$. Hence the paths behave as in fig. 7. The critical point is a stable focus. If $\varrho_1 > \varrho$ the situation is analogous save that we merely conclude that the spirals diverge from the origin. The latter is then an unstable focus.

Suppose in particular $\mu \neq 0$. If $\mu < 0$ then for r small $dr/d\theta$ will steadily decrease and we have the stable focus, while for $\mu > 0$ it will be the reverse with an unstable focus. Thus in the fourth case—real parts of the characteristic roots non-zero—the situation is the same as for the first approximation.

This completes the proof of (3.1).

(5.12) There still remains the following important question to be settled: If $\mu = 0$, to recognize from the given differential system whether the critical point is a stable or unstable focus or a center. The system (5.6) assumes now the form

$$dc_1/d\theta = 0\,,$$

$$\cdots\cdots\cdots\cdots\cdots\cdots\cdots\cdots$$

$$dc_n/d\theta = \varphi_n(c_1, \ldots, c_{n-1})\,,$$

$$\cdots\cdots\cdots\cdots\cdots\cdots\cdots\cdots$$

with the same initial conditions as before. Thus $c_1(\theta) = 1$. Suppose that up to and including φ_{n-1}, but not φ_n, the expressions φ_h, which are polynomials $\psi_h(e^{i\theta}, e^{-i\theta})$ have no constant terms. If $n = \infty$, i.e. if no ψ_h have constant terms then we have a center. If n is finite and δ_n is the constant term of ψ_n the critical point is a focus. In the series (5.5) the first non-periodic term will be $c_n(\theta)$. Thus for ϱ small $r(2\pi, \varrho) - r(0, \varrho)$ will have the sign of δ_n. Hence the focus is stable if $\delta_n < 0$, and unstable if $\delta_n > 0$.

6. *Final remark.* The preceding results have confirmed the property already known that our critical point is stable if the real parts of the characteristic roots are negative, unstable if they are positive, conditionally stable if they are of opposite signs. Now in the applications it is often important to have at hand rapid criteria to detect which one of the three cases arises. This is very easily done in terms of the initial system (1.1).

Let us suppose then that the point $A(x_0, y_0)$ is an elementary critical point. The expansions of X, Y in its neighborhood are

$$X = X_{x_0}(x - x_0) + X_{y_0}(y - y_0) + \ldots\,,$$
$$Y = Y_{x_0}(x - x_0) + Y_{y_0}(y - y_0) + \ldots\,,$$

where $X_x, \ldots,$ are the partial derivatives. The characteristic equation is

$$\begin{vmatrix} X_{x_0} - r, & X_{y_0} \\ Y_{x_0}, & Y_{y_0} - r \end{vmatrix} = r^2 - (X_{x_0} + Y_{y_0})\,r + X_{x_0} Y_{y_0} - X_{y_0} Y_{x_0} = 0\,.$$

Hence for the characteristic roots r_1, r_2:

$$X_{x_0} + Y_{y_0} = r_1 + r_2,\; J = \left|\frac{\partial(X, Y)}{\partial(x_0, y_o)}\right| = r_1 r_2\,.$$

Hence the following comprehensive property:

(6.1) *If* $J(x_0, y_0) > 0$ *and* $X_{x_0} + Y_{y_0} < 0$ *the critical point is stable. If* $J(x_0, y_0) > 0$ *and* $X_{x_0} + Y_{y_0} > 0$ *it is unstable. If* $J(x_0, y_0) < 0$ *the point is conditionally stable (it is a saddle point).*

§ 4. The Index. Application to Differential Equations

7. Let us recall briefly the definitions and simple properties of the index. For full details the reader is referred to Appendix II.

Let $\mathfrak{F}$ be a field defined over a Jordan curve J in the Euclidean plane Π. It is supposed that $\mathfrak{F}$ has no critical points (no vanishing points) on J. Then Index $(J, \mathfrak{F})$ is $1/2\pi$ times the angular variation of the vector $V(M)$ applied at $M \in J$ as M describes J once.

Let the field $\mathfrak{F}$ be defined over a region Ω and let $A \in \Omega$ be at worst an isolated critical point. Let also γ be a small Jordan curve surrounding A but no other critical point and oriented concordantly with the plane Π. Then Index $(\gamma, \mathfrak{F})$ is independent of γ and is by definition Index $(A, \mathfrak{F})$.

Noteworthy properties of the index are:

(7.1) *If two fields* $\mathfrak{F}, \mathfrak{F}'$ *defined over a Jordan curve* J *have no critical points on* J, *and their vectors are never in opposition on* J *then* Index $(J, \mathfrak{F})$ = Index $(J, \mathfrak{F}')$.

(7.2) *The index of a non-critical point is zero.*

(7.3) *If* $\mathfrak{F}$ *is defined in a region* Ω *and the Jordan curve* $J \subset \Omega$ *surrounds a finite set of critical points* $A_1, \ldots, A_s$ *then*

$$\text{Index}\,(J, \mathfrak{F}) = \Sigma\ \text{Index}\,(A_i, \mathfrak{F}).$$

(7.4) *The index according to Poincaré.* In his mémoire [2], p. 29, Poincaré gave a definition of the index of a Jordan curve J which may be formulated as follows. It is supposed that the field $\mathfrak{F}$ crosses a given fixed direction D in at most a finite number of points. By crossing one means definitely that the direction of the vector passes through the direction D and one excludes the places where the vector, say rotating forward just reaches D and then retrocedes. Let p be the number of crossings as the vector V revolves positively and n the number as it revolves negatively. Then Index $(J, \mathfrak{F}) = \dfrac{p-n}{2}$. This definition is highly convenient in the applications, where frequently p and n are easily calculated. That it is equivalent to the one by angular variation is seen as follows: the crossings of D by the vector V count off the multiples of π in the angular variation from D. Hence the total angular variation is $(p-n)\,\pi$ and so the index is $(p-n)\,\pi/2\pi = \dfrac{p-n}{2}$.

8. The application to differential equations is immediate. To the system (1.1) defined over a closed region Ω there corresponds the field $\mathfrak{F}$ of the vectors (X, Y) defined over Ω. Let J be an oriented rectifiable Jordan curve, so that one may integrate along J. Then at once if J contains no critical point:

$$(8.1)\qquad \text{Index } J = \frac{1}{2\pi}\oint_J d \text{ arc tan} \frac{Y}{X} = \frac{1}{2\pi}\oint_J \frac{XdY - YdX}{X^2 + Y^2}.$$

This expression makes it evident that the index varies continuously with J when J is continuously deformed without crossing critical points. Since the integral is an integer its value under the circumstances is constant.

(8.2) Let A be an isolated critical point and J a positively oriented small circle surrounding A so that Index J = Index A. Apply an affine transformation of coordinates, and take (8.1) in the new coordinates. Using the definition of the index of Appendix II (4.5), and referring to Appendix II (4.1) it is readily seen that the value of the index in the new coordinates is the same as in the old. Thus, one may conveniently use (8.1) to calculate Index A in any affine coordinate system.

9. *Index of the elementary critical points.* We may suppose as usual that the critical point is at the origin and that the index is given by (8.1) where J is a small circle of radius r. We first have:

(9.1) *The index of the origin relative to the system* (1.2) *is the same as for its first approximation.*

The components of the vector V for (1.2) and of the vector V^* for its first approximation are respectively

$$X = ax + by + X_2, \quad Y = cx + dy + Y_2,$$
$$X^* = ax + by, \quad Y^* = cx + dy,$$
$$ad - bc \neq 0 .$$

By (7.1) we merely have to show that on J, V and V^* are never in opposition or that we cannot have

$$X + kX^* = 0, \quad Y + kY^* = 0, \quad k > 0,$$

or explicitly that one must rule out relations

$$(9.2) \quad (1 + k)(ax + by) + X_2 = 0, \ (1 + k)(cx + dy) + Y_2 = 0 .$$

As a consequence of (9.2) we find

$$(1 + k)^2 [(ax + by)^2 + (cx + dy)^2] = Z_4 (x, y) .$$

Introducing polar coordinates r, θ we obtain

$$(9.3) \quad (1 + k)^2 [(a \cos \theta + b \sin \theta)^2 + (c \cos \theta + d \sin \theta)^2] =$$
$$= r^2 \alpha_4 (\theta) + r^3 \alpha_5 (\theta) + \dots ,$$

where $\alpha_n (\theta)$ is a form of degree n in $\sin \theta$, $\cos \theta$ and the series at the right converges for r small and any θ, uniformly in θ. As a consequence the right hand side of (9.3) $\to 0$ with r whatever θ. On the other hand since $ad - bc \neq 0$, we can only have $ax + by = 0$, $cx + dy = 0$ if $x = y = 0$. Hence the square bracket at the left $\neq 0$ whatever θ, i.e., on $0 \leq \theta \leq 2\pi$. Thus it is continuous and positive on a closed interval and hence it has a positive lower bound ξ. Since $k > 0$, the left hand side $\geq \xi$ whatever θ. Hence for r sufficiently small the two sides of (9.3) are different. This proves (9.1).

10. We shall now deal directly with the first approximation. Let the reduced forms be those of (2) save that we still use coordinates x, y instead of u, v. Here again each case requires special examination.

(10.1) *Case I.* (*Node.*) Here λ_1, λ_2 are real and of the same sign. Assuming $|\lambda_2| \geq |\lambda_1|$ we have on C:

$$x^2 + y^2 = r^2, X = \lambda_1 x, Y = \lambda_2 y, X^2 + Y^2 = \lambda_1{}^2 x^2 + \lambda_2{}^2 y^2 .$$

It follows that if we vary λ_1, λ_2 continuously, without vanishing, till they both become $+1$ when they are positive, -1 when they are negative, the integral (10.1) will vary continuously also. Since it is an integer it is then constant and so the modifications of λ_1, λ_2 do not change Index A. At the end the vector V will be along the radius AM and always pointing inward or always outward. In both cases the index is unity.

(10.2) *Case II.* (*Node.*) Here $\lambda_1 = \lambda_2 = \lambda$. Instead of the circle C we shall take the curve J indicated in fig. 10: $J = BCDEB'C'D'E'B$.

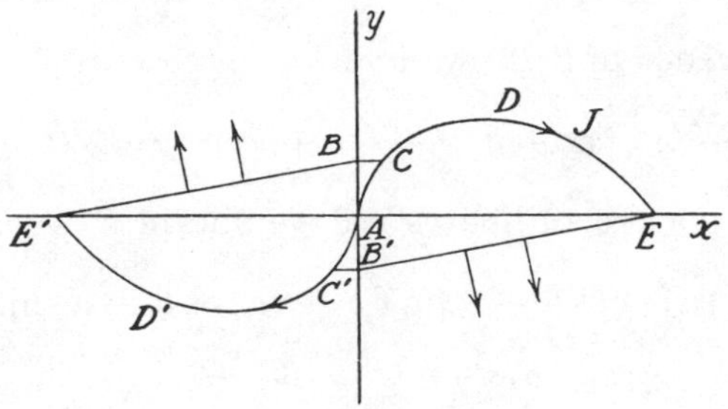

Fig. 10

Here CDE, $C'D'E'$ are portions of paths and the other parts of J are segments. The figure is symmetric with respect to the origin and it is drawn for $\lambda > 0$ (unstable node). It is clear that J is deformable continuously into a circumference without crossing A, and so it is admissible as curve J in (8.1). The angular variation of $V(M)$ along $B'EDCB$ is manifestly π, since the vectors turn continuously forward from $-\pi/2$ to $+\pi/2$. Similarly on the second part of J from B to B'. Hence the total angular variation of $V(M)$ is 2π and the index is again unity. If $\lambda < 0$ all the vectors are reversed and the result is the same.

(10.3) *Case III.* (*Saddle point.*) This time λ_1, λ_2 are real but of opposite signs. We may assume $\lambda_1 < 0$, $\lambda_2 > 0$ and choose an integration curve J which is the arc $BCDEF$ of fig. 11 repeated by symmetry around the axes. Along that arc the angular variation of V is $-\pi/2$, hence its total angular variation is -2π. Hence the index of the saddle point is -1.

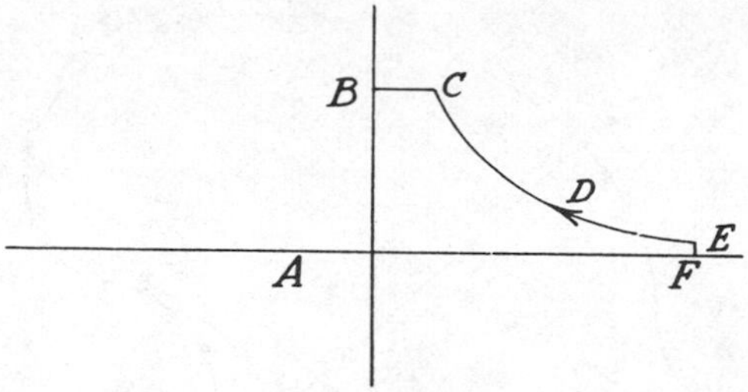

Fig. 11

(10.4) *Cases IV and V (focus and center).* Let the reduced form be this time

$$dx/dt = (\mu + i\omega)\, x, \quad d\bar{x}/dt = (\mu - i\omega)\, \bar{x}\,.$$

The denominator in the integral (8.1) is the squared length of the vector $V(M)$. Its value here for M on the circle of radius one is $\mu^2 + \omega^2$. Hence the same argument as before shows that we may vary μ continuously to zero, and ω continuously to unity without changing Index A. For $\omega = 1$ the vector $V(M)$ is represented in the complex plane by ix, i.e., it is tangent to the circle. As M describes the latter $V(M)$ rotates forward and so the angular variation is 2π. Thus the index of a focus or a center is unity.

To sum up we have proved:

(10.5) THEOREM. *The index of an elementary critical point other than a saddle point is unity; the index of a saddle point is* (—1).

11. We come now to an important proposition due to Poincaré ([2], p. 57):

(11.1) THEOREM. *The index of a positively oriented closed path is unity.*

The following ingenious proof is due to Heinz Hopf.

Let γ be the closed path in the plane Π. Since our system is analytic γ is rectifiable. Take a horizontal tangent to γ as low as possible and let Q be its point of contact. Let arc length s

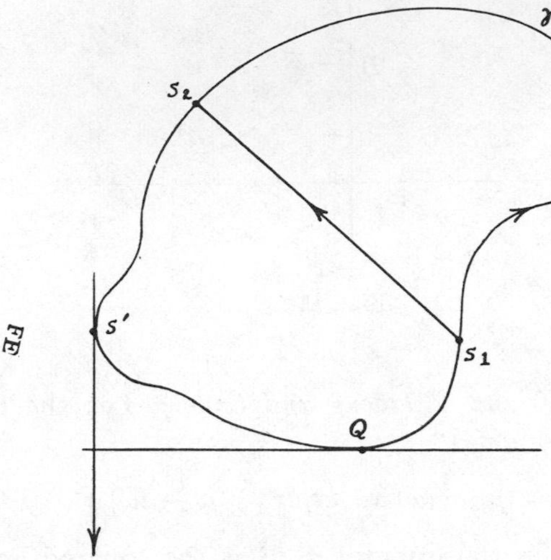

Fig. 12

on γ be counted from Q in the positive direction and let the scale be such that γ is of length unity.

Let now a new plane Π^* be referred to coordinates s_1, s_2. To a pair of points of γ marked by arc lengths s_1, s_2: $0 \leq s_1 \leq s_2 \leq 1$ we assign the point (s_1, s_2) of Π^* and determine at this point a vector V represented by the segment from s_1 to s_2. This defines in the closed triangle ABC a field $\mathfrak{F}$ without critical points. Therefore the angular variation of V around ACB is zero. Now on AC the vector at (s', s') is simply the tangent vector to γ at the associated point s' of γ. Hence the angular variation along AC is 2π Index γ. Along CB and BA it is merely $-\pi$. Hence 2π Index $\gamma - 2\pi = 0$ and this proves the theorem.

(11.2) Corollary. *A closed path must surround at least one critical point.*

(11.3) Corollary. *If a closed path surrounds only elementary critical points they cannot all be saddle points.*

The same argument yields:

(11.4) *Let a positively oriented planar Jordan curve J be rectifiable and possess a continuously turning tangent. If the field*

$\mathfrak{F}$ is tangent to J along J and has no critical point on the curve then Index $(J, \mathfrak{F}) = +1$.

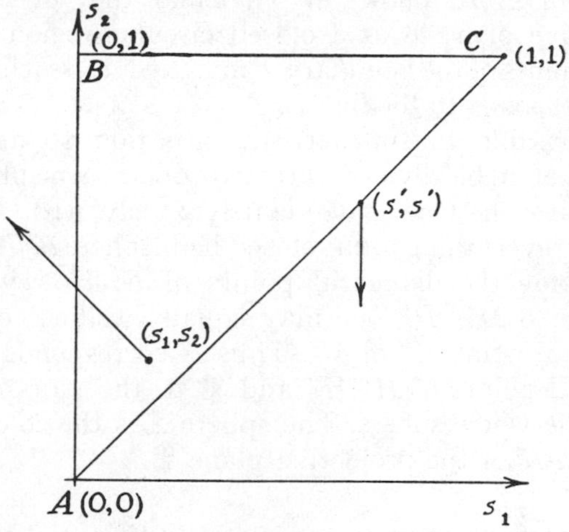

Fig. 13

§ 5. Behavior of the Paths at Infinity

12. Consider now a system

$$dx/dt = P(x, y), \ dy/dt = Q(x, y) \tag{12.1}$$

where this time P and Q are relatively prime polynomials. This last condition is to avoid systems with an infinite number of critical points. We propose to discuss the behavior of the paths far out in the plane. Following Poincaré [2] we shall deal with the problem by completing the plane Π to a projective plane $\mathfrak{P}$ by addition of a certain line, the classical "line at infinity." This process has already been discussed in (I, 3.8). Let $\mathfrak{P}$ be referred to homogeneous coordinates X, Y, Z. If the point M of $\mathfrak{P}$ is not in the line $Z = 0$ we assign to M the cartesian coordinates $x = X/Z$, $y = Y/Z$, and identify M with the point (x, y) of Π. Thus $\Pi = \mathfrak{P} - L$ where L is the line $Z = 0$.

Every line λ: $bX - aY = 0$ of $\mathfrak{P}$ intersects L in a unique point $\lambda^* (a, b, 0)$ and the correspondence $\lambda \leftrightarrow \lambda^*$ is one-one. Since λ determines uniquely the line $bx - ay = 0$ of Π, there is a one-one correspondence between the directions of Π and the points of L. It follows in particular that in the model of the projective plane $\mathfrak{P}$ as a closed circular region Ω with the diametral points of the boundary Γ matched, the set of "matched" points corresponds to the line L.

It is advisable to transfer the situation to an Euclidean sphere S. Let it be divided into two open hemispheres H', H'' by an equator E. One may identify Ω say with $\bar{H}' = H' \cup E$ and transfer everything to the closed hemisphere $\bar{H}'$. The matched points become the diametral points of E. However if $M \in \Pi$ corresponds to $M' \in H'$, one may equally well associate with M the diametral point M'' of M'. Thus Π corresponds to the pairs of diametral points of H', H'' and $\mathfrak{P}$ to the pairs of diametral points of the whole sphere. The sphere S is the so-called *doubly covering sphere* of the projective plane $\mathfrak{P}$.

13. Let us return now to our special differential system which we write

$$Qdx - Pdy = 0\,. \tag{13.1}$$

Let p, q be the degrees of P, Q and set

$$P^* (X, Y, Z) = Z^p P (X/Z, Y/Z),$$
$$Q^* (X, Y, Z) = Z^q Q (X/Z, Y/Z)\,.$$

Let also n be the largest of p and q and consider the differential relation

$$\begin{vmatrix} dX, & dY, & dZ \\ X, & Y, & Z \\ Z^{n-p} P^*, & Z^{n-q} Q^*, & 0 \end{vmatrix} = 0\,. \tag{13.2}$$

At points not on $Z = 0$, i.e. in the Euclidean plane Π, we may take $Z = 1$, $dZ = 0$, $X = x$, $Y = y$, and then (13.2) reduces to (13.1). Therefore (13.2) may be viewed as the extension of (13.1) to the projective plane $\mathfrak{P}$.

Upon expanding the determinant, (13.2) becomes:

$$(13.3) \quad -Z^{n-q+1}Q^* \, dX + Z^{n-p+1}P^* \, dY + (Z^{n-q}XQ^* - Z^{n-p}YP^*) \, dZ = 0 .$$

Since P and Q are relatively prime so are P^* and Q^*. Hence the coefficients of dX, dY have at most a power of Z in common. Now if $p \neq q$ no power of Z divides the coefficient of dZ and (13.3) is to be left as it stands. On the other hand if $p = q = n$ and if $p_n(x, y)$, $q_n(x, y)$, the terms of highest degree in P, Q are such that $xq_n = yp_n$, then Z factors out of every term in (13.3) but no higher power does. Under the circumstances one must cancel out the term Z in (13.3). In one or the other case (13.3) assumes the form

$$(13.4) \quad A(X, Y, Z) \, dX + B(X, Y, Z) \, dY + C(X, Y, Z) \, dZ = 0$$

where A, B, C are forms (homogeneous polynomials) of the same degree.

14. The equivalent cartesian form of (13.4) remains of course (13.1). This means that (13.1) determines a line through each non-critical point M, namely the tangent to the path through M. This is in contrast with (12.1) which defines a definite *ray* through M. The opposite ray corresponds to a change of t into $-t$.

Returning now to the situation in (12) the line through M determines arcs through M', M''. Let us choose a definite time t, and thus have definite rays. Let the directed arc through M' be such that it projects into the ray through M. Then the associated arc through M'' will project into the opposite ray through M. The arcs thus defined determine a vector field $\mathfrak{F}$ on $S - E$, and by continuous extension on the whole of S. The critical points of $\mathfrak{F}$ are associated in pairs with equal index. Such a pair K', K'' gives rise to what we shall describe as a critical point K of the system (13.4), and its index is by definition the common index of K' and K''. The sum of the indices $I(K)$ is half the same as for the sphere, i.e. unity. The critical points are the places where the tangents defined by (13.1) cease to exist, i.e. where $A = B = C = 0$.

To calculate an index $I(K)$ one proceeds as follows: Select say projective coordinates X', Y', Z' such that K is the point $(0, 0, 1)$ then pass to cartesian coordinates $x' = X'/Z'$, $y' = Y'/Z'$. As a result (13.4) will be replaced by a system

$$Q'(x', y')\,dx' - P'(x', y')\,dy' = 0\,. \tag{14.1}$$

One passes now from (14.1) to the associated system in the form (12.1)

$$dx'/dt = P',\quad dy'/dt = Q'\,. \tag{14.2}$$

This amounts to replacing the lines (14.1) through each point by a system of rays. This system has then the same index as one of the points K', K'', i.e. its index is that of K.

It is readily seen that for the critical points in the Euclidean plane Π the procedure just described yields the usual index. For the points at infinity however it yields an index where none was at hand before. In other words the notion of index is thus brought to bear upon the system (13.4) throughout the projective plane $\mathfrak{P}$.

15. As an application let us take the following system discussed by Poincaré ([2], p. 66):

$$\begin{cases} dx/dt = x^2 + y^2 - 1\,, \\ dy/dt = 5\,(xy - 1)\,. \end{cases} \tag{15.1}$$

The right hand sides equated to zero have no common solution and so there are no critical points in the Euclidean plane. In homogeneous form (15.1) becomes

$$\begin{vmatrix} dX, & dY, & dZ \\ X, & Y, & Z \\ X^2 + Y^2 - Z^2, & 5\,(XY - Z^2), & 0 \end{vmatrix} = 0\,,$$

or after expansion

$$-5\,Z\,(XY - Z^2)\,dX + Z\,(X^2 + Y^2 - Z^2)\,dY + \tag{15.2}$$

$$+ (4X^2Y - Y^3 - 5XZ^2 + YZ^2)\,dZ = 0\,.$$

The only critical points are given by

$$Z = 0,\ Y(4X^2 - Y^2) = 0\,.$$

They are thus the three points

$$A\,(1, 0, 0),\ B\,(1, 2, 0),\ C\,(1, -2, 0)\,.$$

Since the system has no finite critical points in the Euclidean plane it possesses no closed paths in that plane. Let us now consider the nature of the critical points.

Critical point A. The appropriate cartesian coordinates are $y = Y/X$, $z = Z/X$. Making $X = 1$, $dX = 0$, $Y = y$, $Z = z$ in (15.2) we find

$$z(1 + y^2 - z^2)\,dy + (4y - y^3 - 5z^2 + yz^2)\,dz = 0\,.$$

The critical point is the same as for the pair of equations

$$dy/dt = -4y + 5z^2 - yz^2 + y^3\,,$$
$$dz/dt = z + z(y^2 - z^2)\,.$$

As the characteristic roots are -4, 1 the point A is a saddle point, and its index is -1.

Critical point B. The transformation

$$X \to X,\ Y \to Y + 2X,\ Z \to Z$$

sends B to the point $(1, 0, 0)$. The same process applied to the transformed equation shows that B has the same behavior as for

$$dy/dt = 4y,\ dz/dt = 5z\,.$$

As the characteristic roots are 4, 5 the point B is a node, its index being $+1$,

Critical point C. The transformation is now $X \to X$, $Y \to Y - 2X$, $Z \to Z$, but otherwise everything is as for B and C is also a node, its index being again $+1$.

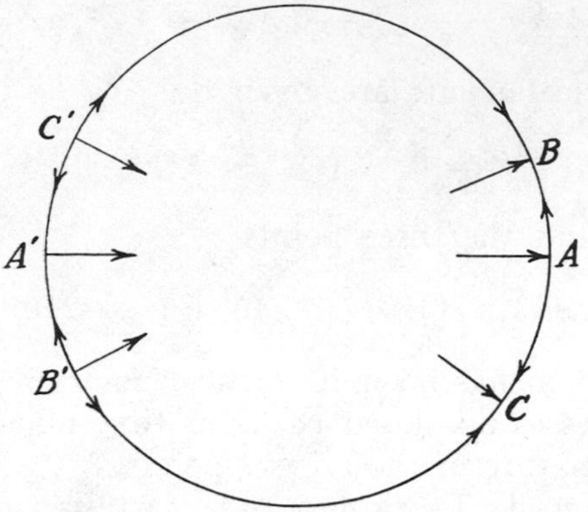

Fig. 14

The sum of the indices is $+1$ as it should be.

As a complementary and useful observation let us point out that (15.2) is satisfied by $Z = 0$. Hence the line at infinity consists of arcs of paths.

The preceding information will now be utilized to construct the phase-portrait. To that end let the projective plane $\mathfrak{P}$ be

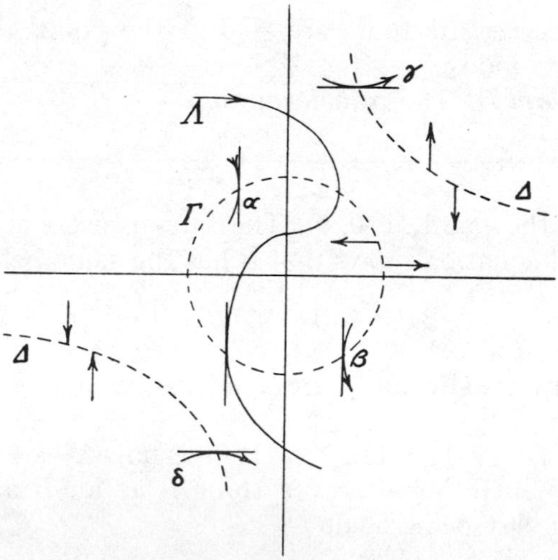

Fig. 15

identified with a closed circular region Ω with diametral points of the boundary Γ matched. Let us mark on Γ diametral points A, A'; B, B'; C, C' for each critical point. Thus A, A' are saddle points and the rest are nodes. Since AB, AC and $A'B'$, $A'C'$ are parts of paths, they are on paths tending to or away from the two saddle points. Referring to (15.1) $dx/dt > 0$ outside the circle $x^2 + y^2 = 1$. Hence the approach to A and removal from A' are as shown in fig. 14. Since the other critical points are

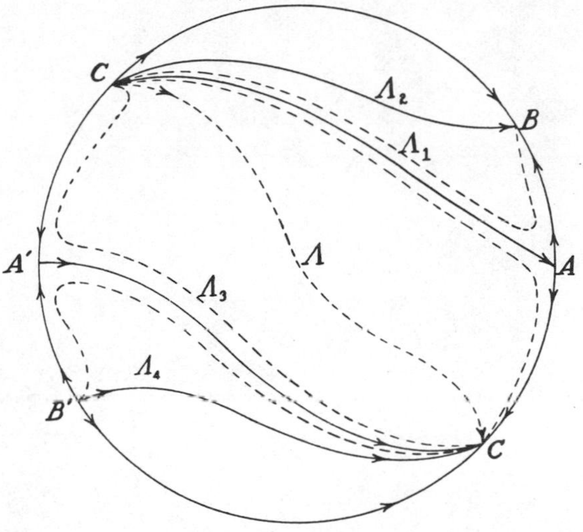

Fig. 16

nodes the paths tend to or away from them as shown in the figure.

Let Γ denote the circle $x^2 + y^2 = 1$ and Δ the hyperbola $xy = 1$. From (15.1) we learn that along a path x decreases inside Γ and increases outside Γ, while y decreases outside Δ and increases in its two interior regions. This situation is described by horizontal and vertical pointers in fig. 15. Observe also that Γ is the locus of the points where the tangents to the paths are vertical and Δ the locus of the points where they are horizontal. The only crossings of Γ, Δ by paths consistent with these various properties are as indicated by the arcs α, β, γ, δ in fig. 15.

If we recollect that within Γ the slopes dy/dx of the paths are always ≥ 0, we find that the paths crossing Γ have the general form of Λ in fig. 15. Returning now to the circular region of fig. 14 we find that Λ assumes the general form indicated in fig. 16. From this follows that the two paths issued from the two saddle points other than the equator are of the types Λ_1, Λ_2 of fig. 16. The other basic types are then represented by Λ_3, Λ_4.

CHAPTER X

Two Dimensional Systems *(Continued)*

In the present chapter one will find a description of the general critical points of analytical systems, and of the limiting sets of their paths as $t \to +\infty$. The groundwork for this study was laid by Poincaré but his results were considerably extended, refined and given more precision by Bendixson.

A full treatment is also given of the behavior of the paths near a critical point with a single non-zero characteristic root, a topic already dealt with by Bendixson. Finally the chapter concludes with a consideration of structural stability, an important concept due to Andronov and Pontrjagin, but fully treated for the first time by De Baggis.

References: De Baggis [1]; Bendixson [1]; Coddington, Levinson [1]; Lefschetz [4]; Niemitzki-Stepanow [1]; Poincaré [2].

§ 1. General Critical Points

1. Consider again an autonomous analytical system

$$dx/dt = X(x, y), \quad dy/dt = Y(x, y), \tag{1.1}$$

and suppose that it has an isolated critical point O which one may take as the origin. We will assume that X, Y are holomorphic in a closed circular region Ω of center O and that O is the only critical point in Ω. Thus X, Y vanish simultaneously in Ω solely at the point O.

We will first prove several preliminary properties.

(1.2) *Let $f(x, y)$ be a real function holomorphic in Ω with $f(0, 0) = 0$. If the radius σ of Ω is sufficiently small then the curve $f(x, y) = 0$ has in Ω at most a finite number of real branches and they are of one of the two types*

(1.2a) $$x = 0\,,$$

(1.2b) $$y = a_k\, x^{k/n} + a_{k+1}\, x^{(k+1)/n} + \dots\,,$$

where k, n are positive integers and the coefficients a_j are all real. Moreover for σ small enough the real branches intersect one another in Ω only at the origin.

By the Weierstrass preparation theorem

$$f \equiv x^m\,(y^p + a_1(x)\,y^{p-1} + \dots + a_p(x))\cdot E(x, y)\,,$$

where the parenthesis is a special polynomial in y and E is a unit (see I, 14.1). Both are real when f is real (I, 14.2).

The solutions of $f = 0$ in Ω, for σ small enough, are thus $x = 0$ if $m > 0$ and those of

(1.3) $$y^p + a_1(x)\,y^{p-1} + \dots + a_p(x) = 0\,.$$

Now the theory of the Puiseux series (see notably Picard [1], vol. II, p. 350) is fully applicable to (1.3) and so the solutions in question are of the type (1.2b) with the a_j real or complex. The real solutions correspond evidently to systems (1.2b) with the a_j all real.

Suppose now that there are two real branches. If one is $x = 0$ and the other is (1.2b) they clearly intersect only at the origin in Ω. Suppose that there are two branches, one of them (1.2b) and the other similar with k', n', a_k' in place of k, n, a_k. The intersections of the two correspond to the solutions in Ω of a relation

$$x^{r/nn'}\,(\beta_0 + \beta_1\, x^{1/nn'} + \dots) = 0$$

where the β_j are not all zero. By a well known property of analytic functions the solutions $\neq 0$ are bounded away from zero and hence correspond to points outside Ω for σ small. This completes the proof of (1.2).

Let A and B be the end-points of an arc λ. We shall conveniently use the designations (AB), $[AB)$, $(AB]$, $[AB]$ for the open arc λ, the are λ closed at A and open at B, the arc λ open at A and closed at B, the closed arc λ.

We shall also say with Poincaré that a differentiable arc λ of one of the above four types, and situated in the field of operation of the system (1.1), is *without contact*, whenever λ contains no critical point, and the unique path γ through any point M of λ is never tangent to λ at M.

(1.4) *Let* $\lambda = [AA')$ *and* $\mu = [BB')$ *be two arcs without contact intersecting at most in* A (*then* $A = B$) *and disposed as in fig.* 1 *so that* A *and* B *are on the same path* γ. *Given any point* $M \in \lambda$ *let the path* $\gamma\,(M)$ *through* M, *followed forward first meet* μ *at a point* M'. *Then* $M \to M'$ *defines a topological mapping* φ *of* λ *on a subarc* $[BA'')$ *of* $[BB')$. *Moreover if* $B' \neq A'$, *then the assign-*

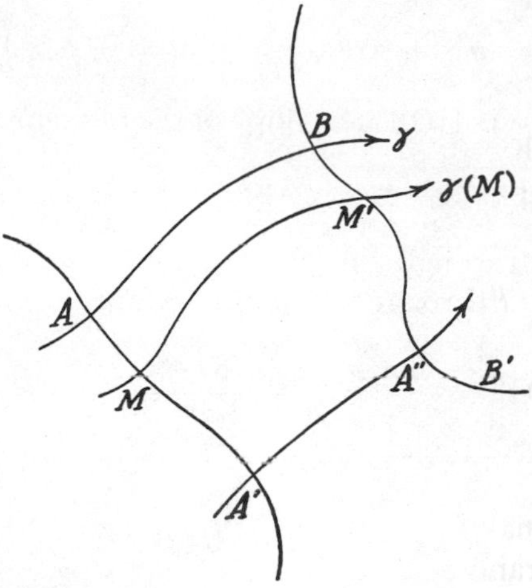

Fig. 1

ment $A'' = \varphi\, A'$ *makes of* φ *a topological mapping* $[AA'] \to [BA'']$. *The same properties hold if the paths are followed backward instead of forward, and likewise even if* $A = B$ *provided that* A *is not a critical point.*

If diam AB is sufficiently small this property is merely a special case of (II, 17.3). If AB is large one decomposes it into a finite or countable set of consecutive small segments (countable set if $A = B$, non-critical) and applies the same property to each segment in turn, resulting in a mapping such as φ.

Consider now the curve

$$rdr/dt = xX + yY = Z\,(x, y)\,. \tag{1.5}$$

Let α be a branch of $Z = 0$. Take a circle C_ϱ of radius ϱ and center O contained in Ω.

(1.6) *If ϱ is sufficiently small the branch α intersects C_ϱ in a single point A and the arc $(OA]$ of the branch is an arc without contact.*

Let the coordinates be so chosen that α is not tangent to one of the axes at the origin. If α is identified with the branch (1.2b) its tangent at the origin is: $x = 0$ if $k < n$; $y = 0$ if $k > n$; $y = a_k x$ if $k = n$. We are then here in the third case. Hence α has a representation

$$y = mx + ax^p + \dots, \; m \neq 0, p > 1 .$$

At any point (x, y) of $(OA]$ the slope of the tangent to the branch is

$$\mu = m + pax^{p-1} + \dots ,$$

while the path through the point has the same slope μ' as the circle of center O through the point. Hence

$$\mu' = -\frac{x}{y} = \frac{-1}{m + pax^{p-1} + \dots} ,$$

and therefore

$$\mu\mu' = -1 + O(x^{p-1}) .$$

Thus for ϱ small the path through any point of α within or on C_ϱ and the branch α are nearly orthogonal and so the part of α in the closed interior of C_ϱ is without contact.

Suppose that α contains several disjoint arcs within C_ϱ. Then as ϱ decreases C_ϱ would sometime become tangent to α. Hence the path through the corresponding point of α would become tangent to α, whereas, as we have seen, they are nearly orthogonal. Thus α has a single arc in C_ϱ. Similarly, if this unique arc α intersects C_ϱ, for ϱ arbitrarily small, in more than one point then some C_σ, σ small, would be tangent to α and this is ruled out as before. To sum up when ϱ is sufficiently small α has a single arc in the closed interior of C_ϱ and this arc intersects C_ϱ in a single point.

(1.6a) Returning now to the curve (1.5), we conclude that

if ϱ is small enough each branch of the curve intersects C_ϱ in a single point and in C_ϱ the only point common to all the branches is the origin. Hence the interior of C_ϱ together with C_ϱ is decomposed by the branches of (1.5) into a finite set of triangular sectors in each of which dr/dt has a fixed sign. Since the paths are nearly orthogonal to the branches in C_ϱ, they all cross a given branch in a fixed direction: all inward or outward relative to a given sector.

(1.7) *If a path γ enters [leaves] a sector OAB and does not leave [enter] it through one of the sides OA, OB, then as $t \to +\infty$ [as $t \to -\infty$] γ must tend to the vertex O of the sector.*

The second case: $t \to -\infty$, is reducible to the first by the change of variable $t \to -t$, so that we only need to consider

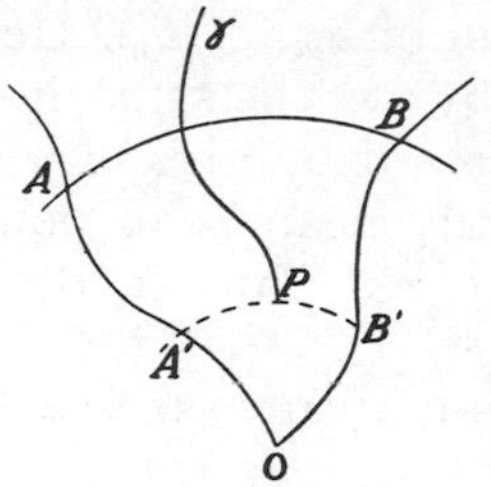

Fig. 2

the first case where γ enters the sector. We must show that $r(t) \to 0$ along γ as $t \to +\infty$. If $r_0 = \inf r(t)$ on $\bar{\gamma}$ in the sector then r_0 is reached on $\bar{\gamma}$ in the sector and we must show that $r_0 = 0$. Suppose $r_0 > 0$ and let $[A'B']$ be the arc of the circle $r = r_0$ comprised in the sector. Since $\bar{\gamma}$ is compact it has a point P on $[A'B']$. Since P is an ordinary point there is a path δ (not a point) through P. Upon constructing the path-rectangle whose axis is an arc λ of δ containing P, it is seen that $\gamma \to P$ in the sector is only compatible with $\gamma = \delta$. Thus γ must leave the sector if P is A' or B'. Hence P is between A' and B'. Since however $dr/dt < 0$ or > 0 at P, γ must cross $A'B'$ at P. Hence if $r_0 \neq 0$ it is not inf $r(t)$ on γ and therefore $r_0 = 0$, proving (1.7).

Let r, θ denote polar coordinates. Then:

(1.8) *Let the path γ tend to the origin so that $r(t) \to 0$, either as $t \to +\infty$ or as $t \to -\infty$. Then along γ, $\theta(t)$ tends to a finite or* *infinite limit, i.e.* $\gamma \to 0$ *in a fixed direction or else it spirals around* O. (Bendixson [1], p. 34.)

We deduce at once from (1.1)

$$(1.9) \quad r^2\, d\theta/dt = xY - yX = r^{m+1}\,(a_{m+1}(\theta) + r a_{m+2}(\theta) + \ldots)$$

where $a_h(\theta)$ is a form of degree h in $\sin\theta$, $\cos\theta$. Suppose first that $xY - yX \not\equiv 0$. Thus $a_{m+1}(\theta) \neq 0$. Let $r(t) \to 0$ as $t \to +\infty$. The case $t \to -\infty$ may be taken care of by the change of variable $t \to -t$. To prove that $\theta(t)$ has a finite or infinite limit we merely need to show that when $\theta(t)$ remains bounded it cannot have two distinct limiting values θ_0, θ_1. Suppose that this is the case. One may select a θ_2 between θ_0 and θ_1 and such that $a_{m+1}(\theta_2) \neq 0$. Thus for r below a certain ϱ, $d\theta/dt$ for $\theta = \theta_2$ will have the fixed sign of $a_{m+1}(\theta_2)$. Hence the ray L: $\theta = \theta_2$ will always be crossed by γ in the same direction. This is however ruled out since L is crossed by arcs of γ arbitrarily near O and joining points arbitrarily near θ_0 to points arbitrarily near θ_1 in both possible directions. Thus in the present case $\theta(t)$ has a limit.

If $xY - yX = 0$ then $d\theta/dt = 0$, $\theta = C$ and (1.8) obviously holds.

§ 2. Local Phase-Portrait at a Critical Point

2. We shall now examine the various possible dispositions of the paths in the individual sectors and then combine the sectors to obtain the complete disposition around the critical point O.

Take then a sector OAB and suppose $dr/dt < 0$ in the sector. Thus the side AB is crossed inward by the paths. There are then various possibilities depending upon the modes of crossing of the other two sides. Each case must be examined separately. The radius ϱ of the circle C_ϱ is supposed throughout to be small enough to have the subdivision into sectors of (1.6a).

Case I. The sides OA and OB are both crossed outward. A path γ through a point M of $(OA]$ when followed backward necessarily goes farther from O throughout the triangle and must cut the arc $[AB]$ at a point N. By (1.4) $N \to M$ defines a topological mapping of a subarc $[AC)$ of $[AB]$ onto $[AO)$. Similarly the paths through $(OB]$ when followed backward define a topological mapping of $[BO)$ onto a subarc $[BD)$ of $[BC)$. Since the paths through the subarc $[CD]$ of (AB) must remain in the sector, by (1.7) they all tend to O as $t \to +\infty$. We thus obtain the disposition of fig. 3 with the typical paths γ, δ, ε. For evident reasons the paths crossing $[CD]$ form a system referred to as a *fan*.

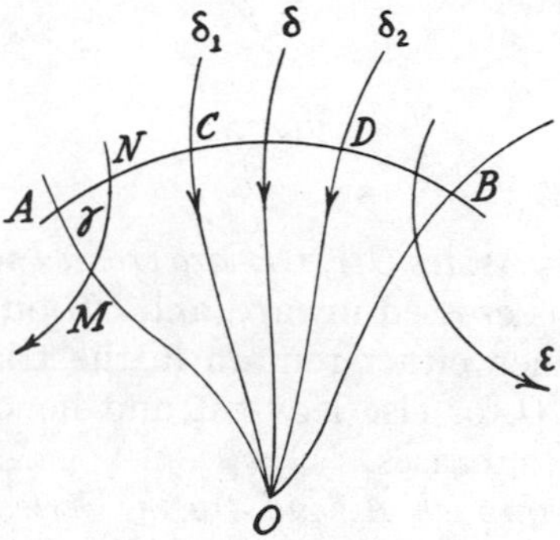

Fig. 3

As a limiting case we might have $C = D$, so that the fan consists of a single arc. This is shown in fig. 4.

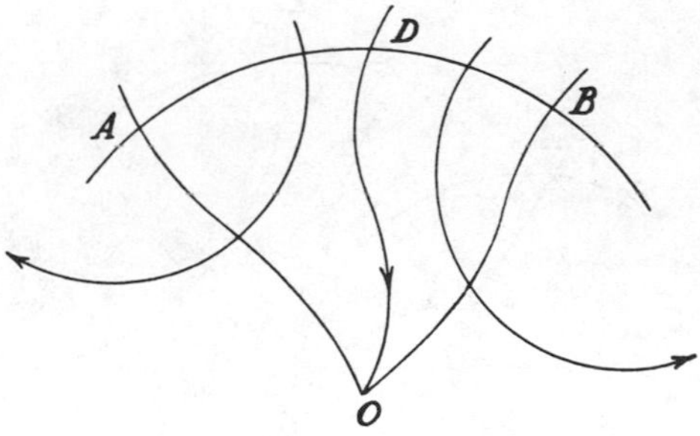

Fig. 4

Case II. The sides OA and OB are both crossed inward. Thus the whole periphery of the triangle, O excepted, is crossed inward. By (1.7) all the paths considered must tend to O, giving the disposition of fig. 5.

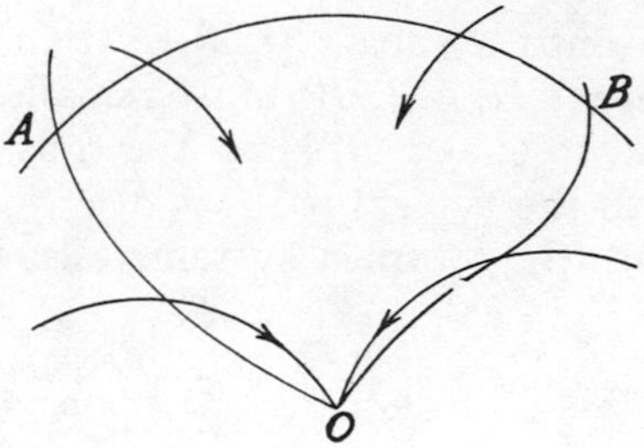

Fig. 5

Case III. The two sides OA, OB are crossed in opposite manner. Assume that OA is crossed inward and OB outward. The path δ through A may then either remain in the triangle, and then it can only tend to O, or else leave it and hence cross (OB). This gives rise to two subcases.

(a) *The path δ through A tends to the vertex O in the triangle.* As under Case I, the paths through the points of $(OB]$ give rise to a topological mapping of $(OB]$ onto a subarc $(CB]$ of $[AB]$ (fig. 6). The paths through $[AC]$ form a fan. The paths γ, ε, ζ exhibit the three types of paths in relation to the triangle.

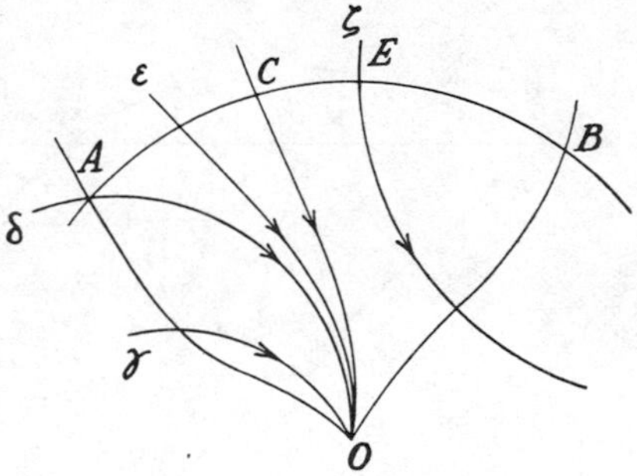

Fig. 6

(b) *The path δ through A crosses (OB) and this no matter how small the radius ϱ of the circle C_ϱ.* We have then the situation of fig. 7, with only two typical paths: γ and ε.

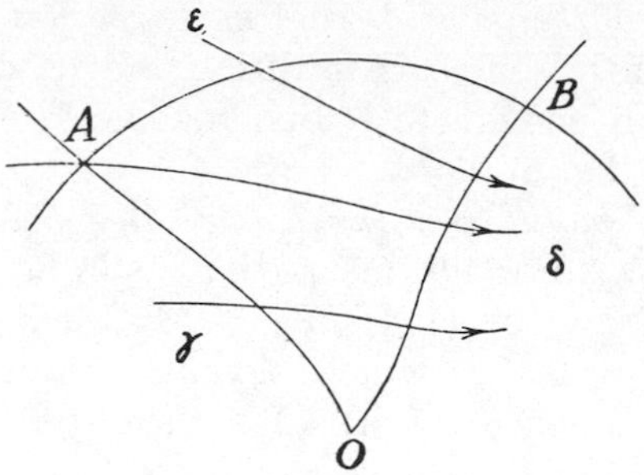

Fig. 7

If the crossings at OA and OB are reversed there will result cases $\text{III}_1 a, \ldots$, derived from IIIa, ..., in the obvious way.

It has been assumed so far that (AB) is crossed inward by the paths. If the crossing is outward the change of time variable $t \to -t$ will preserve the paths but make them cross (AB) inward. The resulting configurations are the same as before but with all the arrows reversed. We shall refer to them as I′, ...

Suppose now that the curve $Z = xX + yY = 0$ has no real branch through the origin. Thus in the vicinity of the origin dr/dt has a constant sign. Let us suppose first $dr/dt < 0$ so that in the vicinity of O, r is decreasing along any path γ. Consequently along γ the coordinate r tends to a limit r_0 and $r = r_0$ is a closed path of diameter $\leq \varrho$. Since this contradicts: $dr/dt < 0$ for $r < \varrho$, the only possibility is $r_0 = 0$, i.e. γ tends to O.

Suppose that there is a ray L through the origin which is not a path. Then as regards the open sector of angle 2π, bounded by L the situation is that of Case III. We have now the two possibilities corresponding to IIIa and IIIb which we consider in turn.

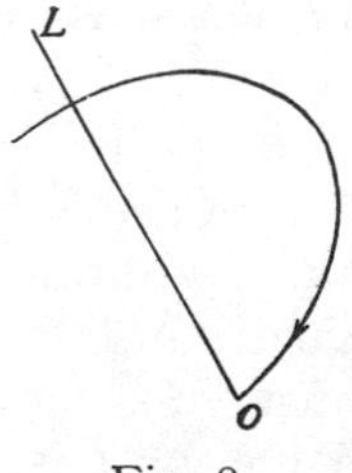

Fig. 8

Case IV. One of the paths δ crossing L reaches the origin in the open sector. Then by our earlier discussion it will be seen that all the paths near enough to O behave like δ. Thus the point O is a *stable node* (fig. 8).

Case V. Every path δ crossing L, say at a point M_1, crosses it again at a point M_2. For another path δ' as in fig. 9, the crossings will be M_1', M_2' and by (1.4) $M_1' \to M_2'$ defines a topological mapping $[OM_1] \to [OM_2]$, which shrinks the interval OM_1. It follows that δ, and hence every path sufficiently near the origin O

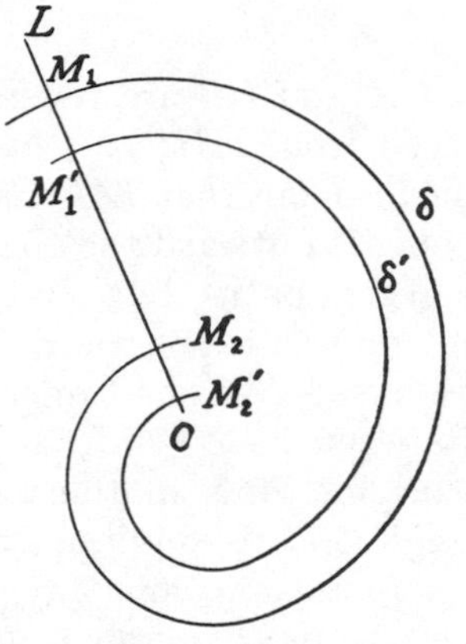

Fig. 9

is a spiral tending to O. Thus O is a *stable focus*. Exceptionally $M'_1 = M_2'$ throughout OM_1 and O is a *center*.

We have in addition

Case VI. $dr/dt = 0$, hence all paths are circumferences of center O. The origin O is then again a center.

If every ray issued from O is a path, we have actually Case IV and O is a node.

Finally if $dr/dt > 0$ near O we have reversal of arrows and Case IV': unstable node, and Case V': unstable focus.

(2.1) *Local phase-portrait.* To obtain the full configuration of the paths around the critical point O (the local phase-portrait around the point) one must combine adjacent sectors of the various types described in the various admissible ways.

Let us observe at the outset that if all the sectors around O are of types IIIb (fig. 7) or III_1b limit (rotation reversed) or the same with III′ instead of III, then all the paths near O are spirals and we merely have a stable or unstable focus. As this case will arise otherwise anyhow we may leave it out of consideration at present. On the other hand, a succession of sectors such as IIIb or III_1b not surrounding fully the point O, has no topological effect on the configuration of the paths. Thus the situation of fig. 7 and related types need not be considered here. Upon matching the other types in the various admissible ways, and adding to them the focus and center we obtain the following list:

I. *Fan.* This may be *attractive* or *repulsive* accordingly as the paths all tend to the critical point or all away from it (fig. 10). As a limiting case we have the *nodes.*

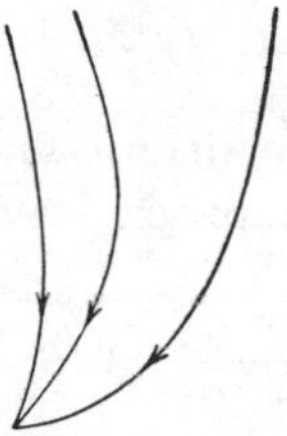

Fig. 10

II. *Hyperbolic sector* (fig. 11).

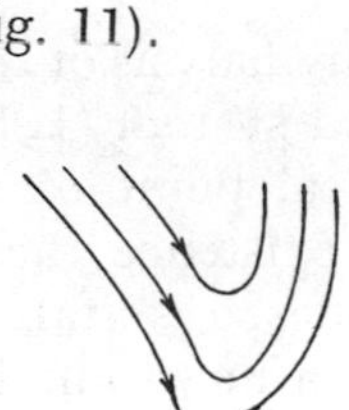

Fig. 11

III. *Elliptic sector* (fig. 12). This is the only truly new type.

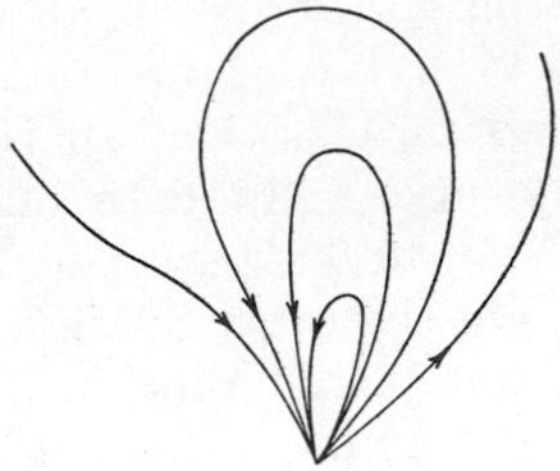

Fig. 12

IV. *Focus.*

V. *Center.*

(2.2) *Existence of the various types.* All the types except III are known to occur as critical points of linear equations. Regarding III the following very simple type was suggested by Gomory. Take the system

$$dx/dt = x, \; dy/dt = -y$$

with a saddle point at the origin and no other critical point. The transformation by reciprocal vectors

$$x \to \frac{x}{x^2 + y^2}, \; y \to \frac{y}{x^2 + y^2}$$

will replace the system by an analytical system with a single critical point at the origin which has four sets of elliptic sectors one in each quadrant.

(2.3) *Remark.* While our analysis of the critical point has been developed for an analytical system (1.1) it is manifestly applicable to an isolated critical point of a system which merely satisfies the fundamental existence theorem provided that the curve $xX + yY = 0$ divides a suitably small circular neighborhood of the origin into a finite number of triangular sectors such as those considered above, in each of which dr/dt has a fixed sign, and similarly for the curve $Yx - Xy = 0$ and the sign of $d\theta/dt$. One must also assume that the branches of the curves clustering around the point possess continuously turning tangents in the given neighborhood of the origin.

(2.4) *Index.* Let us suppose that the critical point O consists of σ hyperbolic sectors, ν elliptic sectors and the rest fans. Let $\alpha_1, \ldots, \alpha_\sigma$ be the positive angles of the tangents to the hyperbolic sectors, $\beta_1, \ldots, \beta_\nu$ and $\gamma_1, \ldots, \gamma_\varrho$ the same for the elliptic sectors and the fans. Thus

$$\Sigma\alpha_h + \Sigma\beta_j + \Sigma\gamma_k = 2\pi .$$

Let us draw now a small circuit γ which coincides in a hyperbolic

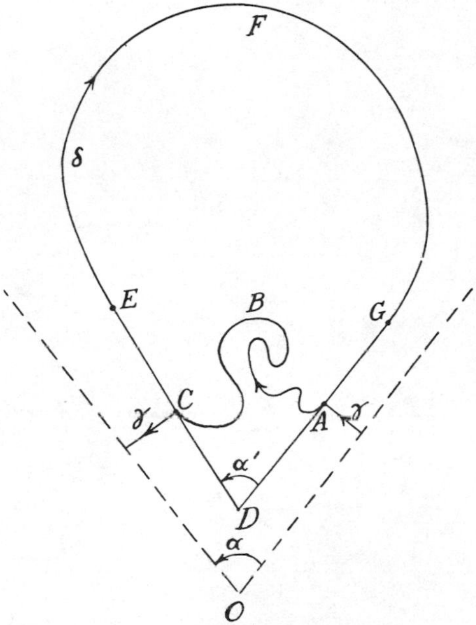

Fig. 13

sector with most of a "hyperbolic" arc such as ABC of fig. 13, and in an elliptic sector sector with most of an oval such as HKL of fig. 14.

Let us carry out the construction of fig. 13. The tangents at A and C meet at D and EFG is a circular arc tangent to the lines DA and DC. Consider now a vector distribution tangent throughout to $\delta = ABCEFGA$ and coincident with the vectors of our system (1.1) along ABC. Since δ is negatively oriented its index is -1 (IX, 11.4). The angular variation V_1 along ABC

is that of our field and along $CEFGA$ it is $W_1 \doteq -(\alpha + \pi)$ ($\doteq$ means "almost equal"). Hence $V_1 \doteq -2\pi - W = -2\pi - (-\alpha - \pi) = \alpha - \pi$, i.e. V_1 is almost $\alpha - \pi$.

For the elliptic sector the situation is very similar. The arc MN is circular and the circuit $\varepsilon = MNHKLM$ is positively oriented. Its index, relative to a tangent vector field, is thus $+1$. The angular variation V_2 along HKL is that of our field, and along $LMNH$ it is $W_2 \doteq \pi - \beta$. Hence $V_2 \doteq 2\pi - W_2 = 2\pi - (\pi - \beta) = \beta + \pi$, i.e. V_2 is almost $\beta + \pi$.

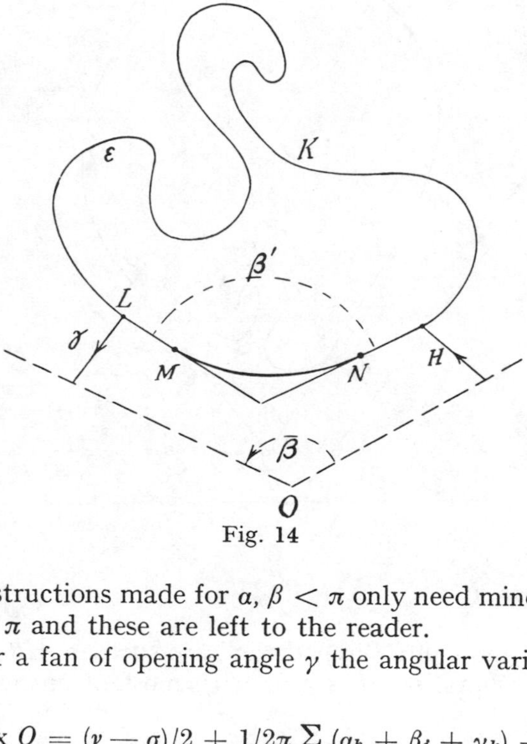

Fig. 14

Our constructions made for $\alpha, \beta < \pi$ only need minor changes if α or $\beta \geq \pi$ and these are left to the reader.

Since for a fan of opening angle γ the angular variation $\doteq \gamma$, we have:

$$\text{Index } O = (\nu - \sigma)/2 + 1/2\pi \sum (\alpha_h + \beta_j + \gamma_k) + \xi ,$$

where ξ is arbitrarily small. Since ξ is zero or a multiple of 1/2 it can only be zero. Since the sum is merely 2π we arrive at the following formula due to Bendixson ([1], p. 39):

$$(2.5) \qquad \text{Index } O = 1 + (\nu - \sigma)/2$$

Since the index of a focus or center is unity, (2.5) holds even in these cases. Hence it holds without exception.

An obvious corollary of (2.5) is:

(2.6) *The numbers* ν *of elliptic sectors and* σ *of hyperbolic sectors have the same parity.*

(2.7) *Stability.* It is at once apparent that if there is a hyperbolic sector or an elliptic sector the critical point is unstable. Thus:

(2.8) *A n.a.s.c. for stability is that the point be a stable node, a stable focus or a center. In the first two cases the stability is asymptotic.*

(2.9) *Limit-cycles. Separatrices.* These exceptional paths will play an important role later (see 8) and in general when one wishes to discuss the full phase-portrait. A limit-cycle is a closed path which is not a member of a continuous family of closed paths, i.e. an isolated closed path. Roughly speaking a separatrix is a true path (not a critical point) behaving topologically abnormally in comparison with neighboring paths. Explicitly a separatrix is a path which is either a limit-cycle or else a path terminating or beginning on the projective plane with a side of a hyperbolic sector. Thus the four arcs of a saddle point A which tend to A belong to separatrices.

3. The local phase-portrait may be completed by information regarding the mode of approach of the paths to the origin. Suppose in fact that

$$(3.1)\quad X = X_m + X_{m+1} + \dots, \quad Y = Y_m + Y_{m+1} + \dots,$$

where X_h, Y_h are forms of degree h in x, y and one of X_m, Y_m is not identically zero.

(3.2) *If the form* $U_{m+1} = xY_m - yX_m$ *is not identically zero then the directions of approach are among those represented by* $U_{m+1} = 0$, *so that their number is at most* $2m + 2$ *and they are opposite in pairs.*

The value of $d\theta/dt$ is given by (1.9) with

$$a_{m+1}(\theta) = \frac{U_{m+1}(x, y)}{r^{m+1}} \not\equiv 0.$$

In addition

$$(3.3) \qquad r\,dr/dt = r^{m+1}(\beta_{m+1}(\theta) + r\,\beta_{m+2}(\theta) + \ldots),$$

where β_h is like a_h in (1.9). Along the path γ tending to the origin replace the variable t by $\tau = \int_{t_0}^{t} r^{m-1}\,dt$. As a consequence instead of $t \to \pm\infty$, τ may tend to a finite value τ_1. (We shall see however that $\tau \to \pm\infty$ as before.) The new system for γ is now

$$(3.4) \qquad \begin{cases} d\theta/d\tau = a_{m+1}(\theta) + r a_{m+2}(\theta) + \ldots, \\ dr/d\tau = r\{\beta_{m+1}(\theta) + r\,\beta_{m+2}(\theta) + \ldots\}. \end{cases}$$

Let r, θ be considered in (3.4) as cartesian coordinates. The critical points of the new system are of two types. First the images of the former critical points other than the origin. These critical points are at a positive distance $\geq \varrho$ from the θ axis. Then there are new critical points on $r = 0$ (the θ axis) located at the discrete places where $a_{m+1}(\theta) = 0$, i.e. at the θ places corresponding to the directions represented by $U_{m+1} = 0$.

Now if $f(\tau)$ is any solution of

$$(3.5) \qquad d\theta/d\tau = a_{m+1}(\theta),$$

then $(\theta = f(\tau), r = 0)$ is a solution of (3.4). Since there is a (non-point) solution of (3.4) starting at any non-critical point $(\theta_0, 0)$ $(a_{m+1}(\theta_0) \neq 0)$, every arc of the axis $r = 0$ is a path of (3.4). It follows that as $r \to 0$ along γ this path can only tend to infinity or else to a critical point such as $(\theta_1, 0)$ $(a_{m+1}(\theta_1) = 0)$, so that at the same time $\tau \to \pm\infty$. Thus, as stated earlier, as $t \to \pm\infty$ along paths γ for which $r \to 0$, $\tau \to \pm\infty$ likewise. Going back to the initial system it means that the directions of approach to the origin are restricted to those represented by $U_{m+1} = 0$ and this proves (3.2).

4. One may actually proceed a step further. Let ξ be a simple root of $a_{m+1}(\theta) = 0$, i.e. let $l = y\cos\xi - x\sin\xi$ be a simple factor of $U_{m+1}(x, y)$. Let also $\beta_{m+1}(\xi) \neq 0$, i.e. l is not a factor of $V_{m+1} = xX_m + yY_m$. Then $A: r = 0$, $\theta = \xi$ is an elementary

critical point of (3.4). Since the path $r = 0$ tends to A in a fixed direction, A is not a focus nor a center. Hence there are at least four directions of approach to A, and at least two distinct from the directions on $r = 0$. These two are opposite, hence one is in the part $r > 0$ of the plane and it corresponds to at least one path of (1.1) tending to the origin in the direction $l = 0$. Since the same situation will hold regarding $\xi + \pi$, paths approach the origin along l in both possible directions. Thus:

(4.1) *If $l = \lambda x + \mu y$ is a simple factor of U_{m+1} and does not divide V_{m+1} then paths approach the origin along l in both possible directions.*

§ 3. The Limiting Sets of the Paths as $t \to \pm \infty$

5. To study these limiting sets it is convenient to close the Euclidean plane at infinity by a point, thus turning it into a sphere. Analytically this is done by applying the standard transformation

$$x' = 1/x, \; y' = 1/y$$

which reduces the infinite region to the origin.

The reason for giving up the projective plane for the present is that the argument will rest heavily upon the Jordan curve theorem. As a matter of fact instead of passing to the sphere in the above manner, one could replace the projective plane by its doubly covering sphere.

A certain number of preliminary properties are needed.

(5.1) *Let the Jordan curve J on the Euclidean plane Π or on the sphere S possess an open analytical arc λ. Let a path γ cross λ at a point P where γ and λ are not tangent. Then, at P, γ crosses from one of the two regions bounded by J into the other.*

Let P be taken as origin for local coordinates x, y and let x, y be so chosen that at P neither γ nor λ are tangent to one of the axes.

Since λ is analytical, a neighborhood λ' (an arc) of P in λ has a holomorphic representation

$$x = \varphi(u), \; y = \psi(u) \tag{5.2}$$

where $\varphi(0) = \psi(0) = 0$ and owing to the choice of axes both

$\varphi'(0)$, $\psi'(0) \neq 0$. One may eliminate therefore u from (5.2) and obtain say an analytic representation for a suitable λ' of the form

$$f(x, y) = y - g(x) = 0 . \tag{5.3}$$

The condition of no tangency at P means

$$X(P) \neq 0, \; Y(P)/X(P) \neq g'(0) . \tag{5.4}$$

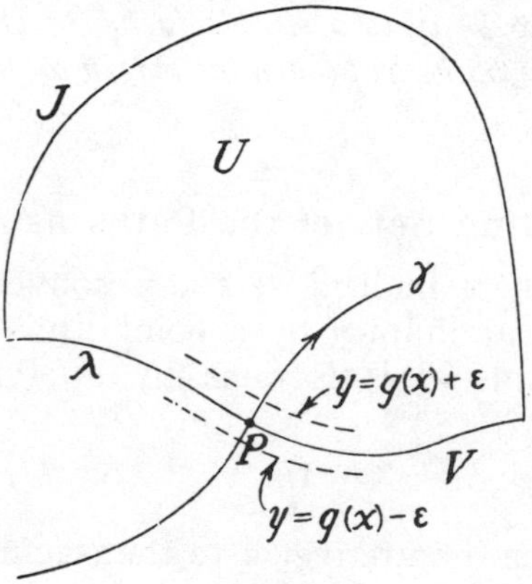

Fig. 15

Let now U, V be the two components of $\Pi - J$ and $S - J$ as the case may be, which have J as common boundary. Consider also for x, y small the two sets

$$\left.\begin{aligned} W_1&: \; y = g(x) + \varepsilon \\ W_2&: \; y = g(x) - \varepsilon \end{aligned}\right\} 0 < \varepsilon < \eta .$$

The two sets W_i are connected and together with λ they make up a full neighborhood of P in Π or S. Hence say $W_1 \subset U$, $W_2 \subset V$.

Let us follow now γ through P. We have on γ:

$$\begin{aligned} df/dt &= f_x X + f_y Y = - g'(x) X + Y \\ &= X(Y/X - g'(x)) . \end{aligned}$$

Hence $(df/dt)_P \neq 0$. It follows that along γ, f is monotonic through P and so varies say from $-$ to $+$, i.e. along γ one passes from W_2 to W_1, that is from V to U, and this proves (5.1).

6. We shall now discuss some simple properties of path-rectangles (IX, 1.4). We have at once from their definition:

(6.1) *The paths which traverse a path-rectangle all describe it in the same direction.*

A noteworthy *path-rectangle* (generalization of the earlier configuration of the same name) is associated with a closed arc without contact λ. Let $2a$ be its length, A its midpoint, s the arc length counted positively along λ in a certain direction. If P is any point of λ and γ the path through P, let γ be made to correspond to the solution $(x(t), y(t))$ of (1.1) such that $(x(0), y(0))$ is the point P. With f as in (5) and changing if necessary f into $-f$, we can find a $\tau > 0$ such that on any γ the

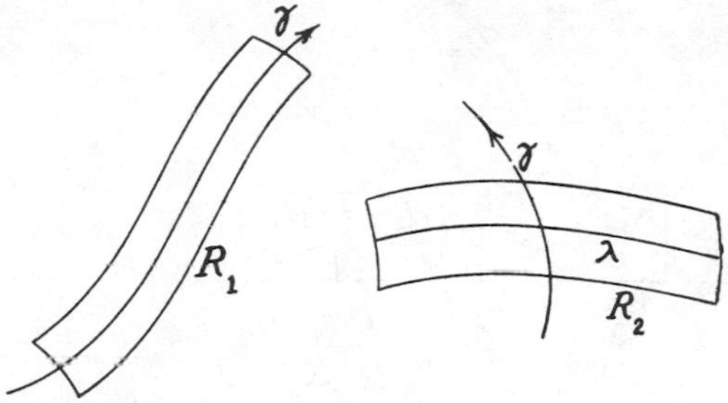

Fig. 16

function $f \geq 0$ for $0 \leq t \leq \tau$, and $f \leq 0$ for $-\tau \leq t \leq 0$. Consider now s, t as rectangular coordinates of an Euclidean plane Π, and let R^* be the rectangle in Π determined by $|s| \leq a$, $|t| \leq \tau$. Define a mapping φ of R^* whereby the point $M^*(s, t)$ goes into the point $M(x(t), y(t))$ of the path γ issued from the point P of λ at a distance s from A. The mapping φ is manifestly continuous and one-one. Therefore since R is compact, φ is topological. We shall refer to the image $R = \varphi R^*$ as a *path-rectangle of median line* λ.

This mild generalization of the earlier path-rectangles has all their properties. The distinction occurs primarily in the applications. The earlier path-rectangle (R_1 in fig. 16) always has a preassigned axis but may be made very thin around that axis. The present type (R_2 in fig. 16) has a preassigned basic line but may be very thin around that line, i.e. in the transverse direction relative to the paths.

(6.2) LEMMA. *If a path γ, followed as t increases, has with an open arc without contact λ three consecutive crossings M, M_1, M_2, then M_1 is between M and M_2 on λ.*

If the lemma is incorrect, then M_2 is between M and M_1 or M between M_2 and M_1. The first case corresponds to fig. 17, and the second to the same figure followed backward. Taking fig. 17 choose at all events λ so small that there is a path-rectangle related to it as in the figure. Let J be the Jordan curve MSM_1M_2M. Then J does and does not separate Q from R, a violation of (5.1) and the lemma follows.

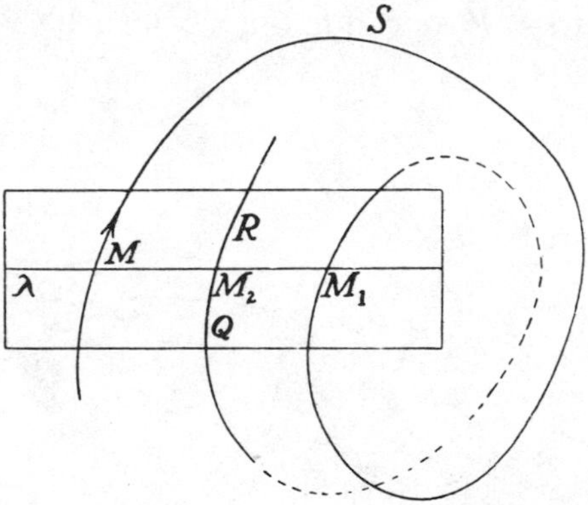

Fig. 17

7. We come now to the limiting sets. Let γ be a path, $\gamma^+(M)$ the subset of γ consisting of M and of all points of γ traversed after M, $\gamma^-(M)$ the analogue referring to the points traversed before M. We refer to $\gamma^+(M)$ and $\gamma^-(M)$ as the *positive* and *negative half-paths determined by* M. The closures $\bar{\gamma}^+(M)$, $\bar{\gamma}^-(M)$ give rise through their intersections to two new sets

$$\Lambda^+(\gamma) = \bigcap_M \bar{\gamma}^+(M), \quad \Lambda^-(\gamma) = \bigcap_M \bar{\gamma}^-(M)$$

called the *positive* and *negative limiting sets* of γ. (They are the ω and α sets of G. D. Birkhoff.)

Since $\gamma^+(M)$, $\gamma^-(M)$ are connected, so are their closures. The latter being also compact and non-empty, we have since we are on the sphere or the projective plane (I, 4.6):

(7.1) *The limiting sets* $\Lambda^+(\gamma)$, $\Lambda^-(\gamma)$ *are compact, connected, and non-empty.*

(7.2) *If* γ *is a critical point, then it coincides with its limiting sets.*

Suppose that P is an ordinary point of say $\Lambda^+(\gamma)$. Thus there is a path δ through P. Construct a path-rectangle Ω with P on its axis and let σ be a small segment transverse to δ at P. Since $P \in \Lambda^+(\gamma)$, there are points of $\gamma^+(M)$ arbitrarily near P and the arcs of $\gamma^+(M)$ through them will contain subarcs in Ω. In particular, these arcs will all cross σ in the same direction and will meet it in points $R_1, \ldots, R_s, \ldots$, encountered in that order as $\gamma^+(M)$ is described forward from M. By virtue of Lemma (6.2) the R_s are in the same order on σ. Hence they can have only one limit-

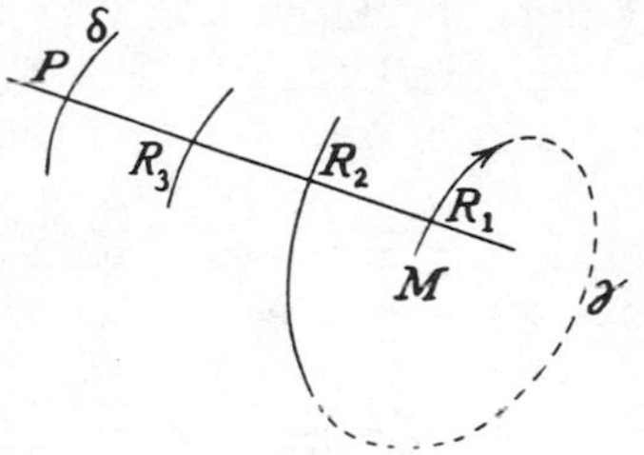

Fig. 18

point on σ and since P is manifestly such a point, $R_s \to P$. Moreover, on σ the points R_s are all on the same side of P. Or explicitly:

(7.3) *If* P *is an ordinary point of* $\Lambda^+(\gamma)$, δ *the path through* P, σ *a suitably small segment transverse to* δ *at* P, *then* $\gamma^+(M)$ *followed forward meets* σ *in successive points* $R_1, R_2, \ldots$, *which are ranged in the same order on* σ, *and tend to* P *on one side of* P *on* σ.

The notations being as before, let P' be a point of $\Lambda^+(\gamma)$ in Ω and δ' its path. Since δ' meets Ω it crosses σ in a point P'' which is a condensation point of the R_s. Hence $P'' = P$, $\delta' = \delta$, and P' is in the axis λ of Ω. Since Ω is a neighborhood of P, we have:

(7.4) *Corresponding to any open arc* λ *of a path* $\delta \subset \Lambda^+(\gamma)$ *and point* P *of* λ, *there is a neighborhood* U *of* P *such that* $U \cap \Lambda^+(\gamma) \subset \lambda$.

(7.5) *A path which meets a limiting set is contained in that set.* It is sufficient to consider $\Lambda^+(\gamma)$, and to show that if it meets the path δ then it contains δ. When δ is merely a critical point this is obvious so this case may be dismissed. Let $E = \delta \cap \Lambda^+(\gamma)$. Since δ is a Jordan curve or an arc, if $E \neq \delta$ there is a boundary point P of E in δ. Since E is a closed subset of δ, P is in E and hence in $\Lambda^+(\gamma)$. Identify now P with the point thus designated before. The arc δ^* of δ in Ω (the axis of Ω) consists of condensation-points of the arcs of $\gamma^+(M)$ through the points $R_q, R_{q+1}, \ldots$, for q above a certain value. Hence $\delta^* \subset \bar{\gamma}^+(M)$ whatever M, and therefore $\delta^* \subset \Lambda^+(\gamma)$. Thus P is an interior point and not a boundary point of E. Therefore $E = \delta$, $\delta \subset \Lambda^+(\gamma)$.

The possibility that $\delta = \gamma$ is not excluded. In that case P must be one of the points R_q. This can only occur when the R_q all coincide with P and γ is closed. Conversely γ closed implies that $\Lambda^+(\gamma) = \Lambda^-(\gamma) = \gamma$ and that all the R_q coincide with P. Notice also that a n.a.s.c. for $\delta = \gamma$ is that the two intersect, i.e., that $\Lambda^+(\gamma)$ meet γ. Since the properties of $\Lambda^-(\gamma)$ follow from those of $\Lambda^+(\gamma)$ by changing t into $-t$, we may state:

(7.6) *A n.a.s.c. for γ to be closed is that one of $\Lambda^+(\gamma)$, $\Lambda^-(\gamma)$ meet γ, and then both coincide with γ.*

(7.7) *Path-polygon.* Let E^2 be a two-cell whose boundary consists of a finite number of separatrices and critical points. This boundary is called a path-polygon. It will be required in a moment.

§ 4. The Theorem of Bendixson

8. We are now in position to give a complete description of the limiting sets and of the related behavior of the paths.

(8.1) THEOREM. *The limiting sets of a path γ fall under the following mutually exclusive categories:*

(a) *$\Lambda^+(\gamma)$ consists of a single point A which is critical and with increasing time γ spirals toward A or else tends to A in a fixed direction.*

(b) *$\Lambda^+(\gamma)$ is a closed path δ and either $\gamma = \delta = \Lambda^+(\gamma)$ or else with increasing time γ spirals towards δ on one side of δ.*

(c) *γ is contained in a 2-cell E^2 whose boundary is $\Lambda^+(\gamma)$ which is now a path-polygon. The path γ spirals towards $\Lambda^+(\gamma)$ and any*

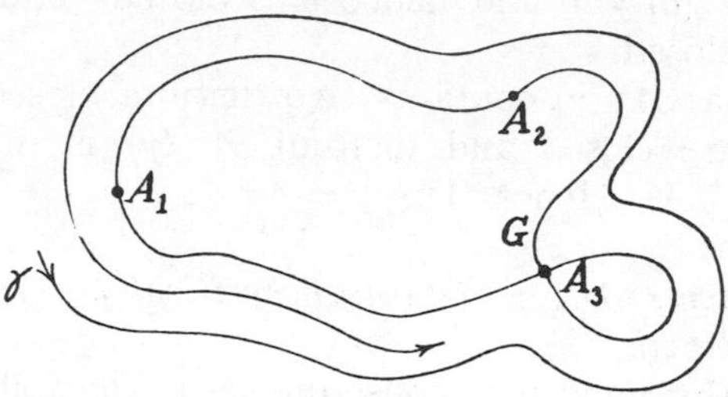

Fig. 19

path which starts from a point of E^2 sufficiently near $\Lambda^+(\gamma)$ spirals towards the latter in the same direction as γ. This last property holds also for (a) *and all the paths starting sufficiently near A, and likewise for* (b) *and those starting sufficiently near δ on the same side as γ.*

(d) *$\Lambda^-(\gamma)$ is of the same nature as $\Lambda^+(\gamma)$ save that the spiralling and tending to a limit occur as $t \to -\infty$.* (Bendixson [1], Ch. I.)

Fig. 19 illustrates the case of a graph and the related behavior of γ.

Part (8.1d) is obvious so that it is sufficient to examine the other parts.

9. *Suppose first $\Lambda^+(\gamma)$ finite.* Since it is connected, it must consist of a single point A. The path through A is in $\Lambda^+(\gamma)$, hence it is A itself. Thus A is a critical point. Combining this with (1.8) we find that under the circumstances the behavior of γ and $\Lambda^+(\gamma)$ conforms with the theorem.

It may be observed that obviously when γ tends to a critical point A then $\Lambda^+(\gamma)$ consists of A alone. Similarly when γ tends to A with $t \to -\infty$, then $\Lambda^-(\gamma) = A$.

(9.1) *Suppose now that $\Lambda^+(\gamma)$ contains no critical point.* Since the set is not empty it contains ordinary points and the paths through them. Let δ be one of these paths. Suppose $\Lambda^+(\delta) - \delta$ non-empty and let P be a point of the set. By hypothesis the path δ_1 through P is distinct from δ. Since P is in the closure of subsets of δ it is in $\bar{\delta}$ and hence in the closed set $\Lambda^+(\gamma)$ which contains δ. Since $P \in \Lambda^+(\delta)$ there are points of δ, i.e. of $\Lambda^+(\gamma)$, arbitrarily near P and not in δ_1. As this contradicts (7.4), we must have $\Lambda^+(\delta) = \delta$ and hence δ is closed. Thus every path in $\Lambda^+(\gamma)$ is closed.

Suppose that $\Lambda^+(\gamma)$ contains two distinct closed paths δ, δ'. Since δ, δ' are closed and disjoint $\Lambda^+(\gamma)$ is not connected, contrary to (7.1). Hence $\Lambda^+(\gamma) = \delta$.

To sum up:

(9.2) *Whenever $\Lambda^+(\gamma)$ contains no critical point, it consists of a single closed path.*

A noteworthy incidental consequence is the following proposition due to Poincaré:

(9.3) THEOREM. *A closed region Ω which is free from critical points and contains a half-path contains also a closed path.*

At the cost of changing t into $-t$ we may suppose that the half-path is $\gamma^+(M)$. Then $\Lambda^+(\gamma) \subset \bar{\gamma}^+(M) \subset \Omega$. Hence $\Lambda^+(\gamma)$ contains no critical point and so it is a closed path in Ω.

10. Continuing with the analysis of the limiting sets let us take any closed path δ and examine its relation to the paths issued from the neighboring points. If $\gamma = \delta$ it is already known (7.6) that $\Lambda^+(\gamma) = \delta$, so we suppose $\gamma \neq \delta$. Choose any point P of δ and a segment σ transverse to δ at P and situated in a certain line L. Referring to (II, 14.8), it is known that starting from any point $M \in \sigma$ and following $\gamma^+(M)$ we first encounter L at a point M_1 and $M \to M_1$ defines a topological mapping φ of the segment σ into another segment σ_1 in L. The point P is a fixed point of φ. Further properties will now be brought out.

Since $\gamma^+(M)$ cannot cross δ, the point M_1 is on the same side of δ as M, i.e. on the same side of P as M in L, and it is an analytical function of M (II, 18.3). This implies that for σ sufficiently small, either $M = M_1$ for all M or else only for $M = P$. In the first case all the paths meeting σ on the side considered are closed; in the second none except δ is closed. Suppose we have the second situation and for one M the point M_1 is between M and P as in fig. 20. Then the same situation will hold for all points between P and M, for otherwise by a well-known continuity argument φ would possess a fixed point between P and M. Under the circumstances, all the paths meeting σ on the same side as M will spiral towards δ with increasing t. Let $M, M_1, M_2, \ldots$ be the successive intersections

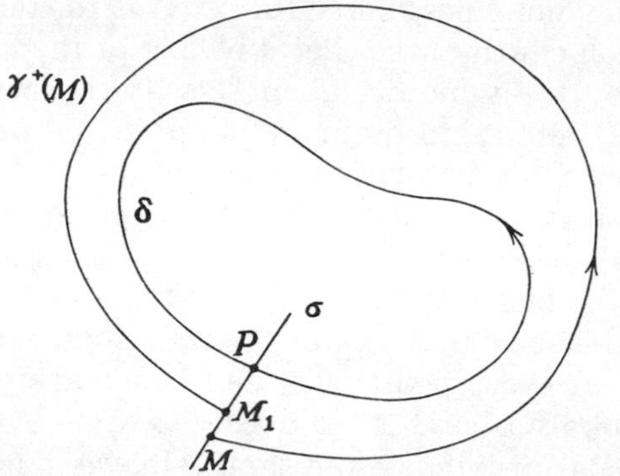

Fig. 20

of any one of them, say γ with σ. By the above $M, M_1, M_2, \ldots$ are met in that order on σ and they are all on the segment MP. Hence $\{M_n\}$ has a limit P' on $\overline{MP}$. Since P' is a fixed point of φ we must have $P' = P$, and so $\{M_n\} \to P$. Since this occurs on any segment transverse to δ, as a point follows $\gamma^+(M)$ with $t \to +\infty$, the point tends to δ. This means that $\Lambda^+(\gamma) = \delta$, and that γ spirals forward around δ towards δ. Moreover all the paths issued from points near enough to δ on the same side as γ behave likewise.

If the point M_1 in the above argument is such that M is between P and M_1, then following $\gamma^-(M_1)$ backward, we would find that γ and all the paths on the same side issued from points near enough to δ spiral towards δ as $t \to -\infty$. In this case $\Lambda^-(\gamma) = \delta$.

When the paths spiral towards [away from] δ on a given side with increasing time, then they do the opposite with decreasing time, and so they are then orbitally stable [unstable] on that side. Notice that the behavior on both sides of δ need not be the same. If δ is orbitally stable [unstable] on both sides, then it is orbitally stable [unstable] in our usual sense. If it does not behave alike on both sides, then it is orbitally conditionally stable.

To sum up, then, when δ is closed the only paths γ which have condensation points on δ, i.e. whose limiting sets meet δ, are δ itself or those which spiral towards or away from δ. In the first case $\Lambda^+(\gamma)$, in the second $\Lambda^-(\gamma)$ is δ itself. The behavior relative to δ is clearly in accordance with the theorem.

11. The remaining possibility for $\Lambda^+(\gamma)$ is to consist of critical points and non-closed paths. Let δ be one of the latter. Now we may prove by the same arguments as at the beginning of (9) that if $\Lambda^+(\delta)$ contains an ordinary point then δ must be closed. Since the δ here considered is not closed its $\Lambda^+(\delta)$ consists only of critical points, and since $\Lambda^+(\delta)$ is connected it consists of a single critical point, say A. The rest of the argument will be based upon the behavior of γ and δ in the vicinity of A.

We shall first show that δ cannot spiral around A. For γ passes arbitrarily near every point of δ, and hence arbitrarily near A. Hence if δ spirals around A, so does γ (3, case V) and therefore $\Lambda^+(\gamma) = A$ contrary to assumption. Thus δ tends to A in a definite direction (1.5).

Take on δ a point P arbitrarily near to A. Since γ passes arbitrarily near P on one side of δ and does not tend to A, δ contains one side of a hyperbolic sector centered at A and so it is a separatrix. Let δ_1 be the separatrix which contains the other side of the sector. As we follow δ_1 away from A we shall reach a critical point A_1 (which may be A itself), then leave it with a separatrix δ_2, etc. Since γ must return to the neighborhood of δ, we shall ultimately return to δ_1 after having described in the process the full set $\Lambda^+(\gamma)$, which is thus a path-polygon Π. Part of the argument of (10) is applicable here to show that γ and any path passing sufficiently near Π on the same side as γ all spiral around Π in the same direction and it need not be repeated. Let U be the component of the complement of Π in the sphere containing γ. Since the possible "loops" in Π are all outside of U, U is a two-cell. This completes the proof of Bendixson's theorem.

(11.1) *Remark.* The proof of the theorem applies with hardly any modification to a basic system **(1.1)** within a closed region Ω where: (a) the system satisfies the conditions of the fundamental existence theorem; (b) in Ω there are at most a finite number of critical points around each of which the property described under **(3.3)** holds. In particular, Poincaré's theorem **(9.3)** on the existence of a closed path is valid without any reference to **(3.3)** (see **34**).

§ 5. Some Complements on Limit-Cycles

12. Suppose that γ, δ are closed paths bounding an annular region free from critical points or other closed paths. We say then that γ, δ are *consecutive.*

(12.1) *Two consecutive closed paths γ, δ cannot both be stable or unstable on the sides facing one another.* (Poincaré.)

That is to say, if say γ is interior to δ then it is not possible to have γ stable [unstable] outside and δ stable [unstable] inside. Suppose, in fact, the assertion false. Replacing if need be t by $-t$, we may dispose of the situation so that both γ, δ are unstable on the sides facing one another. Choosing now paths γ', δ' in the annular region respectively very near γ, δ and suitable transverse segments, we will have the configuration of fig. 21. Let U be the inner annular region bounded by arcs of γ', δ' together with the segments MM', NN'. Any path starting from a point of the boundary of $\bar{U}$ remains in $\bar{U}$. Since $\bar{U}$ is free from critical points it must contain a new closed path ε, and ε must surround γ, else it would bound a region free from critical points. Since this contradicts the assumption that γ, δ are consecutive (12.1) is proved.

(12.2) *Example.* Consider the general equation in polar coordinates

$$dr/d\theta = rf(r^2) \tag{12.3}$$

where f is a real polynomial. Setting $\theta = t$, and pass˙ng to rectangular coordinates, we obtain the system

$$\begin{cases} dx/dt = -y + xf(x^2 + y^2) \\ dy/dt = x + yf(x^2 + y^2) \end{cases} \tag{12.4}$$

which is of the basic type (1.1). The behavior of the paths is more conveniently investigated, however, in the polar form. Since only the positive roots of $f(z)$ matter, let us set $f(z) = z^k g(z)h(z)$ where $g(z) > 0$ for $z > 0$ and $h(z)$ has only positive roots It is also not a genuine restriction to assume $h(0) > 0$, since this may be achieved by the unimportant change of θ into $-\theta$.

Now if α is a root of $h(z)$ the circle $C_\alpha : r^2 = \alpha$ is a closed path. There are two possibilities:

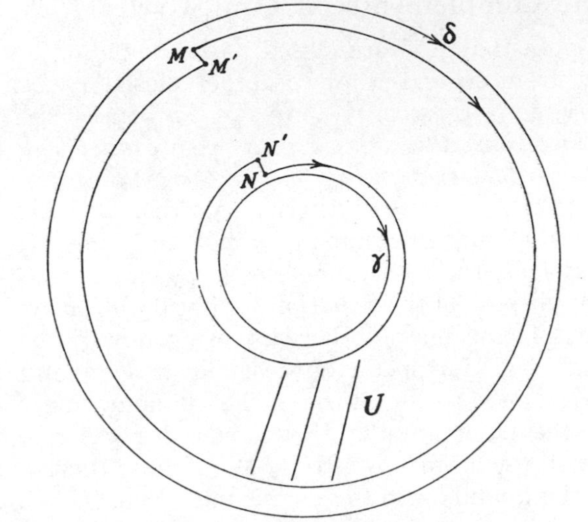

Fig. 21

(a) *The root α is of odd order.* Then $h(r^2)$ changes its sign as r^2 crosses α. Suppose that it goes from — to +. Since dr/dt changes sign from — to + as r^2 crosses α any half-path near C_α inside the circle will spiral away from C_α, and likewise for any half-path near C_α outside the circle. Hence C_α is unstable (on both sides). If the change of sign of $h(z)$ is in the opposite direction, then C_α is orbitally stable.

(b) *The root α is of even order.* Then $h(r^2)$ does not change sign as r^2 crosses α. If $h(r^2) \geq 0$ near α then along a path near C_α, r^2 must increase and so C_α is orbitally stable inside but unstable outside. If $h(r^2) \leq 0$ near α, C_α is orbitally stable outside and unstable inside. Thus C_α is semi-stable in both cases.

If α, β are consecutive roots of $h(z)$, then along any path γ in the region between C_α and C_β the sign of dr/dt is fixed and so r^2 is monotone increasing or decreasing. Hence γ spirals away from one of the two circles and towards the other. Thus the only closed paths in the finite plane are the circles C_α. Beyond the last circumference C_α the paths spiral to infinity if the leading term of $h(z)$ is positive, away from infinity in the contrary case.

The origin is manifestly a focus. The paths passing near the origin spiral away from the origin since $g(0)h(0) > 0$. Thus the origin is an unstable focus.

There is no difficulty in writing down the general solution of (12.3) and it is a simple matter to verify the preceding properties by means of the solution.

§ 6. On Path-Polygons

13. Let us consider again a path-polygon Π with the 2-cell Ω, which it bounds. We suppose also that the two sides that termi-

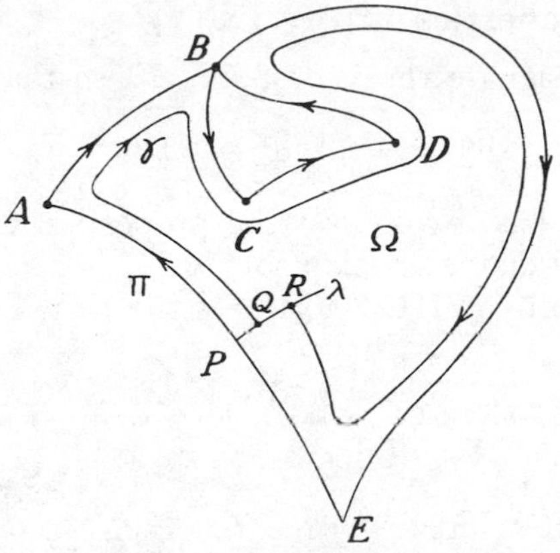

Fig. 22

nate at a vertex A always determine, in an evident sense, a hyperbolic sector centered at A and contained in Ω. Take now any point P on a side δ of Π and draw in Ω a transverse arc λ to δ at P. A very elementary argument will then show that the paths originating very near P in Ω intersect λ in points which tend monotonically towards P or away from P. If the second case holds we replace t by $t' = -t$ and the first case is then obtained. As before we may show that whichever of the two possibilities takes place, it will take place for all the paths sufficiently near Π.

(13.1) *The paths γ passing sufficiently near Π in Ω either all spiral toward Π or away from Π, or else again they are all closed.*

In the first case Π is the $\Lambda^+(\gamma)$, in the second case the $\Lambda^-(\gamma)$ of all such paths.

Let Q, R be two consecutive crossings of λ by γ with Q arbitrarily near P. Then a slight deviation from the proof of (IX, 11.1) will show that Index $(\gamma \cup RQ) = 1$, it being assumed that γ is oriented concordantly with Ω. We may take γ so close to Π (Q so close to P) that there are no critical points between $\gamma \cup RQ$ and Π. Applying now (IX, 7.3) we have:

(13.2) *The 2-cell bounded by a path-polygon contains critical points, and the sum of their indices is $+1$.*

§ 7. Some Properties of div (X, Y)

14. Let V designate the vector (X, Y) so that

$$\text{div } V = \partial X/\partial x + \partial Y/\partial y . \tag{14.1}$$

This divergence has two interesting properties. If γ is a closed path one of the characteristic roots λ_1, λ_2 is zero. Let λ be the other root. Then by (VIII, 1.9) if τ is the period:

$$\lambda = \frac{1}{\tau} \oint_\gamma \text{div } V \cdot dt . \tag{14.2}$$

Hence this property:

(14.3) *If the time average of the divergence on the closed path γ is negative [positive] γ is orbitally asymptotically stable [unstable].*

15. A second property of the divergence is the following:

(15.1) *Criterium of Bendixson. If* div V *has a fixed sign (zero excluded) in a closed two-cell Ω then Ω contains no limit-cycle nor even an oval going from and to a critical point.*

For suppose that there is a limit-cycle γ in Ω and let it bound a region $S \subset \Omega$. Applying Green's theorem we have

$$\iint_S \text{div } V \, dxdy = \oint_\gamma (X\, dy - Y\, dx) = 0 .$$

Hence div V cannot have a fixed sign in S, nor *a fortiori* in Ω.

The case of the oval was pointed out to us by Coleman. If δ is an oval in Ω going to and coming from a critical point A, then

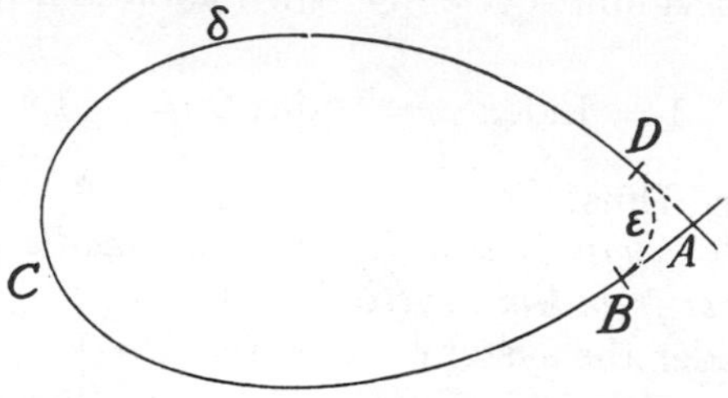

Fig. 23

one may "round-off" δ near A so as to produce a continuously turning tangent on the resulting closed curve δ_1, and this by an arc of arbitrarily small length ε. Let $\alpha = \sup \sqrt{X^2 + Y^2}$, $\beta = \inf |\text{div } V|$ in Ω, and let S, S_1 be the regions bounded by δ, δ_1. Assuming for convenience div $V > 0$ in Ω — the case div $V < 0$ would be treated in the same way—we have

$$0 < \iint_{S_1} \text{div } V\, dxdy = \int_{\varepsilon} V ds < \alpha\, \varepsilon\,.$$

Therefore if S, S_1 denote also for convenience the areas of the two regions $0 < \beta\, S_1 < \alpha\, \varepsilon$. Since the middle term $\to \beta S$ and the last term $\to 0$ as $\varepsilon \to 0$, we have a contradiction, proving the criterium.

16. *Applications.* (16.1) Consider the two equations

$$dx/dt = ax + by + [x, y]_2,\ dy/dt = cx + dy + [x, y]_2\,.$$

Here

$$\partial X/\partial x + \partial Y/\partial y = a + d + [x, y]_1\,.$$

If $a + d \neq 0$ in a certain neighborhood U of the origin, the divergence will have the sign of $a + d$. Hence U can contain no limit-cycle of the system. Now $a + d$ is the sum of the characteristic roots of the origin. Hence if the origin is a node it has a certain neighborhood free from limit-cycles. This result is of course fairly evident otherwise even for a focus. Notice that if the origin is a saddle point this is also the case. For if U is a neighborhood of the origin O containing no other critical point and if γ could be a limit-cycle in U then γ would have to surround O. Then

$$1 = \text{Index } \gamma = \text{Index } O = -1$$

a contradiction. Thus:

(16.2) *If the origin is a node, focus or saddle point it has a neighborhood free from limit-cycles.*

(16.3) Consider the system

$$dx/dt = y, \ dy/dt = x(a^2 - x^2) + by, \ ab \neq 0 .$$

There are three critical points: the origin O and the points $A(a, 0)$, $A'(-a, 0)$.

Origin. The characteristic roots are the solutions of the equation

$$r(r - b) - a^2 = 0 .$$

Hence they are real and of opposite signs. Thus the origin is a saddle point.

Points A and A'. The equation of the characteristic roots, the same for both, is

$$r(r - b) + 2a^2 = 0 .$$

The roots are real and of the sign of b if $b^2 \geq 8a^2$, complex with real parts of the sign of b if $b^2 < 8a^2$. Hence A and A' are stable [unstable] nodes if $b^2 \geq 8a^2$ and $b < 0$ [$b > 0$] and stable [unstable] foci if $b^2 < 8a^2$ and $b < 0$ [$b > 0$].

No oval can proceed from and back to the origin since $\text{div } V = b \neq 0$. Fig. 24 has been drawn for the case when A and A'

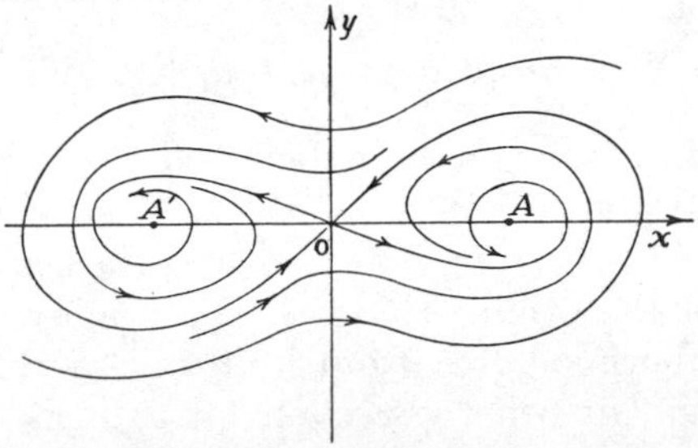

Fig. 24

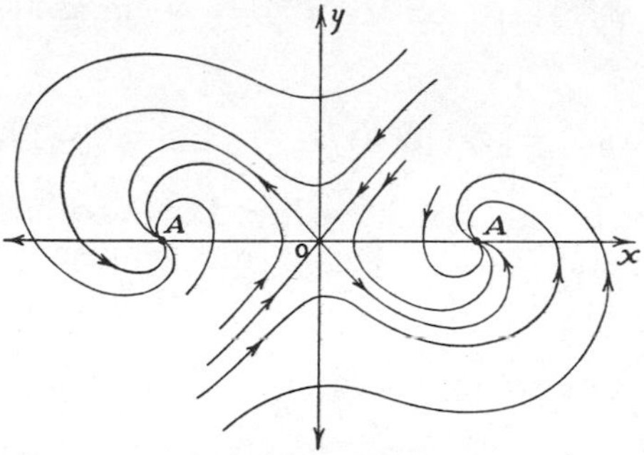

Fig. 25

are stable foci and it depicts adequately the corresponding phase-portrait not too far from the segment $|x| \leq a$ of the x axis. If they are unstable foci one merely has to reverse the arrows. Fig. 25 corresponds in the same manner to A, A' stable nodes and again one merely reverses the arrows if they are unstable.

§ 8. Critical Points with a Single Non-Zero Characteristic Root

17. While in theory one may apply our general method to this case it will be found more expedient to have recourse to a related but more direct procedure.

When there is a single non-zero characteristic root one may reduce the system to the form

$$(17.1) \qquad dx/dt = [x, y]_2, \; dy/dt = \lambda y + [x, y]_2, \; \lambda \neq 0 .$$

We propose to determine the local phase-portrait of this system. Our goal is the proof of the following theorem due to Bendixson ([1], pp. 45–58). Our method however will differ widely from Bendixson's.

(17.2) THEOREM. *The local phase-portrait of an isolated critical point with a single non-zero characteristic root is of one of the following three types: node, saddle-point (four separatrices), two hyperbolic sectors and a fan (three separatrices). The corresponding indices are* 1, —1, 0 *so that they may serve to distinguish the three types.*

Replacing in (17.1) t by $|\lambda|^{-1}t$ the system assumes one of the two forms:

$$(17.3) \qquad dx/dt = [x, y]_2, \; dy/dt = \pm y + [x, y]_2 .$$

The system with $-y$ may be reduced to the other by the change of time parameter $t \to -t$. Hence it is sufficient to discuss the system

$$(17.4) \qquad dx/dt = [x, y]_2, \; dy/dt = y + [x, y]_2 .$$

The change of coordinates $x + y \to x$, $y \to y$ reduces (17.4) to the form

$$(17.5) \quad dx/dt = y + f(x, y), \; dy/dt = y + g(x, y); \; f, g = [x, y]_2$$

and this is the system studied below.

18. We first look for the directions of approach of solutions to the origin. We shall refer to a corresponding branch as a *TO-curve*—a term patterned after Niemitski-Stepanov [1] who refer to such a branch as "*O*-curve". As simple a method as any is to pass to polar coordinates. We find at once

$$(18.1) \quad \begin{cases} dr/dt = r \sin\theta\,(\sin\theta + \cos\theta) + r^2\,[\sin\theta, \cos\theta]_3 + \cdots, \\ d\theta/dt = \sin\theta\,(\cos\theta - \sin\theta) + r\,[\sin\theta, \cos\theta]_3 + \cdots. \end{cases}$$

The θ axis is thus a solution, its arcs between critical points are separatrices and so the images of *TO*-curves can only tend to such critical points. They correspond to $\sin\theta = 0$, $\theta = k\pi$, and $\cos\theta = \sin\theta$, $\theta = \pi/4 + k\pi$. This gives four possible directions for *TO*-curves, the two along the x axis and the two along the first bisector.

19. Take first $\theta = 0$. Expanding the right hand sides of (18.1) around the r, θ origin Ω yields

$$(19.1) \quad \begin{cases} dr/dt = ar^2 + r\theta + \cdots, \\ d\theta/dt = r + \theta + \cdots. \end{cases}$$

Thus Ω is a Bendixson critical point and we cannot say anything about its *TO*-curves, or rather the corresponding curves in the (x, y) plane. The same remarks hold for the critical point $\theta = \pi$.

20. Consider now the direction $\theta = \pi/4$. To see what happens in this case we make the transformation $\theta = \pi/4 + \theta^*$, and expand in powers of r, θ^*. There results

$$(20.1) \qquad \begin{cases} dr/dt = r + [r, \theta^*]_2, \\ d\theta^*/dt = r - \theta^* + [r, \theta^*]_2 . \end{cases}$$

For this last system the origin Ω^* has the characteristic roots $1, -1$ and so it is a saddle point. Since $r = 0$ is evidently a solution of (18.1), its arcs through Ω^* are *TO*-curves for it. Hence in the part of the plane $r > 0$ there is one more *TO*-curve for Ω^* and it is the image of a unique *TO*-curve Δ_1 in the (x, y) plane tending to the origin along the bisector L: $x = y$ in the first quadrant. The change of variables $y = 3\pi/4 + \theta^*$, discloses the presence of a similar *TO*-curve Δ_2 tending to the origin along L in the third quadrant.

21. We now take up the study of the system (17.5). The Weierstrass preparation theorem yields

$$y + f(x, y) = \{y - A(x)\}E_1(x, y),$$
$$y + g(x, y) = \{y - B(x)\} E_2 (x, y),$$

where by identification we find

$$E_i (0, 0) = 1; A, B = [x]_2 .$$

Hence $E (x, y) = E_2/E_1$ is a unit such that again $E (0, 0) = 1$. Upon changing the time unit according to $dt \to E_1\, dt$, (17.5) is replaced by the simpler (geometrically) equivalent system

$$(21.1) \quad dx/dt = y - A (x), \; dy/dt = \{y - B (x)\} E (x, y) .$$

The curves

$$(21.2) \qquad \Gamma_V: y = A(x); \; \Gamma_H: y = B(x) ,$$

will play an all important role in our discussion. They are both tangent to the x axis at the origin. Γ_V is the locus of the points where the tangents to the paths are vertical, and Γ_H the same where the tangents are horizontal. In the first and fourth quadrants Γ_V and Γ_H pass below L. The Δ_i jointly divide the plane into two regions R_1 below and R_2 above. We shall deal directly with the phase-portrait in R_1. The same for R_2 is obtained through the change of variables $x \to -x$, $y \to -y$ in (21.1), and applying the R_1 treatment to the new system. For the present we call Γ_V, Γ_H the branches of (21.2) to the right of the y axis, i.e. in R_1.

Observe that $\Gamma_V \neq \Gamma_H$. For if they were the same (19.1) would yield

$$dy/dx = E\ (x, y)\ ,$$

and hence the origin would not be a critical point. Thus Γ_H is either above or below Γ_V.

22. *Suppose first Γ_H above Γ_V.* From the signs of dx/dt, dy/dt one infers at once that unless Γ_H and Γ_V are both above the x axis, the disposition is that of fig. 30: R_1 is a nodal sector. In the exceptional case there arises the possible alternative indicated in fig. 26 and which might give rise to an elliptic sector. However along Ox the slope of the tangents to the paths is

$$dy/dx = B(x)/A(x)\ .$$

Since Γ_H is above Γ_V which is above Ox for $x > 0$ and small, we have here $B(x) > A(x) > 0$. Hence the slope in question > 1 and so the tangents in question make an angle $> \pi/4$ with Ox. Now under the situation of fig. 26 the angle would be nearly zero. Hence this case is eliminated and fig. 30 for R_1 prevails in all cases.

Suppose now that Γ_V is above Γ_H. This time the phase-portrait in R_1 upon heeding the signs of dx/dt and dy/dt is in accordance with fig. 27 with two hyperbolic sectors, save that we have indicated a separatrix S. However *a priori* a fan could arise in place of S, and also (in R_1) right below Δ_1. They are both eliminated by the same argument as follows. First since Γ_V is above Γ_H, $C = A - B > 0$ for x positive and small. Suppose that we

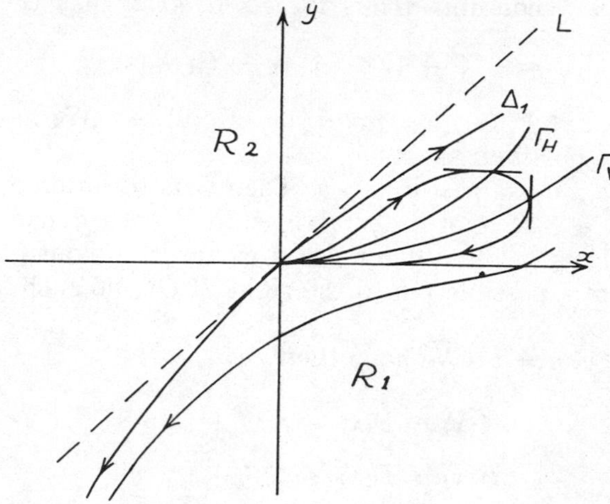

Fig. 26

had a fan and let $y(x)$, $y(x) + \varepsilon(x)$, $\varepsilon > 0$, correspond to two of its *TO*-curves with $\varepsilon(x)$ of at least the same order in x as y. From

$$\frac{dy}{dx} = \frac{(y - B)\,E}{y - A} \tag{22.1}$$

and denoting "approximately equal" by $\doteq$ we have

$$\frac{d\varepsilon}{dx} \doteq \varepsilon \cdot \frac{\partial}{\partial y}\frac{(y - B)\,E}{y - A} \tag{22.2}$$

$$\doteq \varepsilon \left\{\frac{-CE}{(y - A)^2} + \frac{y - B}{y - A}\frac{\partial E}{\partial y}\right\}.$$

Now we may write

$$E = 1 + g(x) + (y - A)\,h(x,y)\,,$$

where g, h are non-units. Hence

$$\frac{d\varepsilon}{dx} \doteq \frac{\varepsilon\,(-C + (y - A)\,(y - B)\,m\,(x, y)\,)}{(y - A)^2}.$$

where m is a non-unit. Thus for $\varepsilon > 0$, $d\varepsilon/dx$ has the sign of

$$\varphi = -C + (y - A)(y - B)\, m(x, y) .$$

Let $A = ax^p E_1(x)$, $B = bx^q E_2(x)$, $E_i(0) = 1$. We now distinguish three possibilities:

I. $p < q$, *or else* $p = q$, $a > b$. Then C is of order p and one of $y - A$, $y - B$ is of order at least p. Hence φ has the sign of $-C$. Thus $d\varepsilon/dx < 0$ for $\varepsilon > 0$. Since this contradicts the structure of a possible fan to the right of Oy, no such fan exists in this case.

II. $p = q$, $a = b$. We have then

$$A(x) = D(x) + a'x^s + \ldots ,$$

$$B(x) = D(x) + b'x^s + \ldots ,$$

$$a' \neq b',\ D = ax^p + \beta x^{p+1} + \ldots + \gamma x^{s-1}$$

so that $s > p$ and of course $p \geqq 2$, since Γ_H, Γ_V are tangent to Ox. Making the regular change of variables $y = z + D(x)$, $x = x$, we have

$$\frac{dz}{dx} = \frac{(z - b'x^s - \ldots)\, E - D'(x)\,(z - a'x^s - \ldots)}{(z - a'x^s - \ldots)} .$$

Upon applying the Weierstrass preparation theorem to the numerator we find

$$\frac{dz}{dx} = \frac{(z - b'x^s - \ldots)\, E^*(x, z)}{z - a'x^s - \ldots},\quad E^*(0, 0) = 1 . \tag{22.3}$$

Since we have dealt with a regular transformation (22.3) may replace (22.1) as regards the search for the fan. Here s is the analogue of p before and since $b' \neq a'$ we are under the preceding case and so the fans are again eliminated.

III. $p > q$. This is only consistent with Γ_V above and Γ_H below Ox, with the possible fan likewise below Ox. Here C is of order q and one of $y - A$, $y - B$ is of order at least q. Hence φ has again the sign of $-C$, which leads once more to the elimination of the fans.

Thus fig. 28 with just the one separatrix in R_1 does represent the situation. One readily verifies that the position of Γ_H, Γ_V relative to the x axis does not change the situation.

Since the situation in R_2 may be reduced to that in R_1 by reversing the positive direction on the axes we conclude that if in R_2, i.e. above L, Γ_H is below Γ_V, R_2 is a nodal sector, while if Γ_H is above Γ_V it consists of two hyperbolic sectors separated by a single separatrix.

23. Let $C(x) = cx^r E(x)$. Taking account of all the preceding remarks we have the following possibilities:

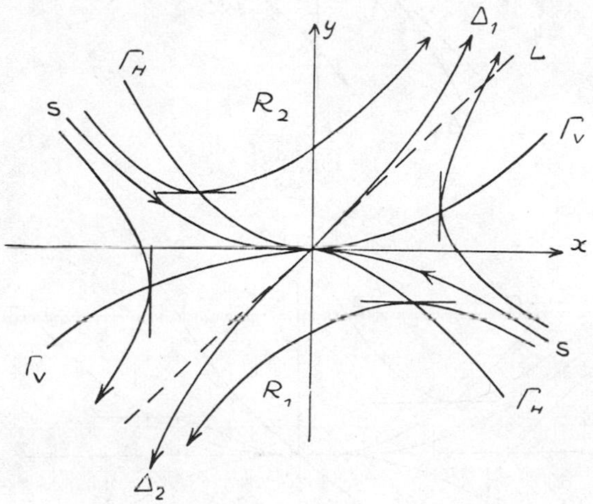

Fig. 27

I. $c > 0$, *r odd*. Then Γ_V is above Γ_H in R_1, below it in R_2 and so we have four hyperbolic sectors separated by four separatrices, i.e. a saddle-point (fig. 27).

II. $c > 0$, *r even*. Here Γ_V is above Γ_H in both sectors and so we have two hyperbolic sectors in R_1 and R_2 is a nodal sector (fig. 28).

III. $c < 0$, *r even*. The same as II with R_1 and R_2 interchanged (fig. 29).

IV. $c < 0$, *r odd*. This time Γ_H is above Γ_V in R_1 and below it in R_2. Hence both sectors are nodal and the total phase-portrait is that of a node (fig. 30).

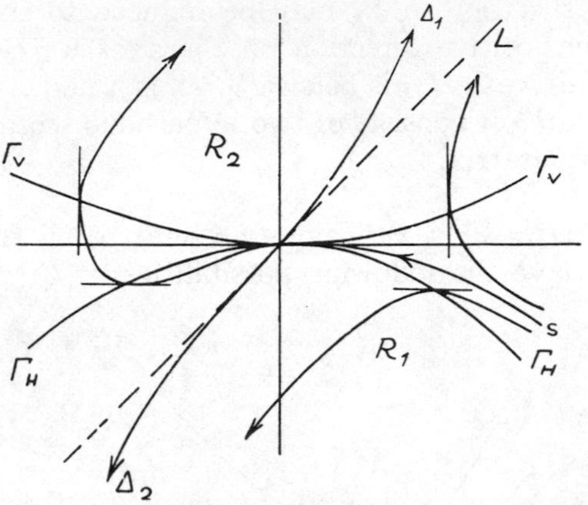

Fig. 28

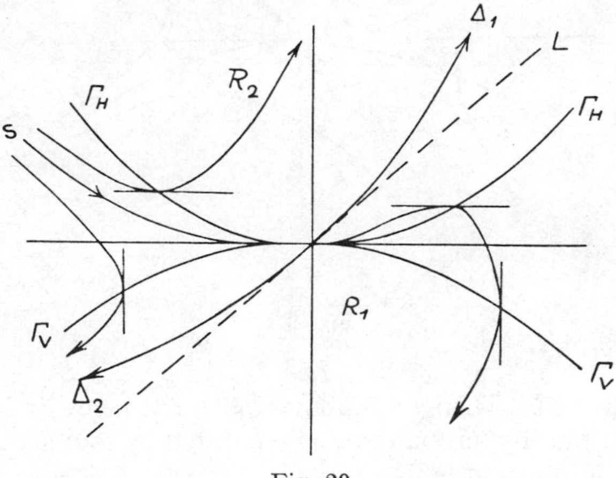

Fig. 29

We have then just the three configurations stated in (17.2) and so this theorem is now proved.

It is interesting to examine the relation of the graphs obtained to the initial variables. Keeping the designation x, y for them let the new, and present variables be written x_1, y_1. Then

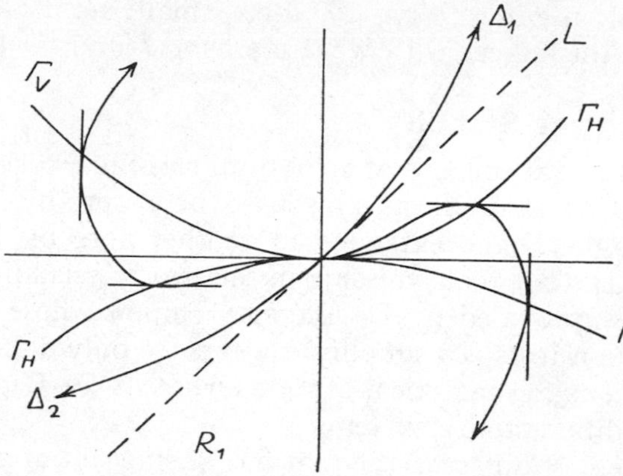

Fig. 30

$x_1 = x + y, y_1 = y$. Thus $x = x_1 - y_1, y = y_1$, the x axis is the same and the old y axis is merely the line L. One may therefore see directly on the graph the relation to the initial axes.

24. *Stability*. A glance at the phase-portraits leads to the conclusion that the only stable system corresponds to fig. 30 with arrows reversed, i.e. to systems reducible to the form

$$(24.1) \qquad \begin{cases} dx/dt = -y + A(x) \\ dy/dt = -y + B(x) \end{cases}$$

where the unit is omitted as not affecting stability. The notations remaining as before they correspond then to r *odd* and c *negative*. The stability is then asymptotic.

It is interesting to compare this stability result with the one obtained from the application of Liapunov's criterium (VI, 16.4). A change of variable must first be applied. Set $x_1 = x - y, y_1 = y$. Thus (24.1) is replaced by

$$dx_1/dt = C(x_1 + y_1) = C(x_1) + y_1 C'(x_1) + \frac{{y_1}^2}{2} C''(x_1) \ldots$$
$$dy_1/dt = -y_1 + B(x_1 + y_1).$$

Since C is of order > 1, according to Liapunov this system is stable if and only if $x_1 C(x_1) < 0$ for x_1 small, i.e. if $c < 0$ and r is odd, and this agrees with what we have found.

§ 9. Structural Stability

25. The general concept of structural stability, together with a statement of results, was presented in a note by Andronov and Pontrjagin [1] and extended in another note by Leontovič and Mayer [1]. The proofs, missing in the two Russian notes, were subsequently provided by De Baggis [1] upon whose work the present section rests. To simplify matters we only discuss necessary conditions for analytical systems whereas De Baggis treats n.a.s.c. for differentiable systems.

In a system of differential equations arising out of a practical problem the coefficients are only known approximatively: various errors are inevitably involved in their determination. The phase-portrait of the system must therefore be such that it is not affected by small modifications in the coefficients. In other words it must be "stable" under these conditions. The general concept behind structural stability is then to look for systems whose phase-portrait is invariant under small variations of the systems. We shall find that this imposes upon our systems restrictions which reduce their "peculiarities" to the simplest possible types.

Be it as it may let us bring a little precision to the general scheme.

Let Ω be a closed 2-cell in the Euclidean plane and let Φ be its boundary. To simplify matters we will suppose that Φ is an analytical Jordan curve. Consider an analytical autonomous system

$$dx/dt = X(x, y), \quad dy/dt = Y(x, y) \tag{25.1}$$

whose domain of analyticity Δ contains Ω. We will also assume that Φ is a curve without contact. Thus the paths along Φ all enter or leave Ω. At the cost of changing t into $-t$, which does not affect the phase-portrait in Ω, we may assume that all the paths enter Ω.

Let $\xi(x, y)$, $\eta(x, y)$ be two analytic functions likewise holomorphic in Δ and consider the system

$$dx/dt = X + \xi = X^*, \quad dy/dt = Y + \eta = Y^*. \tag{25.2}$$

We say that the system (25.1) is *structurally stable in Ω* whenever given any $\varepsilon > 0$ there can be found a $\delta > 0$ such that for all ξ, η such that $|\xi|, |\eta| < \delta$, there exists an ε-topological transformation T of Ω (displacing any point of Ω by a distance $< \varepsilon$) such that $\Omega_1 = T\,\Omega \subset \Delta$ and if γ is a directed path of (25.1) running in Ω then $T\gamma$ is a directed path of (25.2) running in Ω_1. Thus if the system is structurally stable the phase-portraits of (25.1) and (25.2) are qualitatively the same. An immediate consequence of the definition of structural stability is the following comprehensive property:

(25.3) *The following elements of the system* (25.1) *relative to* Ω *are mapped by* T *into the same for* (25.2) *and* $T\,\Omega$: *ordinary point, critical point, condensation point of such points, elementary critical point of a given type, closed path, limit-cycle (stable, unstable, semi-stable), separatrix, continuous family of closed paths, positive or negative limiting set of a path if located in* Ω.

26. We shall now discuss a number of properties of structurally stable systems. In essence these properties imply that they behave in the simplest manner possible.

(26.1) *A structurally stable system in* Ω *has at most a finite number of critical points in* Ω.

Suppose that the system (25.2) is structurally stable in Ω and yet has an infinite number of critical points in Ω. These points will then have a condensation point A in Ω. Let A be taken as origin. Referring to (1.2) X and Y have each a finite number of branches $B_1, \ldots, B_r$ and $B_1', \ldots, B_s'$ issuing from A. As shown in the proof of (1.2) A will be an isolated critical point unless some B_j is a B_k'. Let then $B_1, \ldots, B_\varrho$ be common branches of the two sets. Consider now the system (25.2) with $\xi = \alpha x$, $\eta = \alpha y$, where α is so small that the origin can only arise under T from a point on one of the branches $B_1, \ldots, B_\varrho$. Thus A cannot be an isolated critical point of (25.2), and yet it is such a point. This contradiction proves (26.1).

(26.2) *A structurally stable system in* Ω *has only elementary critical points in* Ω *and none is a center.*

Let this time A be a non-elementary critical point and let it be taken as origin. We have then

$$X = X_2(x, y) + X_3(x, y) + \ldots, \quad Y = Y_2(x, y) + Y_3(x, y) + \ldots,$$

where X_h, Y_h are forms of degree h. Take now in (25.2) $\xi = \varepsilon\lambda x$, $y = \varepsilon\mu y$, with ε arbitrarily small and positive. Then (25.2) will have at A the same type of critical point as the system

$$dx/dt = \varepsilon\lambda x, \ dy/dt = \varepsilon\mu y \ .$$

By choosing λ, μ appropriately A will be a critical point of one of the two types: node, saddle point. Now if the neighborhood U of A is small enough it will only contain a single critical point of (25.2), namely TA. Since A is a critical point of (25.2) $A = TA$. Let T_1 correspond to a choice of λ, μ making of A a node, and T_2 to a choice making of A a saddle point. Then $T_2T_1^{-1}$ is analogous to T and changes a focus into a saddle point in contradiction to (25.3). Hence there are only elementary critical points.

If A is a center in suitable coordinates with $\lambda = \mu = +1$ it becomes a focus for the new system. Since the center is not transformable into this type, it is ruled out. This completes the proof of (26.2).

27. Certain important consequences regarding separatrices and closed paths follow likewise from structural stability.

(27.1) *If* (25.1) *is structurally stable in* Ω *then no separatrix in* Ω *can join two saddle points or go from a saddle point back to the same saddle point.*

Suppose that we have structural stability and that there are two

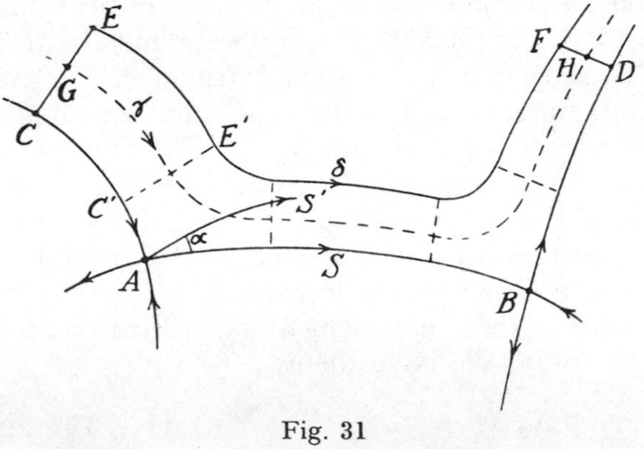

Fig. 31

saddle points A, B joined by a separatrix S. Since we may if need be slightly enlarge Ω we may assume that A, B and S are interior to Ω. We may then draw as in Fig. 31 the various transversals: CE, FD and the dotted transversals all in Ω. Then the path δ issuing from E very near C will be in Ω and behave as shown in the figure. Thus between E and E' it will be a generator of a certain path-rectangle; between E' and the next transversal it will behave as it should near a saddle point (in a hyperbolic sector), etc. A path γ starting between E and C will necessarily behave in similar manner. It is clear that the paths starting from the points of the arc (CE) intersect (DF) in a single point and generate what may be described as an open path-rectangle of which the paths are the generators.

Let now 2ε be less than the least distance from $CEFD$ to $\overline{S}$ and consider the system (27.2) defined by

$$(27.2) \qquad \begin{cases} dx/dt = X \cos \alpha - Y \sin \alpha \\ dy/dt = X \sin \alpha + Y \cos \alpha \end{cases}$$

in terms of a fixed angle α. The vectors of this system are merely the vectors (X, Y) rotated by a fixed angle α. Hence (27.2) has the same critical points as (25.1). We have

$$X \cos \alpha - Y \sin \alpha = X - X (1 - \cos \alpha) - Y \sin \alpha$$

$$X \sin \alpha + Y \cos \alpha = Y + X \sin \alpha - Y (1 - \cos \alpha)$$

Hence here

$$\xi = -X (1 - \cos \alpha) - Y \sin \alpha, \quad \eta = X \sin \alpha - Y (1 - \cos \alpha).$$

It follows that by taking α small enough one may make ξ, η arbitrarily small. Let α be chosen so small that T is an ε-transformation, where T is the topological mapping which exists since we have structural stability. Under T the points A, B remain fixed. Hence they are saddle points of (25.2) and $S' = TS$ must join the two. However owing to the nature of the new field S' behaves as shown in the figure and must cross a certain γ. It enters thus the closed path rectangle $R : GEFHG$, which does not contain B. Since S' joins A to B it must leave R. Owing to the orientation of the vectors of the new field S' cannot leave R

through GH. Hence it can only leave R through the other three sides whose points are at a distance $> 2\varepsilon$ from S. Since $S' = TS$ and T is an ε-transformation, this is ruled out—a contradiction proving (27.1) when the two saddle points are distinct. When they coincide the same treatment applies provided that one takes care to make the whole construction of Fig. 31 outside the loop now formed by the separatrix S.

(27.3) *If the system* (25.1) *is structurally stable in* Ω, *it has at most a finite number of closed paths in* Ω *and they are all limit-cycles.*

Suppose that there is an infinite collection $\{\gamma_\lambda\}$ of closed paths and let $A_1, \ldots, A_r$ be the critical points, whose number as we have seen is finite. Every γ_λ will contain a certain number of A_j in its interior. Hence there is an infinite collection of closed paths, and we will suppose for convenience that it is $\{\gamma_\lambda\}$ itself, whose interiors U_λ contain the same set $A_1, \ldots, A_s$ of critical points. Let $\gamma_\lambda \neq \gamma_\mu$. Since their interiors U_λ, U_μ both contain $A_1 \ldots, A_s$ and since $\gamma_\lambda, \gamma_\mu$ are disjoint one of U_λ, U_μ contains the other. Hence the set $\{U_\lambda\}$ may be ranged in linear order according to inclusion. That is to say we define $\lambda > \mu$ if $U_\lambda \subset U_\mu$, and this makes of $\{\lambda\}$ what is known as a *linearly ordered* set. Consider the closures $\{\bar{U}_\lambda\}$. They are all subsets of the compact set Ω and any finite subcollection $\{\bar{U}_{\lambda_1}, \ldots, \bar{U}_{\lambda_q}\}$, $\lambda_1 < \lambda_2 < \ldots < \lambda_q$, has a non-empty intersection, namely $\bar{U}_{\lambda_q}$. Hence $B = \bigcap \bar{U}_\lambda$ is a non-empty connected closed bounded set (I, 4.6). There are two possibilities:

(a) *B consists of a single point.* Referring to the proof of (I, 4.6) every neighborhood of B contains a $\bar{U}_\lambda$ and hence a γ_λ. Hence, referring to (IX, 5) B is a center, which is excluded by (26.2).

(b) *B consists of more than one point.* Since B is connected it consists of an infinite number of points. The boundary F of B is likewise infinite. For if B has no interior points this is clear. If it has such a point say Q then on each ray through Q there is a point of F and so F is again infinite. It follows that F contains an ordinary point P. Let δ be the path through P, and σ a line through P transverse to δ and containing no critical point. An infinity of γ_λ pass arbitrarily close to P and hence intersect σ. It follows that if M is a point of σ near enough to P, the path γ_M through M will intersect σ again. If M' is its first intersection beyond

$M, M \to M'$ defines an analytical transformation expressed say by an analytical function $u_1 = \varphi(u)$, where u is a coordinate on σ. Regarding δ there are three possibilities.

I. *One of* $\delta^+(P)$ *or* $\delta^-(P)$ *cuts the boundary* Φ *of* Ω. This is ruled out since δ is interior to a γ_λ which is interior to Ω.

II. *Say* $\delta^+(P)$ *tends to a critical point* C. Then $\delta^-(P)$ must tend likewise to a critical point C_1 and conversely. Since there is a path γ_λ arbitrarily close to C or C_1, they are saddlepoints. Thus δ joins two saddlepoints, in contradiction to (27.1).

III. δ *is a closed path.* This is then the only possibility left. It is then known that $\varphi(u)$ is analytical at P.

Since $\varphi(u) - u = 0$ for an infinite number of values of u near $u(P)$, $\varphi(u) = u$. Hence every path through a certain interval of σ containing P is closed. Hence P is exterior to some γ_1 which is ruled out since $P \varepsilon B$. Thus this case is also excluded and (27.3) is proved.

28. We come now to the particular property which requires the full weight of structural stability.

(28.1) *The characteristic number* $\lambda(\gamma)$ *of a limit-cycle of a structurally stable system is non-zero. Hence* γ *is stable or unstable on both sides.*

We have seen (14.2) that

$$\lambda(\gamma) = \frac{1}{\tau}\int_0^\tau \operatorname{div}(X, Y)\, dt,$$

where τ is the period associated with γ. Consider now with Gomory the function $F(x, y)$ defined as follows: — Let $\xi(t)$, $\eta(t)$ be the solution of (25.1) such that $\xi(0) = x$, $\eta(0) = y$. Then $\xi(\tau) - x$, $\eta(\tau) - y$ are analytic functions of x, y in a narrow band Δ, zero on γ but not identically zero near γ, since γ is a limit-cycle. Hence

$$F(x, y) = (\xi(\tau) - x)^2 + (\eta(\tau) - y)^2$$

is analytic in Δ, zero on γ but not identically zero. Now the partials of F are not all identically zero on γ. For otherwise the Taylor expansion of F about some point P of γ is $\equiv 0$ and hence $F \equiv 0$ about P and hence in Ω. Let G be a partial such that

$G = 0$ on γ but G_x and G_y are not both $\equiv 0$ on γ. Thus $S = G_x^2 + G_y^2 \neq 0$ on γ. Following now Schiffer consider the system (25.2) defined by $\xi = aG\, \partial G/\partial x$, $\eta = aG\, \partial G/\partial y$, where a is so small that the associated transformation T is an ε-transformation. In fact T is the identity on γ. The new $\lambda(\gamma)$, say $\lambda(\gamma, a)$ is given by

$$\lambda(\gamma, a) = \frac{1}{\tau}\int_0^\tau \operatorname{div}(X + \xi, Y + \eta)\, dt = \frac{1}{\tau}\int_0^\tau \operatorname{div}(aGG_x, aGG_y) dt =$$

$$= \frac{a}{\tau}\int_0^\tau S dt \neq 0\,.$$

Hence by taking a first positive then negative the limit-cycle will first be unstable then stable, in manifest contradiction to (25.3). This proves (28.1).

29. Consider now an autonomous system such as we have discussed, and suppose that in the associated closed region Ω the system has:

(a) at most a finite number of critical points, all elementary and none a center;

(b) no separatrix joining two saddle points;

(c) at most a finite number of closed paths, each a limit-cycle with a non-zero characteristic number.

A system possessing the three properties (a, b, c) will naturally be referred to as *general*. And now we may state:

(29.1) THEOREM. *A n.a.s.c. for a system to be structurally stable is that it be general.*

We have already proved necessity. For the much more difficult proof of sufficiency the reader is referred to the work of De Baggis [1], where the argument is in fact carried out under much broader conditions.

We have already proved necessity. For the much more difficult proof of sufficiency the reader is referred to the work of De Baggis [1], where the argument is in fact carried out under much broader conditions.

For recent work on structural stability see also Peixoto [1].

§ 10. Non-analytical Systems.

30. In the present chapter and in the preceding one the discussion has been restricted to analytical systems. The motivation was to obtain the greatest geometric yield for the least possible "analytical" price. It remains true, however, that in a large number of problems arising in applications the systems are far from analytical. A considerable interest is therefore attached to the question of how far one may deviate from analytics and still preserve most of the attractive geometric features. This program has been carried out quite completely in an extensive paper by Frommer [1]. The most important problem and the one solely discussed by Frommer and also practically exclusively in the sequel, refers to the disposition of the paths around the critical points.

In the following no attempt will be made to go too far, but only to an extent which may be described by comparatively simple restrictions.

31. Consider then a system (1.1) in which X and Y are now of class C^1 in a certain bounded region Ω. In this region it will be supposed that the system has only isolated critical points. The class C^1 hypothesis guarantees that a unique path passes through each point of Ω. (II, 2.1, 5).

Given a critical point O in Ω let it be taken as origin so that $X(0, 0) = Y(0, 0) = 0$. Let also r, θ be polar coordinates relative to O. It will be assumed first of all, that

$$X = X_m(x, y) + r^{m+1}f(x, y)$$
$$Y = Y_m(x, y) + r^{m+1}g(x, y)$$

where X_m, Y_m are forms of degree $m \geqq 1$ and f and g are of class C^1. In polar coordinates the basic system, with an appropriate time, becomes

$$\text{(31.1)} \qquad \begin{cases} \dfrac{dr}{dt} = r\,[h_{m+1}(\theta) + r\,\varrho\,(r, \theta)] \\[2mm] \dfrac{d\theta}{dt} = k_{m+1}(\theta) + r\,\omega(r, \theta) \end{cases}$$

where h_{m+1}, k_{m+1}, ϱ, ω have the period 2π in θ,

$$(31.2)\qquad \begin{cases} r^{m+1}\, h_{m+1}\,(\theta) = x\,X_m + y\,Y_m \\ r^{m+1}\, k_{m+1}\,(\theta) = x\,Y_m - y\,X_m, \end{cases}$$

and ϱ, ω are of class C^1 and bounded for r small enough and θ arbitrary. Evidently h_{m+1} and k_{m+1} are forms of degree $m+1$ in $\sin\theta$, $\cos\theta$. Much will depend upon the nature of these forms.

Notice at the outset that we cannot have $h_{m+1} = k_{m+1} \equiv 0$. For then (31.2) yields: $X_m = Y_m = 0$. One must distinguish two cases accordingly, as $k_{m+1}(\theta)$ is or is not identically zero.

I. *Suppose first* $k_{m+1}(\theta) \equiv 0$, or which is the same $xY_m - yX_m \equiv 0$. Then (31.1), with a new time, can be replaced by

$$(31.3)\qquad \begin{cases} \dfrac{dr}{dt} = h_{m+1}\,(\theta) + r\,\varrho\,(r, \theta) \\[2ex] \dfrac{d\theta}{dt} = \qquad\qquad \omega\,(r, \theta). \end{cases}$$

Since $r = 0$ is not a solution, there is a path of the system (31.3) passing through every point of the θ axis except possibly through the points for which $h\,(\theta) = 0$. The number of roots, mod 2π, of this equation is finite. In the x, y plane there is a unique path tending to the origin tangentially to any ray which does not correspond to a zero mod 2π of $h\,(\theta)$. Let these exceptional rays be referred to as *singular* and let $\theta = \theta_0$ be a singular ray. If $\omega\,(0, \theta_0) \neq 0$, the point $(0, \theta_0)$ is not critical for (31.3) and so there is a single path tending to it, hence, a single path in the x, y plane tangent to the ray $\theta = \theta_0$. What happens in connection with the singular rays depends upon the nature of the function $\omega\,(r, \theta)$ (for r small) and requires further assumptions. We shall not proceed further in this direction and to simplify matters we assume that $\omega\,(0, \theta) \not\equiv 0$ when $h\,(\theta) = 0$. Thus there are no singular rays and in the x, y plane there is a unique path tangent to every ray. That is, the critical point is a node.

32. II. *Suppose now that* $k_{m+1}(\theta) \not\equiv 0$, or equivalently, that $xY_m - yX_m \not\equiv 0$. Also, at first, let $k_{m+1}(\theta)$ have real roots.

We are dealing now with the system (31.1). The points

$(0, \theta_1)$, $(0, \theta_2)$ . . ., where the θ_i are the roots of $k_{m+1}(\theta)$, are now critical points. The rays $\theta = \theta_i$ in the x, y plane will be called *critical rays* and their number is finite. In the r, θ plane the (open) intervals of the line $r = 0$ between consecutive critical points are complete paths. One operates, of course, solely in the upper half-plane $r \geqq 0$.

With the critical points as centers, draw semi-circles in $r \geqq 0$ of the same radius r_0 with the partial bands of $0 \leqq r \leqq r_0 n$ exterior to the semi-circles (fig. 32). Let n be so small that neither the partial bands nor the semi-circles contain other critical points than those on the r axis. This is always possible because the situation repeats itself under a translation $\theta \to \theta + 2\pi$. The partial bands actually contain path-rectangles

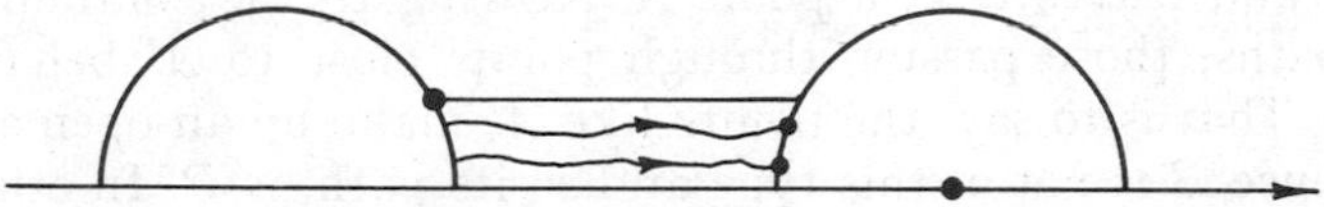

Fig. 32

relative to the paths along $r = 0$. Therefore, all that is required is to study the behavior of the paths within the semi-circles.

We shall now confine our attention to what happens in a semi-circle Γ of center P $(0, \theta_0)$, $k_{m+1}(\theta_0) = 0$ (fig. 33). Let it bound a region Φ. At the cost of replacing t by $-t$ we may assume that the path along $r = 0$ enters Φ at the point A. It follows by continuity that the neighboring paths in $r > 0$ likewise enter the region Φ. There are now two possibilities: either all these neighboring paths leave Φ or else some remain in it. These then, must necessarily $\to P$. Now if a path through a point A_1 quite near A on $\Gamma \to P$ the same obviously holds for all the paths through the points of the arc $(A\,A_1)$ of Γ. It follows that there is a maximal arc $A\,B$ such that all the paths through points of the arc $A\,B \to P$. If $B = A_0$ all the paths $\to P$ form a fan. If $B = A$ all the paths through points very near A on Γ escape from the circle. Let B be $\neq A$, A_0 and suppose first that the path through B is not tangent to Γ on

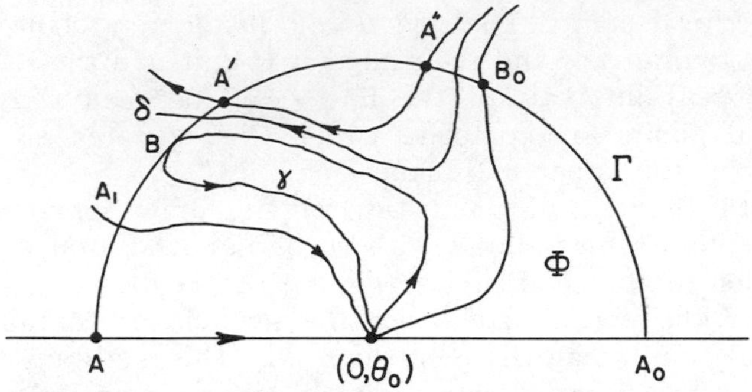

Fig. 33

the side of Φ. If A' is near B outside the arc $A\ B$ its path enters Φ at A' and leaves it at a point A''. Owing to the continuity of the paths, those passing through points close to A' behave likewise. That is to say, the points like A' make up an open arc of Γ. Hence B is not of this type and so its path $\to P$. In other words, there is a closed arc AB such that the paths through their points all tend to P. That set of paths is imaged in the xy plane into a fan whose arcs are all tangent to the same critical ray $\theta = \theta_0$.

Suppose now that the path γ through B is tangent to Γ at B. This can only happen after the manner of fig. **33**. For in the contrary case γ extended beyond B would meet other paths very near B. There are now two possibilities: (a) the path traced backward from B leaves the region at a certain point B'; we then replace in what follows B by B'; (b) γ traced backward from $B \to P$, thus constituting a loop (fig. **33**). This loop cannot contain any closed path since it would have to contain a critical point other than P. Therefore, all the paths within the loop are likewise loops from P to P, and the loop encloses an elliptic sector. This is imaged in the xy plane into an elliptic sector whose ovals have the same tangent ray $\theta = \theta_0$. It is clear that the paths like δ give rise to a hyperbolic sector bounded by the path through B and the one at the other end B_0 tending also to P. We may now treat the arc B_0A_0 of Γ like Γ itself, etc.

Observe now that each sector cuts out on the semi-circle Γ

an open or a closed interval, such as AB or BB_0, etc. The consecutive intervals never both correspond to fans. Hence one of the two is traversed by the paths in both directions: entering and leaving Φ. From this we conclude that *the number of sectors in Φ is finite*. For in the contrary case, some point Q of Γ would have, in any small interval surrounding Q, an interval $\gamma\delta \subset \alpha\beta$ on one side of Q corresponding to an elliptic or hyperbolic sector. Hence $\alpha\beta$ would be traversed, to one side of Q by paths entering and paths leaving the region Φ. This contradicts, however, the continuity of the vector (X, Y) along Γ, and so our assertion follows.

If $k_{m+1}(\theta)$ has real roots the sectors around all the critical points in the r, θ plane, together with the intermediary bands, give rise in the xy plane to a set of triangular sectors which combine, as in the analytical case, to produce the three fundamental types of sectors: elliptic, hyperbolic and fan.

Suppose now that $k_{m+1}(\theta)$ has no real roots. Thus the origin in the xy plane is the center of a circle image of a band. One may then paraphrase the treatment of (2, case V). Referring to fig. 9, the transformation $M_1 \to M'_1$ will no longer be analytic. Let u be a real parameter on the line L with u, $\varphi(u)$ the values at M_1 and M'_1. The zeros of $\varphi(u) - u = 0$ will correspond to periodic solutions with spirals between any two consecutive solutions. The periodic solutions may very well converge to the critical point O, make up continuous subsystems, etc.

33. *Application to quasi-linear systems.* Consider a quasi-linear system of class C^1

$$(33.1) \qquad \begin{cases} \dfrac{dx}{dt} = ax + by + r^2 f(x, y) \\ \dfrac{dy}{dt} = cx + dy + r^2 g(x, y) \\ \qquad ad - bc \neq 0. \end{cases}$$

To avoid complications we will assume that the matrix $\begin{pmatrix} a & b \\ c & d \end{pmatrix}$ has the normal form diag (λ, μ). It is readily shown by the argument at the end of (32) that λ, μ complex lead to a center or a focus. Let the discussion be restricted then to λ, μ real. We first choose coordinates x, y such that (33.1) assumes the form

$$(33.2)\qquad \begin{cases} \dfrac{dx}{dt} = \lambda x + r^2 f(x, y) \\ \dfrac{dy}{dt} = \mu y + r^2 g(x, y) \end{cases}$$

or in polar coordinates

$$\frac{dr}{dt} = r\{h_2(\theta) + r\varrho(r, \theta)\},$$

$$\frac{d\theta}{dt} = k_2(\theta) + r\omega(r, \theta),$$

$$h_2 = \lambda \cos^2 \theta + \mu \sin^2 \theta,$$

$$k_2 = (\mu - \lambda) \sin \theta \cos \theta.$$

If $\mu = \lambda$, with λ and μ of the same sign, we know (31, I) that the origin is a node with a unique path tangent to every ray at the origin. We may, therefore, assume that $\mu \neq \lambda$. There are then the critical directions given by $\sin \theta \cos \theta = 0$. The critical rays are the four halves of the axes.

We notice that the signs of dr/dt, $d\theta/dt$ around the critical directions are the same as for the first approximation. Therefore, the general behavior is the same as for the linear system

$$(33.3)\qquad dx/dt = \lambda x,\ dy/dt = \mu y$$

save that if one of the semi-axes is a direction of approach, it is not known even in the case of the saddle point, whether or not it is the "basis" of a fan — that is, whether one or many paths tend to that direction. The decision will be made in every case by the general process of (22) as applied in the following manner. Instead of $=$ use $\doteq$ to denote neglecting terms of higher order in r. That is, the decision is made by reference to

$$dr/dt \doteq rh_2(\theta),\ d\theta/dt \doteq k_2(\theta),$$

which is the same as for the first approximation. In other words, the character of the critical point is the same as for the first approximation.

The same result is again obvious when the characteristic roots are complex, save that when they are pure complex we cannot distinguish between stable or unstable focus and center. In other words, *the description of the critical point is once more the same as for the analytical case.*

34. Once the description of the critical points is shown (under definite restrictions) to follow the same general rules as in the analytical case, the extension of the Bendixson theorem offers no difficulty. We merely state that it is applicable to paths remaining within a cell whose closure is contained in the basic region Ω.

CHAPTER XI

Differential Equations of the Second Order

In many applications there arise differential equations of the second order. In comparatively simple cases the equations are linear with constant coefficients

$$(\alpha) \qquad A\, d^2x/dt^2 + B\, dx/dt + Cx = F(t)\ .$$

Two well-known instances are the linear spring and the electric circuit. In the spring x is the displacement, A is the mass of the bob, $-B\, dx/dt$ the resistance of the medium, $-Cx$ the "restoring force" of the spring and $F(t)$ the "forcing" term. In the electric circuit x is the current, A is the inductance, B the resistance, $1/C$ the capacitance and $\int F\ (t)\ dt$ the impressed electromotive force. A particularly important case is when F is oscillatory and one asks then for the oscillatory "response" i.e., for the oscillatory solution of the equation.

Since the equation is of the second order $A \neq 0$ so that one may divide by A. One obtains thus

$$(\beta) \qquad d^2x/dt^2 + f\, dx/dt + gx = e(t)\ .$$

The middle term corresponds to a dissipation of energy and it will generally be referred to as the "dissipative" term. It marks in a sense the deviation from the law of conservation of energy.

Usually linear equations such as (α), (β) arise because this type may be solved "in closed form" and one may readily discuss all the peculiarities of the solution. For this reason physicists and engineers tend to simplify the situation which they face to the point where linear equations with constant coefficients cover the case. Whenever for one reason or another this will not do one comes face to face with non-linear equations.

Very general types are

$$(\gamma) \qquad d^2x/dt^2 + f(x)\, dx/dt + g(x) = e(t)\,,$$

$$(\delta) \qquad d^2x/dt^2 + f(x, dx/dt)\, dx/dt + g(x) = e(t)$$

and these are the types to be discussed in the present chapter. Noteworthy special cases are

$$(\varepsilon) \qquad d^2x/dt^2 + f(x)\, dx/dt + g(x) = 0\,,$$

$$(\zeta) \qquad d^2x/dt^2 + f(x, dx/dt)\, dx/dt + g(x) = 0\,.$$

A very important special case of (ε) is the van der Pol equation

$$(\eta) \qquad d^2x/dt^2 + \mu\,(x^2 - 1)\, dx/dt + x = 0,\ \mu > 0\,.$$

It corresponds to a certain electric circuit with a vacuum tube: the resistance is *negative* for x small and *positive* for x large. Thus the system may emit energy for x small and absorb energy for x large. This may and does mean that there may arise an "internal" oscillation (self-oscillation) and its existence was indeed brought out (plausibly) by van der Pol, then rigorously established by Liénard for a general class (ε) which includes (η). The more general system (ζ) was first dealt with by Levinson-Smith and (δ) by Levinson. Mary Cartwright and D. E. Littlewood have investigated at length the system (γ) with f, e replaced by μf, μe, where μ is a positive multiplier. Systems such as (δ) with f, e replaced by μf, μe have also been discussed by Reuter.

Naturally enough sharp restrictions are generally to be imposed upon the coefficients f, g. One endeavors then to prove that the solutions $(x\,(t), x'\,(t))$ eventually penetrate and remain in a bounded region, and furthermore if $e\,(t)$ has the period τ, that there exists a solution with the same period. This will characterize a large portion of the chapter.

The reader will find that variants of the same equation will be discussed in different places. The justification will always be the introduction of a new method of attack available in the particular case under consideration.

References. Cartwright-Littlewood [1]; Cartwright [1]; Lefschetz [4]; Levinson-Smith [1]; Liénard [1]; van der Pol [1]; Shimizu [1].

§ 1. Non-Dissipative Systems

1. They are the systems without dissipative term

$$(1.1) \qquad d^2x/dt^2 + g(x) = 0 .$$

For simplicity we will assume $g(x)$ analytic for all x. If we introduce

$$(1.2) \qquad G(x) = \int_0^x g(x)\, dx ,$$

then $g(x) = dG/dx$, and so the dynamical system is of the type known as *conservative*, a term frequently applied to (1.1) itself. Notice that G is the so-called potential energy.

The equation (1.1) may be integrated as follows: Multiply both sides by $dx/dt\, dt$ and integrate, thus obtaining

$$(1.3) \qquad 1/2\, (dx/dt)^2 + G(x) = u .$$

In the dynamical interpretation as motion of a particle the first term represents the kinetic energy of the particle and (1.3) expresses the law of conservation of energy as applied to the particle.

A further integration yields

$$(1.4) \qquad t - t_0 = \int_{x_0}^{x} \frac{dx}{\sqrt{2\,(u - G(x))}}$$

which is usually the starting point for discussing the solution of the given equation.

To obtain further insight into the situation, it is better, however, to reduce the system to

$$(1.5) \qquad dx/dt = y,\ dy/dt = -g(x) .$$

Then (1.3) yields as the equation of the paths

$$(1.6) \qquad 1/2\, y^2 + G(x) = u .$$

Consider the curve $\Gamma : y = G(x)$ and suppose that the part of Γ below $y = u$ includes a finite arc AB with its end points $A\ (a, u)$, $B\ (b, u)$. Then $y = \pm \sqrt{2\,(u - G(x))}$ will represent an oval in the strip $a \leq x \leq b$ which is a closed path. We thus have here a continuous family of closed paths depending upon the parameter u. We forego a more detailed treatment and will merely make a certain observation regarding the period. For evident reasons of symmetry, the two halves of the oval, above and below the x axis are described in the same time. Hence the period of the resulting oscillation, obtained from (1.4) is:

$$T = 2 \int_a^b \frac{dx}{\sqrt{2\,(u - G(x)\,)}}$$

and it is not difficult to show that it depends generally upon u, and is not constant. In a noteworthy special case this will be proved explicitly later (see 24.7). When (1.1) is linear, however, it is well known that T is constant.

§ 2. Liénard's Equation

2. The French physicist A. Liénard, in a noteworthy, but not very well known paper [1], investigated at length a very general equation with a dissipative middle term

$$d^2x/dt^2 + f(x)\,dx/dt + g(x) = 0\,, \tag{2.1}$$

to which we shall refer as *Liénard's equation*. Liénard makes a certain number of assumptions, very largely fulfilled in practice, and described below. In point of fact he takes $g(x) = x$. The more general form was first dealt with by Levinson-Smith [1].

If we set

$$F(x) = \int_0^x f(x)\,dx, \; G(x) = \int_0^x g\,(x)\,dx$$

$$y = dx/dt + F(x), \;\; u\,(x, y) = y^2/2 + G(x)$$

then in the "spring" interpretation $y^2/2$ is the kinetic energy, $G(x)$ the potential energy and (2.1) yields, after multiplication by y:

$$du = F dy\,.$$

The integral $\int F dy$ taken along a path is the energy *dissipated* by the system.

Following Liénard we shall give a rather full discussion of (2.1) under the assumptions given below:

I. *f is even, g is odd, $xg(x) > 0$ for all $x \neq 0$; $f(0) < 0$;*

II. *f and g are continuous for all x; g satisfies a Lipschitz condition for all x;*

III. *$F \to \pm \infty$ with x;*

IV. *F has a single positive zero $x = a$ and is monotone increasing for $x \geq a$.*

It may be observed that van der Pol's equation

$$d^2x/dt^2 + \mu (x^2 - 1)\, dx/dt + x = 0; \ \mu > 0\,,$$

so important in vacuum tube circuits, is of the type under consideration.

(2.2) THEOREM. *Under the assumptions I, . . ., IV equation (2.1) possesses a unique periodic solution which is orbitally stable* (Liénard).

In terms of x, y, (2.1) is equivalent to

$$dx/dt = y - F(x),\ dy/dt = -g(x)\,. \tag{2.3}$$

Let us show that:

(2.4) *The system (2.3) satisfies a Lipschitz condition for all (x, y). Hence the existence theorem and in particular path uniqueness hold in full for the system.*

The Lipschitz condition reduces in this instance to

$$\begin{aligned} |y - y_1 - (F(x) - F(x_1))| &+ |g(x) - g(x_1)| \\ &< k\,\{|x - x_1| + |y - y_1|\}\,, \end{aligned}$$

say for $|x|$, $|y| < A$. Since $f(x)$ is continuous it has an upper bound say m in $|x| < A$. The mean value theorem for $F(x)$ yields then

$$|F(x) - F(x_1)| \leq m\,|x - x_1|\,.$$

From condition II we also have

$$|g(x) - g(x_1)| \leq n\,|x - x_1|\,,$$

and from these inequalities to the Lipschitz inequality is but a step.

To prove (2.2), we merely need to show that:

(2.5) *The system* (2.3) *possesses a single closed path which is orbitally stable.*

Observe the following simple properties:

(2.6) *If* $x(t)$, $y(t)$ *is a solution of* (2.3), *so is* $-x(t)$, $-y(t)$. *That is to say, a curve symmetric to a path with respect to the origin is likewise a path.*

(2.7) *The only critical point of the system* (2.3) *in the finite plane is the origin. Hence any closed path must surround the origin.*

(2.8) *The slope of a path* Γ *is given by*

$$\frac{dy}{dx} = \frac{-g(x)}{y - F(x)}.$$

Referring to the form of the curve Δ: $y = F(x)$ (fig. 1), *we see that outside the origin on the* y *axis, the tangents to the paths are horizontal while on* Δ *they are vertical.*

Since $xg > 0$, it is a consequence of (2.3) that y decreases along a path Γ to the right of the y axis and increases to the left of the y axis. On the other hand, from (2.3) follows that x increases when Γ is above Δ, and decreases otherwise.

Notice now that beyond B (fig. 1) Γ can only go down remaining below Δ till it reaches the y axis. For otherwise it would

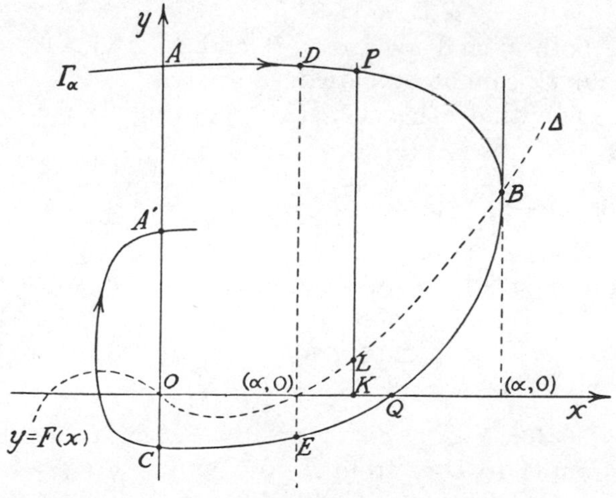

Fig. 1

possess a horizontal tangent with contact to the right of Oy. However, these contacts occur only on $g(x) = 0$, i.e. on $x = 0$ or Oy since g is odd. We conclude then that Γ has the aspect of fig. 1. Let α denote the abscissa of B and write Γ_α for Γ.

If Γ_α is to be closed we must have $OA = -OC$. For suppose that $|OA| \neq |OC|$ and let A_1, C_1 be the symmetrical of A, C with respect to O. By (2.6) the curve symmetric to Γ_α with respect to the origin is a closed path Γ_1 through A_1, C_1. Since the y axis is not tangent to Γ_α, and the segment A_1C_1 crosses Γ_α exactly once and without contact, it follows by (X, 5.1) that A_1 and C_1 are separated by Γ_α. Hence Γ_1 must meet Γ_α, which is ruled out, and so $OA = -OC$.

Conversely, suppose $OA = -OC$. The curve symmetric to the arc AC with respect to the origin is an arc of path joining A to C to the left of the y axis, and so together with AC it makes up a closed path. Thus a n.a.s.c. for a closed path is that $OA = -OC$. Since $u(0, y) = y^2/2$, we may assert:

(2.9) *A n.a.s.c. for Γ_α to be closed is that $u(A) = u(C)$.*

3. The integrals to be written presently are all curvilinear and along Γ_α. Consider

$$\varphi(\alpha) = \int_{ABC} du = \int_{ABC} F dy = u_C - u_A.$$

If $\alpha \leq a$, both F and dy are < 0 and so $\varphi(\alpha) > 0$, $u_C > u_A$. Hence (2.9) Γ_α cannot be closed.

Suppose now that $\alpha \geq a$, i.e., Γ_α is as in fig. 1. We may now introduce

$$\varphi_1(\alpha) = \int_{AD} du + \int_{EC} du, \quad \varphi_2(\alpha) = \int_{DBE} du,$$

so that $\varphi(\alpha) = \varphi_1(\alpha) + \varphi_2(\alpha)$. Along AD or CE we may write

$$du = \frac{F\,dy}{dx}\,dx = \frac{-Fg}{y - F}\,dx.$$

Since $F < 0$ for $x < a$, du is positive when Γ_α is described forward from A to D or from E to C and so $\varphi_1(\alpha) > 0$. On the contrary along DBE we have $du < 0$ and so $\varphi_2(\alpha) < 0$.

The effect of increasing α is to raise the arc AD and to lower CE, thus increasing y for given x. Since for φ_1 the limits of integration are fixed, the result will be a decrease in $\varphi_1(\alpha)$.

Concerning φ_2 we proceed slightly differently. The transformation $X = F(x)$, $Y = y$ is topological to the right of the line DE. The arc DE goes into an arc D^*E^* with end points on the Y axis and $\varphi_2(\alpha)$ is the area between D^*E^* and this axis. Now an increase in α causes D^* to rise say to D_1^* and E^* to be lowered say to E_1^*. The new arc $D_1^*E_1^*$ does not meet D^*E^* and so the effect is an increase in the area between D^*E^* and the Y axis, i.e., again a decrease in $\varphi_2(\alpha)$.

It follows that when $\alpha \geq a$, $\varphi(\alpha) = \varphi_1(\alpha) + \varphi_2(\alpha)$ is monotone decreasing. Notice that $\varphi(\alpha) = \varphi_1(\alpha) > 0$, for $\alpha \leq a$.

We will now show that $-\varphi_2(\alpha) \to +\infty$ with α. Since $du < 0$ on DBE we merely need to show that $-\int du \to +\infty$ along some subarc. Take K in fig. 1 fixed and $\alpha > OK$. We have

$$\int_{PBQ} du = \int_{PBQ} F(x)\, dy < -PK \times KL\,.$$

Since PK is arbitrarily large, the integral $\to -\infty$ and so does $\varphi_2(\alpha)$.

Now φ is monotone decreasing from $\varphi(a) > 0$ to $-\infty$ as α ranges from a to $+\infty$. Hence $\varphi(\alpha)$ vanishes once and only once say for $\alpha = \alpha_0$ and so by (2.9) there is one and only one closed path which is Γ_{α_0}.

When $\alpha < \alpha_0$ we have $\varphi(\alpha) > 0$, while $\varphi(\alpha) < 0$ when $\alpha > \alpha_0$. Hence $u_C > u_A$ for $\alpha < \alpha_0$, while $u_C < u_A$ when $\alpha > \alpha_0$. If A_0, C_0 correspond to Γ_{α_0}, we see then that C is nearer to Γ_{α_0} than A when $\alpha < \alpha_0$. Repeating the same argument, which may be done in view of (2.8), we find that A' is then nearer to Γ_{α_0} than A. The same thing will clearly hold for the same reasons when $\alpha > \alpha_0$. This is what fig. 1 shows. Hence Γ_{α_0} is orbitally stable. Thus Liénard's theorem is proved.

4. What is the effect of omitting condition IV, i.e., of assuming that F ceases to be monotonic? The general aspect of Δ will then be as in fig. 2. It is not difficult to see that we shall still have $\varphi(\alpha) > 0$ for α small and < 0 for α large. Hence it will have

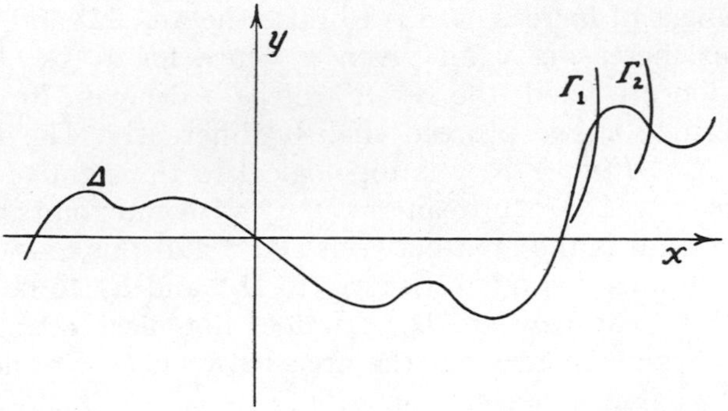

Fig. 2

at least one zero, but may very well have several $a_1, a_2, \ldots$ thus giving rise to a succession of "concentric" closed paths $\Gamma_1, \Gamma_2, \ldots$, and Γ_i will be orbitally stable (unstable) whenever $\varphi'(a_i) < 0$ $[> 0]$. This is proved as at the end of (3). It is readily found that

$$\varphi'(a) = \int_{ABC} f(x)\, dy$$

and thus $\varphi'(a)$ may be calculated directly without passing through the expression for $\varphi(a)$.

§ 3. The Equation of van der Pol: Phase-Portrait

5. The equation of van der Pol

(5.1) $$d^2x/dt^2 + \mu (x^2 - 1)\, dx/dt + x = 0, \mu > 0,$$

is classical. By introducing the new variable

(5.2) $$y = dx/dt + \mu (x^3/3 - x),$$

one may replace (5.1) by the equivalent system

(5.3) $$dx/dt = y - \mu (x^3/3 - x), \quad dy/dt = -x.$$

About all that is known regarding the phase-portrait of this system is that there is a unique stable limit-cycle Γ (2.8) and that the origin is the only critical point. This point is always unstable. In fact the corresponding first approximation is

$$dx/dt = y + \mu x, \; dy/dt = -x .$$

Its characteristic roots are found to be both real and positive when $\mu \geq 2$, both complex with positive real parts when $\mu < 2$. Hence the origin is an unstable node for $\mu \geq 2$ and an unstable focus for $\mu < 2$. From Liénard's theory one knows then that all the paths interior to Γ, and all the paths sufficiently near Γ and exterior to it spiral towards Γ (here in clockwise direction). For the full phase-portrait of the system a discussion of the critical points "at infinity" provides most required information. In point of fact much of it may be gleaned directly from the system (5.3). At the crossings with the y axis the tangents to the paths are horizontal, and at the crossings with the curve

$$\Delta : y = \mu\,(x^3/3 - x)$$

they are vertical. We see also from (5.3) that with increasing time:

x increases above Δ,
x decreases below Δ,
y increases to the left of the y axis,
y decreases to the right of the y axis.

Hence the behavior of the paths at the crossings with the y axis and the curve Δ is as indicated in fig. 3. Remember also that, according to Liénard, the limit-cycle Γ is related to the curve Δ after the manner of fig. 3, and is stable.

The behavior of the paths at infinity is governed by the homogeneous equation

$$\begin{vmatrix} dx & dy, & dz \\ x & y, & z \\ yz^2 - \mu\,(x^3/3 - xz^2), & -xz^2, & 0 \end{vmatrix} = 0 ,$$

which upon expanding the determinant yields

$$xz^3\,dx + z\,\{yz^2 - \mu\,(x^3/3 - xz^2)\}\,dy = \{x^2z^2 + y\,(yz^2 - \mu\,(x^3/3 - xz^2))\}\,dz . \tag{5.4}$$

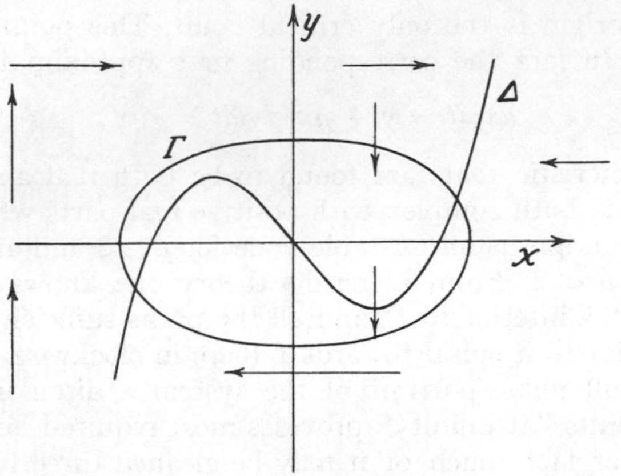

Fig. 3

Since (5.4) is satisfied by $z = dz = 0$, the arcs of $z = 0$ between critical points are paths of (5.1).

Let us now look for the critical points on the line $z = 0$. Consider first the points for which $x \neq 0$. We may thus take $x = 1$, $dx = 0$ and we have to find if the point $y = z = 0$ (point at infinity on the x axis) is critical for

$$z\{yz^2 + \mu(z^2 - 1/3)\}\,dy = \{z^2 + y(yz^2 + \mu(z^2 - 1/3))\}\,dz\,.$$

The "critical point" behavior is governed by the linear approximation

$$z\,dy = y\,dz$$

or with a new t by

$$dy/dt = y,\quad dz/dt = z\,. \tag{5.5}$$

Hence, (IX, 4, first case), (1, 0, 0) is the simplest type of node: paths tending to the node in every direction. In the xy plane the curves tending to infinity in the direction of the x axis will have horizontal asymptotes. Upon referring to the mode of increasing or decreasing of y it will be seen that each curve tends to its asymptote from above in quadrants II, III and from below in quadrants I, IV (fig. 4).

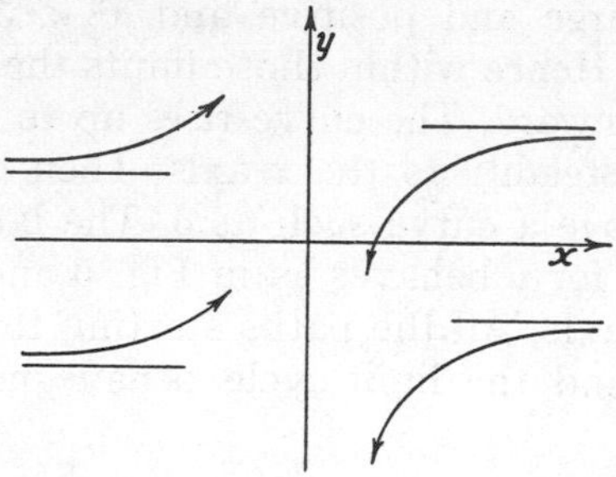

Fig. 4

Assuming similarly $y = 1$, $dy = 0$ we look for the critical points of

$$(5.6) \qquad xz^3dx = (x^2z^2 + z^2 - \mu\,(x^3/3 - xz^2)\,)\,dz$$

and find only the points $x = z = 0$. Thus the only critical points of (5.6) are at infinity on the axes.

The behavior at the origin is the same as for

$$(5.7) \qquad dx/dt = x^2z^2 + z^2 - \mu\,(x^3/3 - xz^2),\; dz/dt = xz^3\,.$$

Since there are no first degree terms the origin is a critical point of higher order. One could apply the analysis of (X, § 1), but it is actually simpler here to proceed directly.

6. Let us examine the behavior of the paths outside of the limit-cycle. Take a path γ beginning at infinity to the left of the y axis (fig. 5). It has a horizontal asymptote and tends to it from above. We have on any path

$$r\,dr/dt = -\mu\,x^2\,(x^2/3 - 1)\,.$$

Hence for $x^2 > 3$, dr/dt is negative and so γ rises but comes nearer to the origin till it reaches the line $x = -3$. On the other hand

$$y' = \frac{dy}{dx} = -\frac{x}{y - \mu\,(x^3/3 - x)}\,,$$

$$-y'' = \frac{y - \mu\,(x^2/3 - x) - x\left\{\dfrac{-x}{y - \mu\,(x^3/3 - x)} - \mu\,(x^2 - 1)\right\}}{(y - \mu\,(x^3/3 - x)\,)^2}\,.$$

Hence for y large and positive and $x^2 < 3$, approximately $y'' = -1/y < 0$. Hence within those limits the concavity of the curve points downward. The curve rises up to the y axis. After that it descends steadily to the x axis. Then it must maintain itself steadily above a curve such as δ. The latter by the same considerations as for γ behaves as in Fig. 5 and so γ returns to the left of the x axis. All the paths starting to the left of the x axis between γ and the limit-cycle behave necessarily like γ,

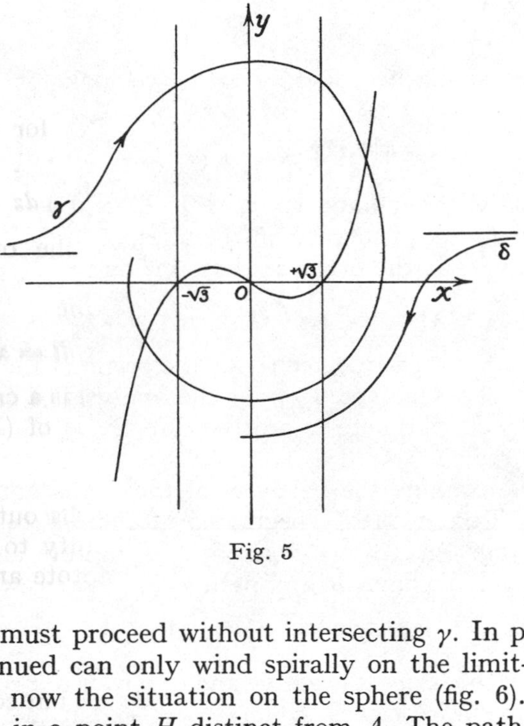

Fig. 5

since they must proceed without intersecting γ. In particular γ itself continued can only wind spirally on the limit-cycle.

Consider now the situation on the sphere (fig. 6). The path δ' cuts OA in a point H distinct from A. The path γ' cuts Δ' and OA at points P and Q. Since no path is tangent to FB nor to OA, $P \to Q$ determines a topological mapping of $[FB]$ onto a segment $[GG_1]$ where G_1 is between G and H. The path ε through G_1 tends to B and it is a separatrix issued from B in the first quadrant.

We shall show by the same argument as for (X, 21.1) that ε is unique. Suppose in fact that there exists a second ε_1 (fig. 7) analogous to ε. If $x(z)$ and $x_1(z)$ are the solutions of (5.6) corresponding to ε and ε_1 then

$$dx/dz = x/z + 1/xz - \mu\,(x^2/3z^3 - 1/z)$$

and similarly for x_1. Hence

$$\frac{d\,(x - x_1)}{dz} = \frac{x - x_1}{z} - \frac{(x - x_1)}{x\,x_1\,z} - \frac{\mu\,(x^2 - x_1{}^2)}{3z^3}.$$

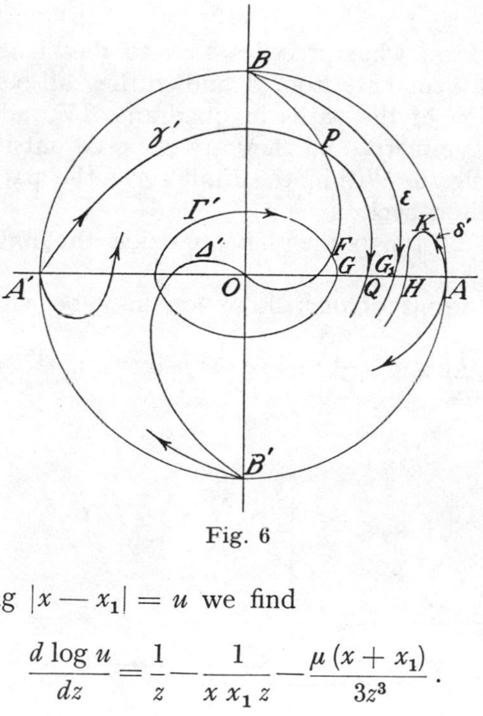

Fig. 6

Hence setting $|x - x_1| = u$ we find

$$\text{(6.1)} \qquad \frac{d \log u}{dz} = \frac{1}{z} - \frac{1}{x\,x_1\,z} - \frac{\mu\,(x + x_1)}{3z^3}.$$

Now both x and x_1 are of order < 1 in z. Hence in (6.1) the last two terms dominate the first and so

$$d \log u/dz < 0$$

for z small. Hence u increases when z decreases. Since this contradicts the disposition of Fig. 7, ε_1 cannot exist and ε is unique.

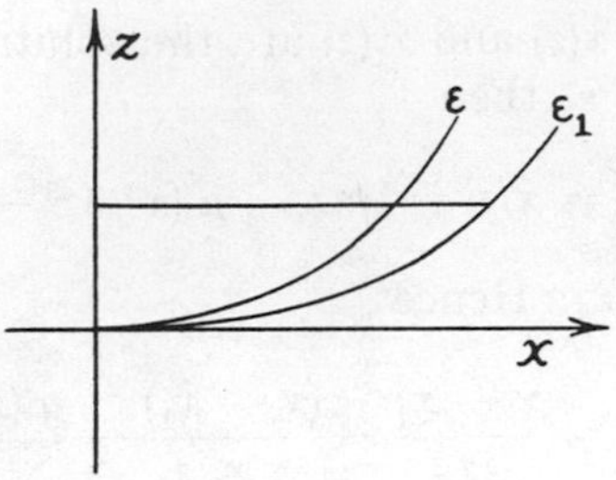

Fig. 7

It follows from what precedes that to the right of BG_1 the paths can only emanate from A and so they all behave like δ'. The disposition of the paths in quadrants IV and III is then obtained by symmetry and they are disposed outside the limit-cycle, as in fig. 6. Within the limit-cycle the paths all spiral toward the limit-cycle.

The paths in the phase-plane, outside the limit-cycle, are indicated in fig. 8.

The main result of our discussion may be summarized as follows:

(6.2) THEOREM. *Every path in the plane (with the origin omitted) tends toward the limit-cycle.*

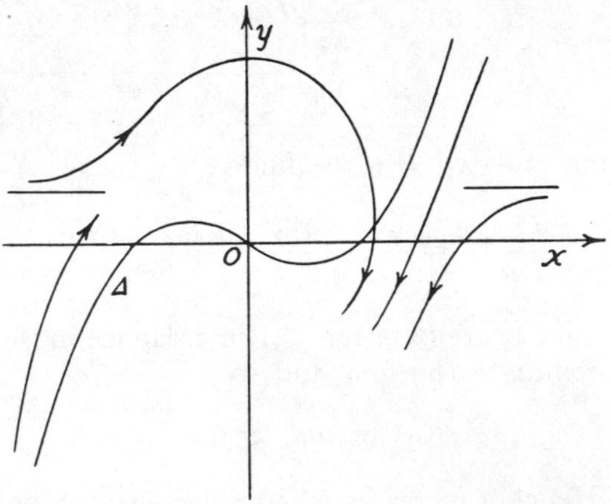

Fig. 8

It may be observed that in the projective plane the critical point B behaves like an ordinary saddle point such that the four paths tending to the saddle point are tangent to one another.

§ 4. The Equation of Cartwright-Littlewood

7. In several noteworthy papers Mary Cartwright and J. E. Littlewood have made an extensive study of the equation

$$d^2x/dt^2 + \mu f(x)\, dx/dt + g(x) = \mu e(t)\,, \tag{7.1}$$

and obtained a great variety of results some of which will be reported here. We shall lean more particularly upon Mary Cartwright [1], although our treatment is often decidedly different. The hypotheses will deviate perhaps somewhat but not significantly from those of the two English authors.

We shall suppose that $\mu > 0$ and endeavor to obtain results valid for the whole μ range.

Guided by the electrical circuit analogy, various restrictions will be imposed on the systems considered here and throughout the remainder of the chapter. By and large it may be said that while they are anything but minimal, they are such as to preserve the essential features of the systems under consideration. In particular we will not impose any symmetry properties upon f and g.

As in (2) we set:

$$F(x) = \int_0^x f(x)\, dx,\ G(x) = \int_0^x g(x)\, dx,\ u = y^2/2 + G(x)\,,$$

and in addition

$$E(t) = \int_0^t e(t)\, dt\,.$$

We make the following explicit assumptions:

I. *For all x : f and g are continuous and g satisfies a Lipschitz condition.*

II. *There exist positive numbers a, α, β such that $f(x) \geq \alpha$ for $|x| \geq a$, and that $g(x) \geq \beta > 0$ for $x \geq a$, $g(x) \leq -\beta$ for $x \leq -a$.*

III. *$E(t)$ is continuous and $|E(t)|$ is bounded for all t.*

Note that in view of our conditions F and G exist and even have continuous derivatives for all x.

We will denote by E the bound of $|E(t)| : |E(t)| \leq E$. We note also that in view of II, $|F(x)|$ and $|G(x)| \to +\infty$ with $|x|$. We may therefore suppose a so large that:

IV. (a) *There exists a positive γ such that:*

$$F(x) - E \geq \gamma \text{ for } x \geq a, \; F(x) + E \leq -\gamma \text{ for } x \leq -a;$$

(b) $G(x) > 0$ *for* $|x| \geq a$.

Finally since g is continuous $|g|$ will have an upper bound $\delta > 0$ in the strip $|x| \leq a$.

We replace the system (7.1) by the equivalent system

$$\text{(7.2)} \qquad \begin{cases} dx/dt = y - \mu\,(F(x) - E(t))\,, \\ dy/dt = -\,g(x)\,. \end{cases}$$

(7.3) *The system (7.2) satisfies a Lipschitz condition for all x, y. Hence the existence theorem and trajectory uniqueness hold in full for the system.*

The proof is the same as for (2.4).

8. (8.1) THEOREM. *There exists a bounded closed 2-cell Ω such that a solution $x(t)$, $y(t)$ starting at time t_0 from any point M_0 in the phase-plane, will after a time T, which generally depends on the solution, enter and remain permanently in Ω.*

Let λ_0 be an upper bound for $|F| + E$ in the strip $|x| \leq a$ and let λ_1 be a positive constant to be specified later. Consider the rectangle R:

$$|x| \leq a, \; |y| \leq \begin{cases} \lambda_0\,\mu + \lambda_1 \text{ if } \mu \geq 1\,, \\ \lambda_0 + \lambda_1 \text{ if } \mu \leq 1\,. \end{cases}$$

(8.2) *A trajectory Γ which remains outside the open rectangle R necessarily spirals around the rectangle and each turn is accomplished in finite time.*

The slope of a trajectory is

$$\text{(8.3)} \qquad m = \frac{dy}{dx} = \frac{-g(x)}{y - \mu\,(F(x) - E(t))}\,.$$

Within the strip and outside R: $|F(x) - E(t)| \leq \lambda_0$, hence $|y - \mu\,(F(x) - E(t)\,)| \geq \lambda_1$. Also $|g(x)| \leq \delta$. Hence

$$(8.4) \qquad |m| \leq \delta/\lambda_1 = m_0\,.$$

Hence Γ must merely cross the strip in one direction or the other. Since above the rectangle and within the strip $y \geq \lambda > 0$, $dx/dt > 0$ and so x increases. Thus the arc such as $\sigma = \zeta\eta$ is

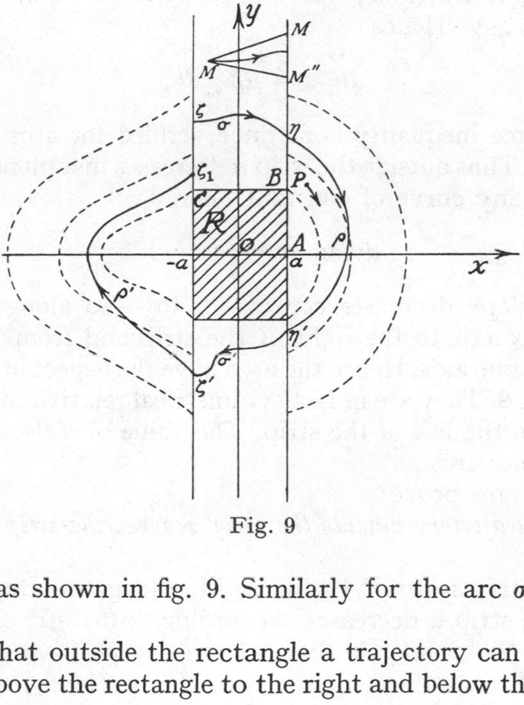

Fig. 9

described as shown in fig. 9. Similarly for the arc $\sigma' = \eta'\zeta'$ of fig. 9.

Notice that outside the rectangle a trajectory can only leave the strip above the rectangle to the right and below the rectangle to the left.

(8.5) *The crossing time of the strip by a trajectory above or below the rectangle has a positive upper bound* τ_1.

Assume say that the crossing is above like σ. On σ

$$dx = \{y - \mu\,(F(x) - E(t)\,)\}\,dt > \lambda_1\,dt\,.$$

Hence

$$t_0 = t(\eta) - t(\zeta) \leq 2a/\lambda_1 = \tau_1\,.$$

9. We shall now discuss the parts of trajectories outside the strip and here we must have recourse to the energy u. We have

$$du = y\,dy + g(x)\,dx, \tag{9.1}$$

and hence from (7.2) along a trajectory

$$du = \mu\,(F(x) - E(t))\,dy\,. \tag{9.2}$$

Now along a trajectory for $x \geq a$: $dy/dt = -g(x) \leq -\beta$, and $F(x) - E(t) \geq \gamma$. Hence

$$du < -\mu\beta\,\gamma\,dt\,, \tag{9.3}$$

and the same inequality is at once verified for arcs to the left of the strip. Thus outside the strip u decreases in a monotone way.

Now on any curve of constant energy

$$dy/dx = -g(x)/y\,.$$

Thus the slope decreases constantly to $-\infty$ along the curve above the x axis to the right of the strip and from $+\infty$ down below the same axis. Hence the arcs have the aspect of the dotted lines in Fig. 9. They are in fact symmetrical relative to the x axis. Similarly to the left of the strip. The value of u decreases from outer to inner arcs.

We may now prove:

(9.4) *A trajectory outside the strip reaches the strip in a finite time.*

Let it start say at P to the right. Along the arc beyond P outside the strip u decreases. According to (9.3) it decreases in time dt by $|du| > \mu\beta\,\gamma\,dt$. Since u remains positive outside the strip it will reach the strip in a time $\tau(P) < (1/\mu\beta\gamma)\,u(P)$, which proves (9.4). This estimate holds of course even when P is on one of the lines $x = \pm a$.

Clearly (8.5) and (9.4) combined imply (8.2).

10. (10.1) *Let a trajectory Γ spiral around the open rectangle and cross one of the lines $x = \pm a$ at two consecutive points ζ, ζ_1, on the same side of the rectangle. Then for λ_1 sufficiently large,—as we assume henceforth—ζ_1 is nearer the rectangle than ζ and $y^2(\zeta)$—*

$y^2(\zeta_1)$ has a positive lower bound μq. Hence, after a certain finite number of turns (after a finite time) Γ must enter the open rectangle.

The proof rests upon an estimate of $u(\zeta) - u(\zeta_1)$. Let us suppose say that ζ is as in fig. 9—the other dispositions would be dealt with likewise. The gain in u along σ is at most

$$|\mu \int_\sigma (F(x) - E(t) \frac{dy}{dx} dx| \leq \mu \int_{-a}^{+a} (|F| + E) \left|\frac{dy}{dx}\right| dx$$

$$\leq 2a\mu \lambda_0 m_0 = \frac{2a\mu \lambda_0 \delta}{\lambda_1}.$$

The same estimate holds for the arc σ' of fig. 9. Hence the total gain in u along these two arcs is at most

$$\frac{4 a\mu \lambda_0 \delta}{\lambda_1}.$$

On the other hand u decreases from η to η', at the same time as y decreases. The decrease in y is at least $2 \lambda_1$, (since the rectangle R has its vertical side $> 2 \lambda_1$). Since for $x \geq a : F(x) - E \geq \gamma$, the loss in u along the arc ϱ is at least

$$|\mu \int_\varrho (F - E(t)) \, dy| \geq 2\mu \lambda_1 \gamma,$$

and equally along ϱ'. Hence

$$u(\zeta) - u(\zeta_1) \geq 4\mu (\lambda_1 \gamma - \lambda_0 a\delta/\lambda_1).$$

This will be positive if we choose, as we may $\lambda_1 > \sqrt{\dfrac{\lambda_0 a \delta}{\gamma}}$.

Under the circumstances

$$y^2(\zeta) - y^2(\zeta_1) = 2(u(\zeta) - u(\zeta_1))$$

$$\geq 8\mu \left(\lambda_1 \gamma - \frac{\lambda_0 a\delta}{\lambda_1}\right) = \mu q,$$

$$q = 8 \left(\lambda_1 \gamma - \frac{\lambda_0 a\delta}{\lambda_1}\right).$$

This proves (10.1).

11. We proceed now with the construction of the closed region Ω. In fig. 10 points such as B, B' or C, C', ..., are symmetrical with respect to Ox. The shaded rectangle is R and the slanted lines have slopes $\pm m_0$. The arcs such as DQD', ..., are parts of curves $u =$ const. The closed region Ω is the interior and boundary of the Jordan curve $FPF'G'SGF$. The open region Σ is the interior of the Jordan curve $EDQD'H'K'RKE$. Thus

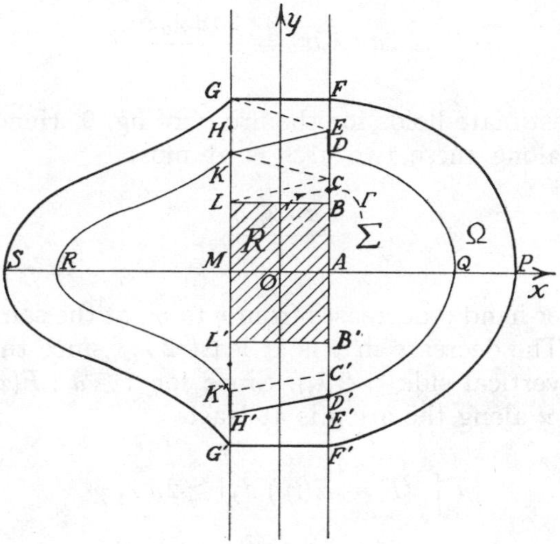

Fig. 10

the *closure* of Σ is in the *interior* of Ω and the boundary of Σ is at a positive distance from the boundary of Ω.

It is already known from (10.1) that every trajectory meets the open rectangle R after a finite time. To prove (8.1) it is thus more than sufficient to prove the following strong result needed later:

(11.1) *Every trajectory Γ leaving the open rectangle R remains in the region Σ.*

If Γ leaves say across BB', as it cannot cross the arc of curve $u =$ const. joining B to B' it remains in Σ outside R or else returns to R (the second possibility alone takes place). Similarly if it leaves across LL'. If Γ leaves across LB then it cannot reach

LC and so it must leave the strip across BC. Since it does not cross the arc DQD' it remains in Σ outside the strip and either returns to R or else re-enters the strip through the interval $(B'C')$. Then either Γ re-enters R or else crosses the strip above $K'C'$ and below $L'B'$. Then Γ re-enters R or else, remaining in Σ, re-enters the strip across (LK). It must then re-enter R or else leave the strip *below its first departure at* T (10.1) and hence across (CB). From this point on the argument just repeats itself and so (11.1) follows. This completes the proof of theorem (8.1).

12. (12.1) THEOREM. *Every trajectory reaches and remains in a region* $|x| < R$, $|y| < R(1+\mu)$ *where* R *is a positive constant independent of* μ. (Cartwright [**1**]).

Since for $\mu < 1$, Ω is already in a region such as stated we may suppose $\mu \geqq 1$. We also notice that at all events from the construction of Ω follows that sooner or later

$$|y| < A + B\mu;\ A,\ B > 0.$$

Hence we merely need to prove:

(12.2) *For* $\mu > 1$, $|x|$ *reaches and remains below a certain bound independent of* μ. (Cartwright [**1**], p. **163**).

Now if (x, y) remains in the rectangle we are through. By (10.1) it reaches the rectangle. Let it leave it say along $x = a$ at time t_0 and with initial value x'_0 for dx/dt. It will turn at a certain time t at which $dx/dt = 0$. Integrating the initial equation (7.1) we find, since $g > \beta$ for $x \geqq a$ and $|E(t)| \leqq E$:

$$-x'_0 + \mu(F(x) - F(a)) + \beta(t - t_0)$$
$$\leq \mu(E(t) - E(t_0)) \leq 2\mu E .$$

Hence

$$\mu(F(x) - F(a)) \leq 2\mu E + x_o' . \tag{12.3}$$

In view of (7.2) the largest value of x'_0 on $x = a$ is of the form $\mu C + D \leq \mu(C + D)$. Hence (12.3) yields for $x \geq a$:

$$F(x) \leq F(a) + 2E + C + D = K > 0 .$$

Since $F(x)$ is monotone increasing and positive for $x \geq a$, our inequality yields $x \leq \varepsilon$, $\varepsilon \geq a$. Similarly for the left of the strip. Thus (12.2) holds and hence also (12.1).

13. A somewhat stronger result may be proved with a slight additional hypothesis. Namely:

(13.1) *If* $f(x) \geq A > 0$, *i.e., if the system is unexceptionally dissipative,* $\mu > 1$ *and* t *is large, then* $|dx/dt| < B$ *where* $B > 0$. (Cartwright [1], p. 184).

We may also say: under the conditions stated the point x, dx/dt of the phase-plane of x and dx/dt penetrates into a square of side $R > 0$, where R does not depend on μ.

For the following proof we are indebted to Gomory. Given any $\varepsilon > 0$ we cannot have $x'(t) > \varepsilon$ for all t above any specific value since $|x(t)|$ is bounded. Hence either $x(t) \to 0$ as $t \to +\infty$, and so the theorem holds, or else $x'(t)$ oscillates indefinitely with increasing t. If t_1 is anyone of its maxima we will have $x''(t_1) = 0$. Hence from (7.1):

$$x'(t_1) = \frac{\mu\, e(t_1) - g\,(x(t_1))}{f\,(x(t_1))}.$$

As $|x|$ is bounded independently of μ so is $g(x)$, say $|g(x)| < M$. Hence since $\mu \geq 1$

$$x'(t_1) < \frac{\mu\, E + M}{\mu\, A} \leq \frac{\mu\, E + \mu\, M}{\mu\, A} = K,$$

where K is independent of μ. Thus $x'(t)$ has a fixed upper bound independent of μ at all these maxima and hence for all t large enough. A similar remark holds for all negative values and (13.1) follows.

§ 5. Applications and Complements

14. *Autonomous oscillations* ($E(t) = 0$). When $E(t) = 0$, equation (7.1) becomes

$$(14.1) \qquad d^2x/dt^2 + \mu\, f(x)\, dx/dt + g(x) = 0$$

which is a generalized form of Liénard's equation. The equivalent system reads

(14.2) $$dx/dt = y - \mu F(x),\ dy/dt = -g(x)\,.$$

We make the following supplementary assumptions to those of (7):

V. $f(0) = \alpha < 0$;

VI. $\gamma(x) = g(x)/x$ *is continuous and* > 0.

Also to simplify matters:
VII. *f and g are analytic for all x.*
Except for analyticity this is still decidedly more liberal than the assumptions made regarding Liénard's equation. The price to pay is the weaker property:

(14.3) *Under our various assumptions the system* (14.2) *has at least one limit-cycle and hence* (14.1) *has at least one periodic solution.*

As already pointed out at the end of § 2 there might well arise several limit-cycles.

The proof of (14.3) is quite simple. Under our assumptions the origin is the only critical point.

The first approximation to (14.2) is

$$dx/dt = y - \mu\,\alpha\,x,\ dy/dt = -\beta x,\quad \beta = \gamma(0).$$

The characteristic equation is

$$r^2 + \mu\,\alpha\,r + \beta = 0\,.$$

The two characteristic roots have therefore positive real parts and so the origin is an unstable node or focus. Thus with increasing time any path Γ will stay out of a suitably small circular region ω surrounding the origin.

On the other hand by (10.1) and (11.1) with increasing time Γ will remain in Σ, and so *a fortiori* in $\Omega - \omega$. Hence Poincaré's theorem (X, 9.3) is applicable here and so (14.1) follows.

15. *Periodic function* $E(t)$. Let $E(t)$ be periodic and of period τ. If $e(t)$ is developable in Fourier series this means that the series has no constant term or again that the mean value of $e(t)$ for a period is zero. We have then:

(15.1) THEOREM. *If $E(t)$ has the period τ there exists a periodic solution with period τ.*

Let $S(t_1)$ denote the transformation whereby if $x(t; x_0, y_0)$, $y(t; x_0, y_0)$ is the solution passing through the point $P_o(x_0, y_0)$ at time t_0 then $S(t_1) P_o = P_1(x_1, y_1)$ where

$$x(t_0 + t_1; x_0, y_0) = x_1, y(t_0 + t_1, x_0, y_0) = y_1 .$$

Thus $S(t_1)$ is a topological mapping of the Euclidean plane Π into itself. Let in particular $S = S(\tau)$.

Consider now any point M of the closed region Ω and the trajectory Γ starting from M at time t_0. According to (10.1) at some time t_1 Γ has a point M_1 within the rectangle R. Since $S(t_1) M$ is in R and $S(t_1)$ is continuous, there is a neighborhood U_1 of M_1 in R. Hence $S^{-1}(t_1) U_1 = U$ is a neighborhood of M such that if $V = U \cap \Omega$ then $S(t_1) V \subset R$. It follows that $S(t) V \subset \Sigma$ for all $t \geq t_1$.

Let $n(V)$ be the least multiple of τ such that $n\tau \geq t_1$. Then $S^\nu V \subset \Sigma$ for all $\nu \geq n$.

Since Ω is compact and $\{V\}$ is an open covering of Ω there is a finite subcovering $\{V_1, \ldots, V_r\}$. Let $N = \Pi n(V_i)$. Then $S^\nu \Omega \subset \Sigma$ for all $\nu \geq N$.

Going back now to the proof of (11.1) we observe that every trajectory issued from a point of the closed rectangle R remains in the open region Σ. Therefore $S(t) R \subset \Sigma$ whatever t. In particular $S^k R \subset \Omega$ whatever k.

Consider now the successive transforms $\Omega_0 = \Omega$, $\Omega_1 = S\,\Omega, \ldots$, $\Omega_k = S^k\,\Omega, \ldots$, and let $Z_0, Z_1, \ldots$, be the closed complements of the unbounded components of the exteriors of Ω_0, $\Omega_0 \cup \Omega_1, \ldots$ Since $S^k R \subset \Omega_0$ whatever k, every Ω_k meets Ω. Referring to Appendix II (9.1) if $J_0, J_1, \ldots$, are the boundary Jordan curves of Ω_0, $\Omega_1, \ldots$, and $H_0, H_1, \ldots$, those of $Z_0, Z_1, \ldots$, then $H_k \subset J_0 \cup J_1 \cup \ldots \cup J_k$. Let $H = H_{N-1}$, $Z = Z_{N-1}$. Since $S^N \Omega_0 \subset \Omega_0$, we have $S^N J_0 \subset \Omega_0 \subset Z$. Hence $S(J_0 \cup \ldots \cup J_{N-1}) \subset Z$ and therefore $SH \subset Z$. It follows that if W is the closed interior of SH then $W \subset Z$. But the closed interior of SH is SZ. Thus $SZ \subset Z$. Since Z is a closed 2-cell, by Brouwer's fixed point theorem S has a fixed point $P(x_1, y_1)$ in Z. Thus if $\Gamma(x(t), y(t))$ is the trajectory starting at P at time t_0 then $x(t_0 + \tau) = x_1$,

$y(t_o + \tau) = y_1$. Since the system (7.2) is now periodic with period τ, the continuation of Γ for time $t \geq t_o + \tau$, is the arc of Γ decribed for $t_o \leq t \leq t_o + \tau$. Hence Γ is closed and $x(t)$, $y(t)$ is periodic. This completes the proof of (15.1).

As an interesting consequence of (10.1) one may assert the following:

(15.2) *A closed trajectory (periodic solution) cannot surround the closed rectangle R. This holds especially for the autonomous oscillation of* (14) *and the forced oscillation of* (15.1).

16. We have already established a boundedness result, (13.1), for a strictly dissipative system. We shall now prove, under additional restrictive hypotheses, the following much more powerful property.

(16.1) CONVERGENCE THEOREM. *Let again as in* (13.1) $f(x) \geq A > 0$ *for all x. By* (13.1) *the solution* $(x(t), x'(t))$ *of* (7.1) *enters a certain square R for large t. Let in addition $g''(x)$ exist and $|g''(x)|$ be bounded in R. Furthermore let $g'(x) \geq B > 0$ for all x. Then for μ sufficiently large all solutions $x(t)$, $x'(t)$ converge to one another as $t \to +\infty$.* (Cartwright [1] p. 186.)

Noteworthy special case: $F(x)$ and $g(x)$ are monotone increasing and $g(x)$ is twice continuously differentiable. Then for μ sufficiently large all solutions converge to one another as $t \to +\infty$. (Cartwright *loc. cit.*)

For all solutions enter a closed square in which $|g''|$ is bounded and under the assumptions $f(x) \geq A > 0$ and $g'(x) \geq B > 0$.

We also note at once the following.

(16.2) COROLLARY. *Let $E(t)$ have the period τ, so that there exists a periodic solution $x(t)$ of period τ. Then under the conditions of the theorem all solutions converge to $x(t)$, $x'(t)$. Hence this periodic solution is stable.*

Proof of theorem (16.1). Let $x(t)$, $x_1(t)$ be any two solutions and set

$$z(t) = x_1(t) - x(t),\ \Delta F = F(x_1(t)) - F(x(t)),$$

$$\Delta g = g(x_1(t)) - g(x(t)).$$

From (7.1) there follows

$$z''(t) + \mu\, d\Delta F/dt + \Delta g = 0. \tag{16.3}$$

Multiplying by $z'(t)$ and integrating from T to t, $t \geq T \geq 0$, we find:

$$1/2\,[z'^2]^t{}_T + \mu\,[z'\Delta F]^t{}_T - \mu \int_T^t z''\Delta F \cdot dt + \int_T^t z' \cdot \Delta g \cdot dt = 0\,.$$

Upon substituting z'' from (16.3) and integrating by parts we find

$$\left[\frac{1}{2} z'^2 + \mu\, z'\Delta F + \frac{\mu^2}{2}(\Delta F)^2 + \frac{z\,\Delta g}{2}\right]_T^t + \int_T^t z^2 \left\{\mu \frac{\Delta F}{z}\frac{\Delta g}{z} - \frac{1}{2}\frac{d}{dt}\frac{\Delta g}{z}\right\} dt = 0\,. \tag{16.4}$$

Denoting by $O(1)$ a function which is bounded in absolute value as $t \to +\infty$ we have from (16.4) and since z, z', x, x', x_1, x'_1 are all $O(1)$:

$$\int_T^t z^2 \left\{\mu \frac{\Delta F}{z}\frac{\Delta g}{z} - \frac{1}{2}\frac{d}{dt}\frac{\Delta g}{z}\right\} dt = O(1)\,. \tag{16.5}$$

And now

$$\frac{d}{dt}\frac{\Delta g}{z} = \frac{d}{dt}\frac{g\,(x+z) - g(x)}{z}$$

$$= \frac{x'(t)}{z}\{g'\,(x+z) - g'(x)\}$$

$$- \frac{z'(t)}{z^2}\{g\,(x+z) - g(x) - z\,g'\,(x+z)\}$$

$$= x'(t)g''\,(x+\theta z) - \frac{z'(t)}{z^2}\left\{\frac{z^2}{2}\,g''\,(x+\theta' z)\right\} = O(1)\,,$$

$$0 \leq \theta,\, \theta' \leq 1\,.$$

Thus say $|d/dt\ \Delta g/z| < a$, $a > 0$, for t large. On the other hand

$$\Delta F/z = f\,(x+\eta z),\ \Delta g/z = g'\,(x + \eta'\, z)\,,$$

$$0 \leq \eta,\, \eta' \leq 1\,.$$

Hence for t large

$$\Delta F/z \cdot \Delta g/z \geq AB .$$

It follows that if $\mu AB > a$, or $\mu > a/AB$, the bracket in (16.5) is positive and $O(1)$. Thus (16.5) yields

$$\int_0^t z^2\, dt = O(1) .$$

It follows that given any $\varepsilon > 0$ there exists a $T > 0$ such that for any $t > T$

$$\text{(16.6)} \qquad \int_T^t z^2\, dt < \varepsilon .$$

(16.7) *As a consequence of* (16.6) $z \to 0$.

For in the contrary case there exists $\varrho > 0$ and a sequence $T < t_1 < t_2 < \ldots \to +\infty$ such that $z^2(t_n) > \varrho$. We may suppose the t_n so spaced that for some t_n' between t_n and t_{n+1} we have $z^2(t_n') < 1/2\, \varrho$ since otherwisc (16.6) would not hold. Let t'_n be the first place above t_n where this occurs and set $\delta t_n = t'_n - t_n$. Thus

$$\int_T^t z^2\, dt > 1/2\, \varrho \sum \delta t_n .$$

Hence $\delta t_n \to 0$ as $n \to \infty$. On the other hand $|z'(t)|$ acquires in δt_n a value $> \varrho/2\delta t_n$ which contradicts $z'(t) = O(1)$. Thus (16.7) holds.

Returning now to (16.4) it yields

$$\text{(16.8)} \qquad \left[z'^2\right]_T^t = -2\left[\mu z' \Delta F + \frac{\mu^2}{2}(\Delta F)^2 + z\frac{\Delta g}{2}\right]_T^t + \mathrm{O}(1)\int_T^t z^2\, dt .$$

Since $z' = O(1)$ and $x_1 \to x$, one may take T so large that at the right $|[\ldots]| < \varepsilon/4$ for any $t > T$, and by (16.7) so that the second term $< 1/2\, \varepsilon$ in absolute value. As a consequence $[z'^2]^t_T < \varepsilon$. Thus $z'^2(t)$ tends to a limit as $t \to +\infty$. Let $z'^2 \to k^2 \neq 0$.

Then $z' \to +k$ or $-k$. For if not there is a sequence $t_1 < t_2 < \ldots \to +\infty$ such that if $z'_n = z'(t_n)$ then say $z'_{2n} \to +k$, and $z'_{2n+1} \to -k$. Indeed take $\{\varepsilon_n\} \to 0$ monotonically, $\varepsilon_n > 0$, and choose z'_1 within ε_1 of $-k$, z'_2 $(t_2 > t_1)$, within ε_2 of $+k$, etc. Since $z(t)$ is continuous it will cross zero say at t'_n between t_{2n} and t_{2n+1}. Thus $z(t'_n) \to 0$. hence $z'^2(t'_n) \to 0$, which contradicts $z'^2(t) \to k^2 \neq 0$ as $t \to +\infty$. Thus the only possibility is say that $z'(t) \to +k$. But then this means that $z(t)$ is monotone increasing as $t \to +\infty$. Since $z(t) \to 0$ this means that $z(t) \to 0$ monotonically from below. This in turn implies that $z'(t) \to 0$. Similarly if $z'(t) \to -k$.

To sum up the only possibility is that $z'(t) \to 0$.

Since both $z(t)$ and $z'(t) \to 0$ as $t \to +\infty$, necessarily $x_1(t) \to x(t)$ and $x_1'(t) \to x'(t)$. This proves our theorem.

§ 6. The Differential Equation

$$x'' + f(x, x')\,x' + g(x) = e(t)\,.$$

17. This type of equation without forcing term was first discussed by Levinson and Smith [1], and with forcing term by Levinson [2]. We shall follow here more closely the latter, and in particular utilize an interesting device due to Levinson.

Contrary to what has been done in § 3, we shall first consider the autonomous system, as the appropriate boundedness theorem is simpler and constitutes an introduction to the other case.

We take then the equation

$$d^2x/dt^2 + f(x, dx/dt)\,dx/dt + g(x) = 0\,, \tag{17.1}$$

or equivalently

$$dx/dt = y,\ dy/dt = -f(x, y)\,y - g(x)\,. \tag{17.2}$$

Setting as before

$$G(x) = \int_0^x g(x)\,dx\,,$$

we impose the following conditions:

I. *Throughout their ranges $f(x, y)$ and $g(x)$ are continuous and each satisfies a Lipschitz condition;*

II. *for x large $g(x)$ has the sign of x and $|g(x)|$ increases monotonically to infinity with $|x|$ and*

$$\frac{g(x)}{G(x)} = O\left(\frac{1}{|x|}\right);$$

III. *there exists an $a > 0$ such that $f(x, y) \geq M > 0$ for $|x| \geq a$ and $f(x, y) \geq -m$, $m > 0$ for $|x| \leq a$.*

Since one may enlarge a at will, also in view of I, II we may have:

(17.3) $|g(x)| \leq \delta$, $\delta > 0$, *in the strip* $|x| \leq a$ *and* $|g(x)| \geq \beta > 0$ *for* $|x| \geq a$.

Let us also show that

(17.4) *The system (17.2) satisfies a Lipschitz condition throughout the phase-plane. Hence the existence theorem and path uniqueness hold throughout the plane.*

Assuming that one remains in a bounded region of the plane: $|x|, |y| \leq A$, we must prove that in that region

$$(17.5) \quad |y - y_1| + |f(x, y)\, y + g(x) - f(x_1, y_1)\, y_1 - g(x_1)| < k\{|x - x_1| + |y - y_1|\}.$$

In the region in question by assumption

$$(17.6) \quad \begin{aligned} &|g(x) - g(x_1)| < \lambda\, |x - x_1|, \\ &|f(x, y) - f(x_1, y_1)| < \mu\{|x - x_1| + |y - y_1|\}. \end{aligned}$$

We also have for (x_1, y_1) in the region $|f(x_1, y_1)| < \nu$.

Now

$$\begin{aligned} |y f(x, y) - y_1 f(x_1, y_1)| &= |y\,(f(x, y) - f(x_1, y_1)) \\ &\quad + (y - y_1) f(x_1, y_1)| \\ &< A\mu\{|x - x_1| + |y - y_1|\} + \nu\, |y - y_1|. \end{aligned}$$

Upon combining with (17.6) we arrive easily at (17.5).

18. (18.1) THEOREM. *There exists a bounded closed 2-cell Ω such that:* (a) *all the paths cross the boundary J of Ω inward;* (b) *all solutions tend to Ω as $t \to +\infty$.*

(18.2) COROLLARY. *There exists a positive R such that for t large every solution* $x(t)$, $y(t)$ *of* (17.2) *satisfies* $|x|, |y| \leq R$.

The only critical points of the system are the places where $g = 0$ on the x axis and they occur only in the interval $|x| < a$. Let us suppose that $xg(x) \neq 0$ for $x \neq 0$, and of course $g(0) = 0$. Thus the origin is the only critical point. Then

(18.3) COROLLARY. *If the origin is the only critical point and it is unstable, the system* (17.2) *possesses at least one stable limit-cycle. Equivalently also the system* (17.1) *has a stable periodic solution.*

For one may in fact apply to Ω with the origin removed the theorem (X, 9.3) of Poincaré.

(18.4) *Application.* To simplify matters let f and g be analytic at the origin and let us suppose that $f(0, 0) = \lambda$, $g = \mu x + \ldots$, and that g has no zeros except the origin for $|x| \leq a$. Then the first approximation is

$$dx/dt = y, \quad dy/dt = -\mu x - \lambda y.$$

The characteristic equation is

$$r^2 + \lambda r + \mu = 0.$$

Hence if $\mu > 0$ and $\lambda < 0$ the origin is an unstable node or focus and there are limit-cycles.

19. The proof of the theorem rests directly upon the construction of Ω.

We shall require the constant energy curves

$$y^2/2 + G(x) = u \tag{19.1}$$

as well as the curve

$$y = -g(x)/M \tag{19.2}$$

taken to the right of the strip $|x| \leq a$. Let us also make an estimate of the slope of the paths in the strip. We find from (17.2)

$$dy/dx = -f - g(x)/y.$$

Hence

$$|dy/dx| \leq m + \delta/|y|. \tag{19.3}$$

Upon taking $|y|$ sufficiently large we shall therefore have in the strip

$$|dy/dx| \leqq 2m = m_0. \tag{19.4}$$

We will set $h = 2m_0 a$.

The region Ω is now constructed as follows. The references are to fig. 11. Take the point 3 so far down that: (a) the point 2

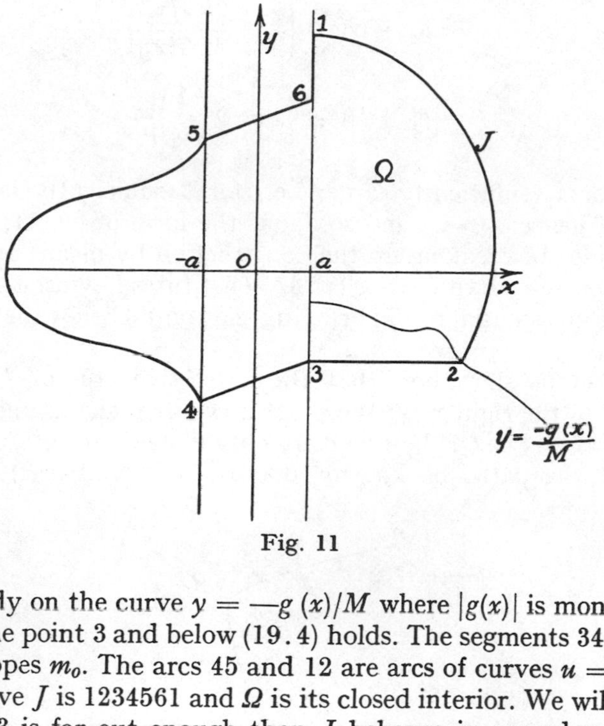

Fig. 11

is already on the curve $y = -g(x)/M$ where $|g(x)|$ is monotonic; (b) at the point 3 and below (19.4) holds. The segments 34 and 56 have slopes m_0. The arcs 45 and 12 are arcs of curves $u =$ const. The curve J is 1234561 and Ω is its closed interior. We will prove that if 3 is far out enough then J behaves in accordance with the theorem. For convenience we will denote by $\doteq$ equality modulo a constant or a quantity that $\to 0$ as $3 \to \infty$. Let us also write x_i, y_i, u_i to denote the values of x, y, u at the point i. We have at once

$$u_1 = u_2 = 1/2\, y_2{}^2 + G(x_2)\,;\; u_3 \doteq 1/2\, y_2{}^2\,;\; u_4 \doteq 1/2\,(|y_2| + h)^2$$
$$= u_5 \doteq 1/2\, y_5{}^2\,.$$

Thus

$$y_5 \doteq |y_2| + h\,.$$

Then

$$u_6 \doteq 1/2\,(|y_2| + 2h)^2 \doteq 1/2\,y_2^{\,2} + 2h\,|y_2|\,.$$

Hence

$$\begin{aligned} u_1 &\doteq u_6 + G(x_2) - 2h\,|y_2| \\ &= u_6 + G(x_2)\left\{1 - \frac{2h}{M}\frac{g(x_2)}{G(x_2)}\right\} \\ &= u_6 + G(x_2)\left\{1 - o\left(\frac{1}{x_2}\right)\right\}. \end{aligned}$$

Thus for x_2 sufficiently large, i.e., for 3 sufficiently far down $u_1 > u_6$, hence $y_1 > y_6$, and so J has the form of fig. 11.

The idea of breaking up the construction by means of energy curves on the curve $y = -g(x)/M$ constitutes Levinson's device. It plays an essential role in bringing the point 6 below the point 1.

20. Let us show now that the successive arcs of J are all crossed in the right way. We must however make a preliminary remark. Let PQ, QR be two consecutive open arcs of J (fig. 12) and let the paths be known to cross (PQ) and (QR) inward.

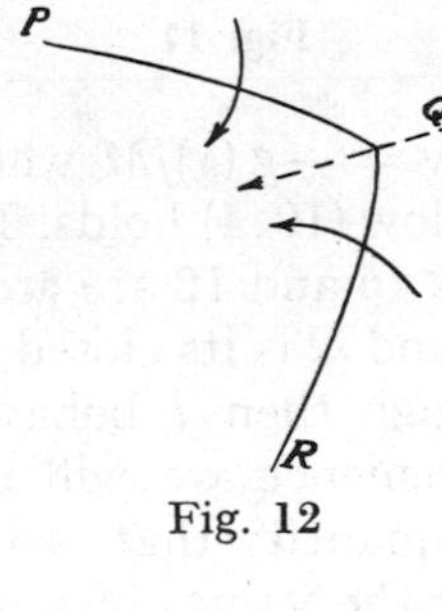

Fig. 12

Then the crossing at Q must be inward also. This is seen at once by reference to a path rectangle with its axis on the path through Q. We will refer to this as "extension by continuity."

We proceed now with the successive arcs of J.

Arcs 12 and 45. We find from (17.2) that along a path

$$du = -fy^2\, dt\,.$$

Since $f \geq M$ for $|x| \geq a$, u decreases steadily for $|x| \geq a$ and so the two arcs in question are both crossed inward.

Arc 23. Along this arc $x' < 0$. On the other hand except at 2:

$$y' = -f(x, y_2)\, y_2 - g(x)$$

$$= f(x, y_2)\,(-y_2) - g(x) > -My_2 - g(x) = 0\,.$$

Hence y' points upward and so the vector (x', y') points inward. At the point 2 the vector points along 23 towards 3. Thus along the whole of [23] the paths enter Ω.

Arcs 34 and 56. Since (19.4) holds and $x' > 0$ [< 0] above [below] the x axis, the pointing is inward along [34) and [56]. Since it is also inward on (45] by continuity it is inward at 4.

Arc 61. Since $x' > 0$ above the x axis the pointing is inward on [61) and by continuity likewise at 1.

Thus J has the proper behavior.

21. Let us suppose that the construction has been carried out for $x_2 = x_2^0$ and that $g(x)$ is monotone increasing for $x > x_2^0$. Let Ω, J correspond to x_2^0 and let Ω_ϱ, J_ϱ correspond to $x_2 = \varrho > x_2^0$ The family $\{J_\varrho\}$ is manifestly continuous and fills up the exterior U of J. Let a path Γ start at some point P of U. As time increases $\varrho(P)$ decreases monotonically. Let it have a lower limit $\sigma > x_2^0$. This means that $\Gamma \to J_\sigma$ from the outside and hence that it spirals around J_σ. In particular very near the arc 12 of J_σ, $du/dt \to 0$ along Γ. This contradicts, however, the result obtained: $du/dt = -fy^2$, $|du/dt| > My^2$, which at points of 12 away from the x axis will be decidedly $\neq 0$. Thus $\sigma = x_2^0$ and Γ enters J. This completes the proof of Theorem (18.1).

22. We shall now take up the equation with forcing term $e(t)$:

$$d^2x/dt^2 + f(x, dx/dt)\, dx/dt + g(x) = e(t)\,, \tag{22.1}$$

or in equivalent form

$$dx/dt = y,\ dy/dt = -f(x, y)\, y - g(x) + e(t)\,. \tag{22.2}$$

The assumptions are the same as in (17) with the sole addition of:

V. *$e(t)$ is continuous and $|e(t)| \leq E$ for all t.*

Properties (17.3) and (17.4) continue to hold with proofs unchanged. We shall also prove:

(22.3) *Theorem* (18.1) *continues to hold.*

We state as:

(22.4) COROLLARY. *If $e(t)$ is periodic and has period τ then there exists a solution of period τ.*

The proof of this corollary is quite simple. If we define the transformations $S(t)$ and S as in (15) then $S(t)\Omega \subset \Omega$, hence $S\,\Omega \subset \Omega$. Hence S has a fixed point in Ω (Brouwer's fixed point theorem) and so, as at the end of (15), there is a periodic solution of period τ.

23. There remains then to prove (22.3). Here again everything will center around the construction of an appropriate Jordan curve J, boundary of the closed 2-cell Ω.

Observe first that the inequality (19.3) still holds provided that δ is replaced by $\delta + E$. Hence (19.4) continues to hold provided that $|y|$ is large enough.

Setting now $b = E/M$ we draw, as in fig. 13, the curve $y = -(g(x)/M) - b$, to the right of the strip $|x| \leq a$. The construction of the curve $J = 1, 2, \ldots, 12,1$ is then started at the point 5 far out on that curve. The arcs 12, 45, 78, 10–11, are arcs of curves $u =$ const.; the arcs 34, 9–10, 12–1 are vertical segments; the arcs 67 and 11–12 have slope m_0; the arc 56 is a horizontal segment; as for 23 and 89 they are arcs of curves $u = \lambda x + \mu$ defined and dealt with below.

Arc 23. The point 2 is determined from 3 by the condition

$$G(x_3) - G(x_2) = 3/2\, b^2 . \tag{23.1}$$

Then λ, μ are determined by the relations asserting that the arc $u = \lambda x + \mu$ joins (x_2, b) to $(x_3, 0)$ or

$$b^2/2 + G(x_2) = \lambda x_2 + \mu,\ G(x_3) = \lambda x_3 + \mu . \tag{23.2}$$

This yields

$$\lambda = \frac{b^2}{x_3 - x_2} . \tag{23.3}$$

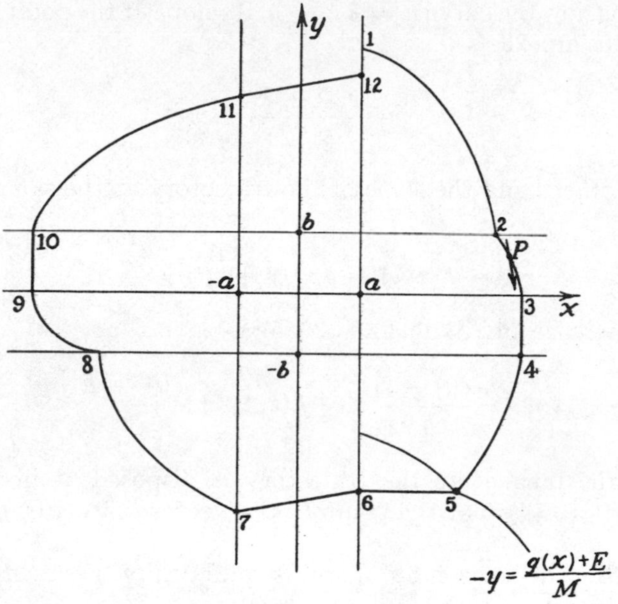

Fig. 13

Now the mean value theorem yields from (23.1)

$$g(x_2') \cdot (x_3 - x_2) = 3/2\, b^2,\ x_2 \leq x_2' \leq x_3 .$$

Hence

$$(x_3 - x_2) = \frac{3}{2}\frac{b^2}{g(x_2')} \leq \frac{3}{2}\frac{b^2}{g(x_2)} . \tag{23.4}$$

From (23.1) follows that $x_2 \to +\infty$ with x_3. For if say x_2 remained $\leq \xi$ then (23.1) would imply

$$G(x) \leq 3/2\, b^2 + G(\xi)$$

as $x \to +\infty$ which is incorrect since $G(x) \to +\infty$ with x.

Since $g(x_2) \to +\infty$ with x_2, hence with x_3, (23.4) implies that $x_3 - x_2 \to 0$ as $x_3 \to +\infty$. It follows then from (23.3) that $\lambda \to +\infty$ with x_3. In particular one may assume that x_3 is large enough i.e., that 5 is far out enough, to have $\lambda > E$.

Taking now the curve $u = \lambda x + \mu$, its slope at the point $P(x, y)$ along the arc 23 is

$$\frac{-g(x) + \lambda}{y}. \tag{23.5}$$

On the other hand the slope of the trajectory at the same point is

$$-f(x, y) - g(x)/y + e(t)/y . \tag{23.6}$$

The difference of (23.6) and (23.5) is

$$-f(x, y) + \frac{e(t) - \lambda}{y} < -f(x, y) + \frac{e(t) - E}{y} < 0 .$$

Hence the tangent to the trajectory is disposed as in fig. 13. Since $x'(t) = y > 0$ at the point P, the vector $(x'(t), y'(t))$ points inward at P.

Arc 89. The reasoning there is practically the same. We impose again

$$G(x_9) - G(x_8) = 3/2\, b^2$$

and take an arc of $u = \lambda_1 x + \mu_1$. We must have

$$b^2/2 + G(x_8) = \lambda_1 x_8 + \mu_1$$

$$G(x_9) = \lambda_1 x_9 + \mu_1 .$$

Hence

$$\lambda_1 = \frac{b^2}{x_9 - x_8} < 0 .$$

We show as before that $x_9 - x_8 \to 0$ with x_9 and hence for x_9 large, i.e., for 5 far out enough, one will have $\lambda_1 < -E$. One shows then as before that along 89 the trajectories point inward.

Arcs 34 and 9–10. These arcs did not appear in fig. 11 and so one must show that they are likewise crossed inward. This is immediate, however, since on 34: $x'(t) < 0$ and on 9–10: $x'(t) > 0$.

Arc 12–1. Assuming that the figure is drawn correctly, i.e., that 12 is below 1, on 12–1: $x'(t) > 0$ and so the crossing is again inward.

For the remaining arcs the proof that the crossing is always inward is the same as in the previous case (see 20) and so it need not be repeated.

All that is now left is to prove:

(23.7) *The point* (12) *is below the point* 1.

We have at once

$$u_1 = u_2 \doteq u_3 \text{ (by 23.1) } \doteq u_4 = u_5;$$

$$u_7 \doteq \frac{(|y_5| + h)^2}{2} \doteq u_8 \doteq u_9 \doteq u_{10} = u_{11} \doteq \frac{y^2_{11}}{2}.$$

Hence

$$y_{11} \doteq |y_5| + h = |y_5| + h' \text{ (h' bounded)};$$

$$u_{12} \doteq 1/2\,(|y_5| + 2h')^2 \doteq 1/2\, y_5{}^2 + 2h'\,|y_5|$$

$$\doteq 1/2\, y_5{}^2 + \frac{2\,h' g(x_5)}{M}.$$

On the other hand

$$u_1 \doteq u_5 = 1/2\, y_5{}^2 + G(x_5).$$

Therefore

$$u_1 - u_{12} = \frac{y_1{}^2 - y_{12}{}^2}{2} \doteq G(x_5)\left\{1 - \frac{2h'\, g(x_5)}{M\, G(x_5)}\right\} \to G(x_5) > 0$$

which becomes unbounded as x_5 increases. Hence if the point 5 is taken sufficiently far out (23.7) will hold and fig. 13 will be correct.

The proof that all trajectories penetrate a certain Ω is the same as in (21). This completes the proof of (22.3).

§ 7. A Special Differential Equation

$$x'' + g(x) = \mu \sin \omega t.$$

24. We shall only consider this equation under the hypothesis

that μ is small and above all that $g(x)$ is an odd polynomial with positive coefficients:

$$g(x) = 2a_2 x + 4a_4 x^3 + \dots + (2n + 2)\, a_{2n+2}\, x^{2n+1}.$$

Thus

$$G(x) = a_2 x^2 + a_4 x^4 + \dots + a_{2n+2}\, x^{2n+2}.$$

The graph of $u = G(x)$ is then of the type of a parabola turned upward. Let (ξ, u) be a point of this parabola. Then one may write

$$u - G(x) = (\xi^2 - x^2)\, H(x),$$

and one finds by direct division that

$$H(x) = a_{2n+2}\, x^{2n} + (a_{2n+2}\, \xi^2 + a_{2n})\, x^{2n-2} + \dots + (a_{2n+2}\, \xi^{2n} + \dots + a_2). \tag{24.1}$$

Consider first the equation

$$d^2x/dt^2 + g(x) = 0, \tag{24.2}$$

or equivalently

$$dx/dt = y,\ dy/dt = -g(x). \tag{24.3}$$

As we have seen a first integral is the closed path

$$y^2/2 + G(x) = u \tag{24.4}$$

and the period of the oscillation corresponding to (24.4) is

$$T = \frac{4}{\sqrt{2}} \int_0^{\xi} \frac{dx}{\sqrt{u - G(x)}}. \tag{24.5}$$

The quantity ξ is the amplitude of the oscillation (24.4). The change of variable $x = \xi \sin\theta$ replaces (24.5) by

$$T(\xi) = \frac{4}{\sqrt{2}} \int_0^{\frac{\pi}{2}} \frac{d\theta}{\sqrt{H(\sin\theta)}}, \tag{24.6}$$

$$H(\sin\theta) = a^{2n+2}\,\xi^{2n}\sin^{2n}\theta + (a_{2n+2}\,\xi^2 + a_{2n})\,\xi^{2n-2}\sin^{2n-2}\theta$$
$$+ \ldots + (a_{2n+2}\,\xi^{2n} + \ldots + a_2)\,.$$

This expression calls for a very simple observation. As ξ increases so do the coefficients of the powers of $\sin\theta$ in $H(\sin\theta)$. Hence the integrand for a given value of θ decreases and so does $T(\xi)$. It is clear that $T(\xi) \to 0$ monotonically as $\xi \to +\infty$. On the other hand as $\xi \to 0$, $T(\xi)$ increases monotonically, its limiting value being $T_0 = 2\pi/\sqrt{2a_2}$. Or explicitly:

(24.7) *As ξ increases from* 0 *to* $+\infty$, $T(\xi)$ *decreases monotonically from* $T_0 = 2\pi/\sqrt{2a_2}$ *to zero. Notice that for the special type of function* $G(x)$ *considered one verifies the assertion of* (1) *regarding the period: it varies here monotonically with the amplitude. Moreover the upper bound* T_0 *of the period depends solely upon the linear term* $2a_2x$ *of* $g(x)$.

Finally since $T(\xi)$ is monotone we have:

(24.8) *Given any* T_1, $0 < T_1 < T_0$, *there exists a solution of* (24.2) *or equivalently of* (24.3) *whose period is* T_1 *and this solution is unique.*

25. Consider now the particular equation

$$d^2x/dt^2 + g(x) = \mu \sin \omega t \tag{25.1}$$

or equivalently the system

$$dx/dt = y,\ dy/dl = -g(x) + \mu \sin \omega t\,. \tag{25.2}$$

We will prove with Shimizu [1]:

(25.3) THEOREM. *Given any* $T_1 < T_0$ *and* $\omega = 2\pi/T_1$, *then if* μ *is sufficiently small there exists a solution of the system* (25.1) *which has the period* T_1.

(25.4) *Remark.* Strictly speaking Shimizu only considers polynomials $G(x)$ of degree four. This enables him to have recourse to elliptic functions. Furthermore he assumes erroneously that the period has an infinite upper bound. In the main however, the crux of our argument is the same as his.

Proof of (25.3). Since $T_1 < T_0$ there exists a unique closed path of (24.3) described in time T_1. Let $x^*(t)$, $y^*(t)$ be the particular solution associated with this path such that $x^*(0) = 0$ and let

$$\eta^* = y^*(0),\ \xi^* = x^*(T_1/4),\ 0 = y^*(T_1/4)\,.$$

By Poincaré's theorem (II, 9.1) for η near η^* and μ small the solution of (25.2) which passes through $(0, \eta)$ at time $t = 0$ is analytic in η, μ and t. Let it be $x(\eta, \mu, t)$, $y(\eta, \mu, t)$. At the first crossing (near $(\xi^*, 0)$) we will have

$$y(\eta, \mu, t) = 0\,. \tag{25.5}$$

The partial $\partial y/\partial t$ for $\eta = \eta^*$, $\mu = 0$ and $t = T_1/4$ is given by

$$(dy/dt)_{T_1/4} = -g(\xi^*) \neq 0\,.$$

Hence (25.5) has a unique solution in t: $1/4\ T(\eta, \mu)$ such that $T(\eta^*, 0) = T_1$ and this solution is valid for $\eta - \eta^*$ and μ small. In particular $T_1 = T(\eta^*, 0)$ is the number that was designated by $T(\xi^*)$ in (24). Since η^* is monotone increasing with ξ^*, we see from (24.7) that $T(\eta^*, 0)$ is monotone decreasing with η^*.

Let now $0 < \eta' < \eta^* < \eta''$ and set $T(\eta', 0) = T'$, $T(\eta'', 0) = T''$. Thus by the remark just made $T'' < T_1 < T'$. One may take μ so small that

$$|T(\eta', \mu) - T'| < 1/2\,(T' - T_1),$$
$$|T(\eta'', \mu) - T''| < 1/2\,(T_1 - T'')\,.$$

As a consequence

$$T(\eta'', \mu) < T_1 < T(\eta', \mu)\,.$$

Hence as μ is kept fixed and η varies from η' to η'', $T(\eta, \mu)$ will cross the value T_1. The corresponding solution

$$x(\eta, \mu, t),\ y(\eta, \mu, t) \tag{25.6}$$

has then the property that

$$x(\eta, \mu, T_1/4) = \xi,\ y(\eta, \mu, T_1/4) = 0\,.$$

Let us show that this solution has effectively the period T_1.

Upon making in (25.2) the change of variables $t' = \pi/\omega - t =$

$T_1/2 - t$ and $y' = -y$, the system remains the same. It has therefore the solution

$$(25.7) \qquad x(\eta, \mu, T_1/2 - t), \; -y(\eta, \mu, T_1/2 - t).$$

But for $t = T_1/4$ this solution assumes the same values ξ, 0 as the solution (25.6). Therefore (25.7) is merely the continuation of (25.6). In particular upon setting $t = 0$ in the two expressions the result must be the same. Comparing the values obtained we find the relations

$$(25.8) \qquad x(\eta, \mu, T_1/2) = 0, \; y(\eta, \mu, T_1/2) = -\eta.$$

Similarly the change of variables

$$x' = -x, \; y' = -y, \; t' = \pi/\omega + t = T_1/2 + t$$

made in (25.2) does not change the form of the system, so that

$$(25.9) \qquad -x(\eta, \mu, T_1/2 + t), \; -y(\eta, \mu, T_1/2 + t)$$

is a solution. Since for $t = 0$ this solution and (25.7) assume the same values 0, η, the solution (25.9) is a continuation of the solution (25.7) and therefore also of (25.6). Comparing the values of (25.6) and (25.9) for $t = T_1/2$ we find in view of (25.8):

$$x(\eta, \mu, T_1) = 0, \; y(\eta, \mu, T_1) = \eta.$$

That is to say after augmenting t by the period T_1 of $\sin \omega t$ the solution (25.6) returns to its initial position. Therefore it has the period T_1 and the theorem is proved.

(25.10) *Subharmonics*. The meaning of this term may be described in the following manner. The solution say of (25.1) whose period is T_1 may be expanded in Fourier series and the series will contain a "harmonic" or terms in $\sin \omega t$, $\cos \omega t$ and generally also the "higher" harmonics or terms in $\sin n\omega t$, $\cos n\omega t$, i.e., of period T_1/n. By contrast there may exist solutions of period nT_1, $n > 1$, and such solutions are known as subharmonics. We prove in effect:

(25.11) *The system* (25.1) *or equivalently* (25.2) *has, for* μ *sufficiently small, a subharmonic solution whose period is* $(2n + 1) T_1$

for all integers n such that $(2\,n+1)\,T_1 < T_0$ *and the subharmonic solutions corresponding to distinct values of n and* μ *small enough are distinct.*

Following once more Shimizu we select this time a solution of (24.3) whose period $T_1{}^* = (2\,n + 1)\,T_1$ and the same reasoning as before yields the desired subharmonic. This subharmonic starts on the x axis very near the point $\xi\,(\,(2\,n + 1)\,T_1)$. Since these points, in finite number are all distinct, so are the various subharmonics under consideration.

§ 8. A Special Differential Equation

$$x'' + f(x)\,x' + g(x) = e(t).$$

26. It is clear that the difficulties attendant upon the construction of the region Ω in § 4 have much to do with the great generality of the basic equation (7.1) under consideration. We have shown in [3] that by imposing decidedly sharper restrictions it was possible to obtain a region Ω which is merely the interior of a large ellipse. See also in this respect a note by Urabe [1] and extensions by Antosiewicz [1] and by Graffi [1]. The following treatment is closely related to our Note [3]. As in our Note no attempt will be made to exploit the method to the full.

We consider then the equation in the title or the equivalent system

$$dx/dt = y - F(x) + E(t), \quad dy/dt = -g(x) \tag{26.1}$$

where the notations are those of (7). Our present assumptions are I and III of (7) and in addition for x large:

IV. $g(x)$ *and* $F(x)$ *have the sign of* x;

V. $g(x) = O\,(|x^{2n+1}|)$, *where n is a positive integer*;

VI. *there exist a number* $b > 0$ *such that* $(b\,g\,(x) - F(x)\,) = o(|x|)$.

(Here $O[o]$ mean: is of order of [less than] for x large.)

Under these conditions we prove:

(26.2) *There exists an elliptical region* Ω *which is entered by all the trajectories of the system.*

We shall be writing certain relations in x and y which are to be taken for $r = |x| + |y|$ large. An equality modulo terms of lower order than the maximum will be denoted by $\doteq$.

Consider then the family of similar ellipses defined by

$$ax^2 - 2\,xy + by^2 = 2\,v,\ a > 0,\ ab > 1\,.$$

Along a trajectory

$$(26.3)\qquad dv/dt = (ax - y)\,dx/dt + (by - x)\,dy/dt$$
$$= (ax - y)\,(y - F(x) + E\,(t)\,) + (x - by)\,g(x)\,.$$

We have now three possibilities:

(a) $x = O(1)$. Then for r large y is large also and so the sign of (26.3) is that of $-y^2$, i.e. it is *negative.*

(b) x becomes large and $y = O(x)$. From (26.3) there follows

$$dv/dt = (ax - y)\,(y - bg(x) - (F - bg(x) + E(t)\,) - (by - x)\,g(x)$$
$$\doteq \{(x - by) - b(ax - y)\}\,g(x) = (1 - ab)\,xg(x)$$

which is again *negative* for x large.

(c) x becomes large and $y > O(x)$. From (26.3) we see that

$$dv/dt \doteq -y\,(y - F(x) + bg(x)\,) \doteq -y^2$$

which is once more *negative.*

Thus in all cases when r and hence v is sufficiently large v decreases along the trajectories and this clearly proves (26.2).

We note the corollaries:

(26.4) *All the solutions penetrate and remain in a bounded 2-cell.*

(26.5) *If* $E(t) = 0$ *and the origin satisfies the condition* $f(0) < 0$ (V *of* 14) *then there exists at least one limit-cycle* (same proof as 14.3).

(26.6) *If* $E(t)$ *has the period* τ *there is a solution with the same period* τ.

§ 9. Certain Periodic Systems Investigated by Gomory.

27. Almost all the equations dealt with so far fall under a common pattern: the oscillations are established by means of the Brouwer fixed point theorem applied to a finite two-cell whose boundary is crossed inward by all trajectories. In his

noteworthy thesis, Gomory [1] succeeded in establishing the existence of oscillations, using primarily the vector-field index but in a different manner. The paper is too long for a full description and so we shall merely present a reasonably extensive résumé.

Basically, Gomory considers a system of the form

$$(27.1) \qquad \begin{cases} \dfrac{dx}{dt} = X(x, y) + E_1(t) \\ \dfrac{dy}{dt} = Y(x, y) + E_2(t) \end{cases}$$

where X, Y are polynomials of the same degree $n > 0$ and E_1, E_2 continuous periodic functions of period $T > 0$. The success of Gomory's method rests essentially upon the fact that X, Y *are* polynomials, for his results are based entirely upon the nature of the critical points at *infinity* of the associated system

$$(27.2) \qquad \frac{dx}{dt} = X(x, y), \quad \frac{dy}{dt} = Y(x, y).$$

The first theorem of Gomory is:

(27.3) THEOREM. *If the critical points at infinity of* (27.2) *satisfy the following conditions*: (a) *they are all elementary*; (b) *their index sum* $\neq 1$; (c) *there are no consecutive saddle points, then the system* (27.1) *has a periodic solution of period* T.

Of course this is an entirely novel periodicity theorem with very simple restrictions upon the polynomials X, Y.

The general method of proof is more or less this. Let x, y, z be projective coordinates with $z = 0$ as the line L at infinity. Each critical point P is enclosed in a rectangular box on one side of L, containing in its interior no other critical point. The boxes are "locally small" and consecutive boxes are joined by an arc without contact not too far from L. If P is a node, the boundaries of its box are such that they are penetrated in the same manner by the trajectories of (27.1). If P is a saddle point, the consecutive box is not, and the boundaries of the two, together with the intermediary arc, are joined into a curve crossed in the same sense by the trajectories. Boundary pieces and joining arcs are united into a *finite* Jordan curve J with the noteworthy property:

(d) *If a trajectory of* (27.1) *passes through a point* Q *of* J *at time* t *it will never pass again through* Q *at any later time.*

Take now, J so large that it contains all the finite critical points of (27.2). It is then shown that on J the two vector distributions (X, Y) and $[X + E_1(t), Y + E_2(t)]$ (the latter for any fixed t) have the same index. By virtue of property (b) it is then found that the transformation $t \to t + T$ has a fixed point inside J. There results a periodic solution, whose trajectory Γ, by property (d), cannot cross J and therefore lies in the finite plane. The solution in question is thus an ordinary periodic solution of period T and the theorem follows.

In the proof, in order to make the system (27.1) operate properly in the whole projective plane, Gomory used the following interesting device. Instead of (27.1) he considers

$$\text{(27.4)} \qquad \begin{cases} \dfrac{dx}{dt} = [X(x, y) + E_1(t)]\, F(x, y) \\[2ex] \dfrac{dy}{dt} = [Y(x, y) + E_2(t)]\, F(x, y) \end{cases}$$

where $F = 1$ within and on a large circle C containing J and $\to 0$ continuously as $x^2 + y^2 \to \infty$ outside C. The system (27.4) has the same trajectories as (27.1) and coincides with it within and on C. Thus one may use (27.4) to calculate indices.

28. The argument proceeds to the study of an equation

$$\text{(28.1)} \qquad \frac{d^2x}{dt^2} + f(x)\frac{dx}{dt} + g(x) = E(t)$$

where $E(t)$ has a continuous derivative and has period T, and f, g are real polynomials. More precisely

$$f(x) = a_0 + a_1 x + \ldots + a_m x^m$$
$$g(x) = b_0 + b_1 x + \ldots + b_n x^n$$

where $m \geqq n > 0$. The basic result is:

(28.2) THEOREM. *If* n *is odd and we do not have both* m *odd and* $b_n > 0$ *then* (28.1) *has a periodic solution of period* T.

While the restriction to polynomials for the functions f, g is rather severe, it is hardly necessary to underscore how far we have come from the other periodicity results discussed in the chapter. Thus, far from being dissipative, $f(x)$ could have any sign for x large.

As a step in the proof one must first analyze the critical points at infinity of the associated system

$$\frac{dx}{dt} = y, \quad \frac{dy}{dt} = -f(x)\, y - g\,(x). \tag{28.3}$$

If $m > n$ this system does not fall under the one already discussed but the direct study of the critical points at infinity is quite feasible.

In the first place there are only two: P (0, 1, 0) and Q (1, 0, 0). In any case, intervals of the line L at infinity: $z = 0$ between critical points are paths.

Consider first the point P. Let $f^H(x, z)$ and $g^H(x, z)$ stand for "homogenized" $f(x)$ and $g(x)$. The (x, z) equations are, with new time t,

$$\frac{dz}{dt} = z\,(f^H + z^{m-n+1}g^H),$$

$$\frac{dx}{dt} = x\,(f^H + z^{m-n+1}g^H) + z^m.$$

From the relation

$$x\frac{dz}{dt} - z\frac{dx}{dt} = z^{m+1}$$

one infers that the infinite line L is the only possible tangent to the paths at the point P. There are basically two sectors made up of the two sides of L. A detailed study, in which one may systematically apply the procedure of (X, 22), leads to the conclusion that either P is a (complicated) node or else one of the sectors is elliptic and the other hyperbolic.

For the point Q (1, 0, 0) the appropriate system is

$$\frac{dz}{dt} = -yz^{m+1},$$

$$\frac{dy}{dt} = -f^H(1, z)\, y - z^{m-n+1}g^H(1, z) - z^m y^2.$$

The lowest degree term in dy/dt is $-a_m y$, and since $a_m \neq 0$ the system is of Bendixson type. The critical point may in fact be any one of the three Bendixson types (X, 17.2): node, half node and half saddle point, saddle point.

The analysis of each situation is made to rest upon the signs of a_m, b_n and the parities of m, n.

Returning now to the initial equation (28.1), or to its equivalent (28.3), Gomory shows that whenever the conditions of (28.2) are fulfilled one may apply a treatment analogous to that of (27) to complete the proof of the theorem.

CHAPTER XII

Oscillations in Systems of the Second Order: Methods of Approximation

We have already discussed in the preceding chapter systems of the second order containing a variable parameter. In the present chapter we shall mainly concentrate on the quasi-harmonic systems

$$(a) \qquad d^2x/dt^2 + \omega^2 x = \mu f(x, dx/dt), \mu f(x, dx/dt, t)$$

where μ is a small parameter, or exceptionally, a large parameter. The term "quasi-harmonic" refers of course to the fact that when $\mu = 0$, (a) becomes the harmonic equation, or equation of harmonic oscillations.

The endeavor will be to obtain approximate solutions of various types. Notable among these is Poincaré's method of small parameters, or as it is also known: method of perturbation —the small term μf being thought of as a perturbation of the harmonic oscillation. Poincaré's method is indeed exact; it offers a solution as a power series in the parameter μ. The "approximation" feature comes in through the fact that in any practical problem one must be content with a few, very few, terms of the series.

We shall examine in some detail the application to autonomous systems or self-oscillations, to Mathieu's equation and to forced oscillations.

We shall also discuss a true approximation—the method of van der Pol and of Krylov-Bogoliubov. Finally we shall study the limiting positions of limit-cycles in systems of Liénard type for large values of the parameter.

References: Andronov-Chaikin [1]; Krylov-Bogoliubov [1]; Lasalle [1]; Minorsky [1]; Poincaré [1]; van der Pol [1], [2]; Stoker [1].

§ 1. Self-Excited Systems

1. The term "self-excited" as applied to an autonomous oscillatory system refers to the fact that an oscillation is induced without any external agency acting upon the system. The basic equation of a self-excited "quasi-harmonic" system is

$$d^2x/dt^2 + \omega^2 x = \mu f(x, dx/dt).$$

With the change in the unit of time $\omega t \to t$ the equation can be given the form

$$d^2x/dt^2 + x = \mu f(x, dx/dt) \tag{1.1}$$

which has the equivalent companion system

$$\begin{cases} dx/dt = y\,, \\ dy/dt = -x + \mu f(x, y)\,. \end{cases} \tag{1.2}$$

The parameter μ is supposed to be small and one may as well assume $\mu > 0$, since if $\mu < 0$ one may merely replace it by $-\mu$ at the cost of replacing f by $-f$.

To simplify matters we will make the following assumptions as to f:

I. *f is a polynomial and its expansion about the origin is of the form $f = ax + by + \dots$, with $b \neq 0$;*

II. *$\varphi(x) = f(x, 0)/x$ has a fixed sign or is identically zero.*

Since for $\mu = 0$ the paths of the system are all closed, one may hope that for μ small some paths will be closed and it is on the discovery of these closed paths that attention will be focused.

At all events the system (1.2) has only one critical point: the origin. The characteristic equation corresponding to the origin is

$$\begin{vmatrix} -r, & 1 \\ a\mu - 1, & b\mu - r \end{vmatrix} = r^2 - b\mu r + (1 - a\mu) = 0\,.$$

When μ is small the characteristic roots are complex and their real part has the sign of b. Thus the origin is asymptotically stable [unstable] if $b < 0$ [$b > 0$]. Therefore there is no closed path arbitrarily near the origin—the origin is certainly not a center since $b \neq 0$.

2. To find the periodic solutions tending to a solution of the harmonic system as $\mu \to 0$, we apply the general method of Poincaré as described in (VIII, §§ 5, 6). The first step is to expand the general solution in series of powers of μ

$$(2.1)\qquad \begin{cases} x = A_0\,(x_0, y_0; t) + \mu\, A_1\,(x_0, y_0; t) + \dots \\ y = A_0'\,(x_0, y_0; t) + \mu\, A_1'\,(x_0, y_0; t) + \dots \end{cases}$$

where the primes designate time derivatives, and x_0, y_0 is the position of x, y at time $t = 0$. For a time $> 2\pi$ and sufficiently small μ the solution (2.1) will be within a small neighborhood of a circumference—a path of the harmonic system for $\mu = 0$:

$$(2.2)\qquad dx/dt = y,\ dy/dt = -x\,.$$

Hence the path will cross the y axis say at $(0, \eta)$. If time be counted from that moment then $x_0 = 0, y_0 = \eta$ and so (2.1) becomes

$$(2.3)\qquad \begin{cases} x = A_0\,(\eta, t) + \mu\, A_1\,(\eta, t) + \dots, \\ y = A_0'\,(\eta, t) + \mu\, A_1'\,(\eta, t) + \dots, \end{cases}$$

where the A_i, A_i' are analytic in η. For $t = 0$ we must have $x = 0, y = \eta$ whatever μ. Hence

$$(2.4)\qquad \begin{cases} A_o\,(\eta, 0) = 0, \dots, A_n\,(\eta, 0) = 0, \dots, \\ A_0'\,(\eta, 0) = \eta, \dots, A_n'\,(\eta, 0) = 0, \dots, \end{cases}$$
$$n = 1, 2, \dots.$$

Since $A_0(\eta, t), A_0'(\eta, t)$ is merely the solution through $(0, \eta)$ at time $t = 0$, of the harmonic system (2.2), we have

$$A_0(\eta, t) = \eta \sin t, \; A_0'(\eta, t) = \eta \cos t,$$

so that (2.3) reads now

$$(2.5) \qquad \begin{cases} x = \eta \sin t + \mu A_1(\eta, t) + \ldots, \\ y = \eta \cos t + \mu A_1'(\eta, t) + \ldots . \end{cases}$$

3. Since the system is autonomous there is no reason to assume that a periodic solution very near the solution $\eta \sin t$, $\eta \cos t$ has exactly the period 2π. Let its period be $2\pi + \tau(\mu)$ where, as we shall see, τ is analytic in μ. It causes the following indirect attack which is simpler than the general method outlined in (VIII, 9.9). If we substitute x, y from (2.5) in μf, and apply (III, 10.5), we obtain for the solution of (1.2) the relations

$$(3.1) \qquad \begin{cases} x = \eta \sin t + \mu \displaystyle\int_0^t f(x, y) \cdot \sin(t - u)\, du \\ y = \eta \cos t + \mu \displaystyle\int_0^t f(x, y) \cdot \cos(t - u)\, du \end{cases}$$

where under the integration signs x, y are the solutions expressed as functions of u.

Expressing the fact that (3.1) has the period $2\pi + \tau$ we obtain the basic relations:

$$(3.2) \qquad H(\eta, \mu, \tau) = \eta \sin \tau + \mu \int_0^{2\pi + \tau} f(x, y) \sin(\tau - u)\, du = 0$$

$$(3.3) \quad K(\eta, \mu, \tau) = \eta(\cos \tau - 1) + \mu \int_0^{2\pi+\tau} f(x, y) \cos(\tau - u)\, du = 0.$$

This is a real analytical system for determining η, τ as functions of μ. Let us suppose that $\eta(0) = a \neq 0$. Then (3.2) yields $\tau(0) = 0$.

It is readily found that the Jacobian

$$\left| \frac{\partial(H, K)}{\partial(\eta, \tau)} \right|_{(a,\, 0,\, 0)} = 0$$

and so we cannot apply the implicit function theorem to obtain solutions $\eta(\mu)$, $\tau(\mu)$ in the neighborhood of $(a, 0, 0)$. However since

$$\left(\frac{\partial H}{\partial \tau}\right)_{(a,\,0,\,0)} \neq 0\,,$$

one may solve (3.2) for τ as a holomorphic function $\tau(\eta, \mu)$ in the neighborhood of $(a, 0)$. We find also directly from (3.2) that $\partial\tau/\partial\eta = 0$ for $\mu = 0$. Hence

$$\tau(\eta, \mu) = B_1(\eta)\,\mu + B_2(\eta)\,\mu^2 + \ldots \tag{3.4}$$

where the $B_h(\eta)$ are holomorphic in a certain circular region $|\eta - a| < \beta$.

It will be convenient to have the value of $B_1(\eta)$ later. We find at once

$$B_1(\eta) = \left(\frac{\partial \tau(\eta, \mu)}{\partial \mu}\right)_{(\eta,\,0)}.$$

Now by differentiating (3.2) as to μ with $\tau = \tau(\eta, \mu)$ there comes:

$$\frac{\partial H}{\partial \mu} = \eta \cos \tau \cdot \frac{\partial \tau}{\partial \mu} + \int_0^{2\pi+\tau} f(x, y) \sin(\tau - u)\, du + \mu[\quad] = 0\,,$$

where the value of the square bracket is omitted since it is not needed. Since $\tau(\eta, 0) = 0$, we have

$$B_1(\eta) = \frac{1}{\eta}\int_0^{2\pi} f(\eta \sin u, \eta \cos u) \sin u\, du\,. \tag{3.5}$$

4. Once $\tau(\eta, \mu)$ is known it is substituted in (3.3) which is then replaced by

$$K(\eta, \mu, \tau(\eta, \mu)) = 0\,. \tag{4.1}$$

The left-hand side is a series in $(\eta - a)$ and μ containing no term independent of μ. Thus

$$K(\eta, \mu, \tau(\eta, \mu)) = \mu\, K_1(\eta, \mu)\,,$$

and the solution $\eta(\mu)$ of (4.1) such that $\eta(0) = a$, will satisfy

$$K_1(\eta, \mu) = 0\,. \tag{4.2}$$

It also yields $\tau(\eta(\mu), \mu)$ as analytic in μ.

In order that (4.2) possess a solution of this nature we require

$$K_1(a, 0) = 0. \tag{4.3}$$

Now

$$K_1(a, 0) = \left(\frac{\partial K(\eta, \mu, \tau(\eta, \mu))}{\partial \mu}\right)_{(a,\,0)}$$

$$= \int_0^{2\pi} f(a \sin u, a \cos u) \cos u \, du,$$

and so we must have

$$\Phi(a) = \int_0^{2\pi} f(a \sin u, a \cos u) \cos u \, du = 0. \tag{4.4}$$

Since f is a polynomial so is Φ. Thus a must be a real root of a certain algebraic equation.

Notice that if a satisfies (4.4) so does $-a$. Therefore $\Phi(a) = a^s \Psi(a^2)$, with Ψ a polynomial. Thus if Φ has any real root a it also has the positive root $|a|$. It will therefore be sufficient to confine our attention in the sequel to the positive roots of $\Phi(a)$.

5. Choosing now a definite positive root a of Φ, we may expand $K_1(\eta, \mu)$ as a power series in $\eta - a$ and μ and we will have

$$K_1(\eta, \mu) = A(a)(\eta - a) + B(a)\mu + \ldots, \tag{5.1}$$

and to discuss (4.2) and stability we require information about $A(a)$. We have

$$A(a) = \left(\frac{\partial K_1}{\partial \eta}\right)_{(a,\,0)} = \frac{1}{\mu}\left(\frac{\partial K(\eta, \mu, \tau(\eta, \mu))}{\partial \eta}\right)_{(a,\,0)}. \tag{5.2}$$

Now

$$\frac{\partial K(\eta, \mu, \tau(\eta, \mu))}{\partial \eta} = \frac{\partial K(\eta, \mu, \tau)}{\partial \eta} + \frac{\partial K(\eta, \mu, \tau)}{\partial \tau}\frac{\partial \tau}{\partial \eta}$$

$$= \cos \tau - 1 - \eta \sin \tau \cdot \frac{\partial \tau}{\partial \eta} + \mu \, [f(x, y)]_{t=2\pi+\tau} \cdot \frac{\partial \tau}{\partial \eta} +$$

$$\mu \int_0^{2\pi+\tau} \left\{ \frac{\partial f}{\partial x}\left(\frac{\partial x}{\partial \eta} + \frac{\partial x}{\partial \tau}\frac{\partial \tau}{\partial \eta}\right) + \right.$$

$$\left. \frac{\partial f}{\partial y}\left(\frac{\partial y}{\partial \eta} + \frac{\partial y}{\partial \tau}\frac{\partial \tau}{\partial \eta}\right)\right] \cos(\tau - u) - f(x, y) \sin(\tau - u) \left.\frac{\partial \tau}{\partial \eta}\right\} du.$$

Hence from (5.2) and since by (3.4) $\frac{\partial \tau}{\partial \eta} = 0$ for $\mu = 0$:

$$A(a) = \int_0^{2\pi} \left(\frac{\partial f}{\partial x}\frac{\partial x}{\partial a} + \frac{\partial f}{\partial y}\frac{\partial y}{\partial a}\right) \cos u \, du\,,$$

where under the integration sign $f = f(x, y)$, and $x = a \sin u$, $y = a \cos u$. Therefore finally

$$A(a) = \Phi'(a)\,. \tag{5.3}$$

Consider now any simple positive root a of Φ, that is to say we suppose that (4.4) holds and that

$$\Phi'(a) \neq 0\,, \tag{5.4}$$

and we have the corresponding development (5.1). Under the circumstances in view of (5.4), the implicit function theorem asserts that (4.2) has a unique analytic solution $\eta(\mu)$ in the neighborhood of $\mu = 0$, such that $\eta(0) = a$. Substituting then in (3.4) we obtain a similar series $\tau(\mu)$ such that $\tau(0) = 0$. The two functions $\eta(\mu)$, $\tau(\mu)$ verify (3.2), (3.3) and hence the corresponding solution $(x(\eta(\mu), \mu, t))$, $y(\eta(\mu), \mu, t))$ represents a closed path of (1.2) which tends to the circle Δ_a: $x^2 + y^2 = a^2$ when $\mu \to 0$. We may also say that $x(\eta(\mu), \mu, t)$ represents the oscillatory solution of (1.1) which tends to $a \sin t$ when $\mu \to 0$. In other words under the circumstances there is a unique periodic solution of the desired type. The circle Δ_a is known as the *generating circle* of the closed path.

We have tacitly assumed the solutions to be real. It is only necessary to observe that the determination of the coefficients of the series involved never makes an appeal to any irrational operations. Hence the coefficients are all real and so are the series.

There remains to discuss the question of stability. Let $\gamma_{\eta,\mu}$

be the path of (1.2) passing through $(0, \eta)$ and let us assume $\eta - \alpha$ and μ small. Under the circumstances the solution $\gamma_{\eta, \mu}$ will cross the positive axis at a point $(0, \eta_1)$ at a time $2\pi + \tau(\eta, \mu)$, where $\tau(\eta, \mu)$ is the solution (3.4) of $H(\eta, \mu, \tau) = 0$. Let

$$\delta(\eta, \mu) = \eta_1 - \eta.$$

We have at once

$$\delta(\eta, \mu) = K(\eta, \mu, \tau(\eta, \mu)) = \mu K_1(\eta, \mu) = (\eta - \alpha) P(\eta) + \mu Q(\eta, \mu),$$

where P, Q are holomorphic in the neighborhood of $(\alpha, 0)$ and $P(\alpha) = A(\alpha) = \Phi'(\alpha) \neq 0$.

We are interested in the behavior of $\gamma_{\eta, \mu}$ relative to the closed path when the latter is very near Δ_α, i.e., when μ is arbitrarily small. At all events let us choose μ which is assumed positive so that $0 < \mu < (\eta - \alpha)^2$. Then for $(\eta - \alpha)$ sufficiently small $\delta(\eta, \mu)$ has the sign of $(\eta - \alpha)\ \Phi'(\alpha)$. Therefore when $\Phi'(\alpha) > 0$, δ is positive for $\eta > \alpha$ and negative for $\eta < \alpha$, while for $\Phi'(\alpha) < 0$ the reverse takes place. Thus when $\Phi'(\alpha) > 0$ if $\gamma_{\eta, \mu}$ starts from a point M outside Δ_α on the positive y axis, it will intersect the latter at a point M' above M (fig. 1). On the other hand if $\gamma_{\eta, \mu}$ starts from M inside Δ_α on the positive y axis, it will intersect

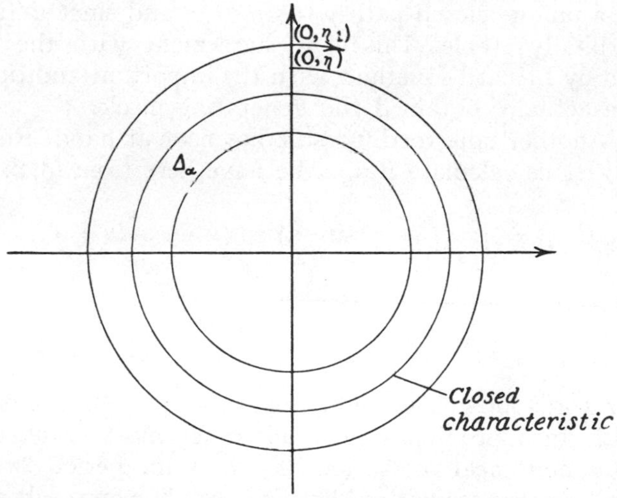

Fig. 1

the latter at a point M' below M. Hence $\Phi'(a) > 0$ implies instability, and clearly when $\Phi'(a) < 0$ we have orbital asymptotic stability.

To sum up we have proved:

(5.5) THEOREM. *The generating circles* Δ_a: $x^2 + y^2 = a^2$ *of the closed paths of* (1.2) *correspond to the real positive roots of* $\Phi(a)$. *If* a *is such a root and* $\Phi'(a) \neq 0$, *then there is one and only one closed path* γ_μ *tending to* Δ_a *when* $\mu \to 0$, *and it is orbitally asymptotically stable or else unstable accordingly as* $\Phi'(a)$ *is negative or positive. Correspondingly there exists a unique oscillatory solution of* (1.1) *(to within an arbitrary phase* θ*) which tends to the harmonic oscillation* $a \sin (t + \theta)$ *when* $\mu \to 0$, *and whose period* $2\pi + \tau(\mu)$ *tends at the same time to* 2π.

(5.6) *Application to van der Pol's equation.* Here $f = (1 - x^2)\, y$ and so

$$\Phi(a) = \int_0^{2\pi} (1 - a^2 \sin^2 u)\, a \cos^2 u \, du = \pi a \left(1 - \frac{a^2}{4}\right),$$

$$\Phi'(a) = \pi \left(1 - \frac{3a^2}{4}\right).$$

Hence there is just one generating circle Δ_2 $(a = 2)$. It is the limit of a unique closed path γ_μ as $\mu \to 0$, and since $\Phi'(2) < 0$, γ_μ is orbitally stable. This is in agreement with the results obtained by Liénard's method, with the important addition that we have actually obtained the generating circle.

(5.7) Another noteworthy result lies near at hand. Referring to (3.4), let us calculate $B_1(\eta)$. We have here from (3.5)

$$B_1(\eta) = \int_0^{2\pi} (1 - \eta^2 \sin^2 u) \cos u \sin u \, du = 0\,,$$

and therefore

$$\tau(\eta, \mu) = \mu^2 \{B_2(\eta) + B_3(\eta)\, \mu + \dots\}\,.$$

In other words for small μ and η near a, $\tau = O(\mu^2)$. One may say that with an approximation sufficient for most practical purposes, the non-linear oscillation has the same period 2π as the limiting harmonic oscillation. This is a well known result already brought out by van der Pol.

§ 2. Forced Oscillations

6. We shall now consider a system

$$(6.1) \qquad d^2x/dt^2 + x = \mu f(x, dx/dt, t)$$

where f has the period 2π in t. The parameter μ is still positive and small. The quantity μf represents then a periodically acting forcing term, hence "forced" oscillations. Any periodic solution must have then a period commensurable with 2π. In point of fact we shall mainly discuss solutions of period 2π. The method followed is that of (VIII, 10).

It is assumed here that $f(x, y, t)$ is a polynomial in x, y, $\sin t$, $\cos t$.

The general solution of (6.1) or of the equivalent system

$$(6.2) \qquad dx/dt = y,\ dy/dt = -x + \mu f(x, y, t),$$

with the initial position ξ, η at $t = 0$ is given by a pair of series

$$(6.3) \qquad \begin{cases} x = A_o(\xi, \eta, t) + \mu A_1(\xi, \eta, t) + \ldots, \\ y = A_o'(\xi, \eta, t) + \mu A_1'(\xi, \eta, t) + \ldots. \end{cases}$$

Substituting in (6.2) and identifying like powers of μ there is obtained the recurrent system

$$(6.4) \qquad \begin{cases} A_o'' + A_o = 0, \\ A_1'' + A_1 = f_1(A_0, A_0', t), \\ \cdots\cdots\cdots \\ A_n'' + A_n = f_n(A_0, A_0', \ldots, A_{n-1}, A_{n-1}', t), \\ \cdots\cdots\cdots\cdots\cdots\cdots \end{cases}$$

with the initial conditions

$$A_o(0) = \xi,\ A_o'(0) = \eta;\ A_n(0) = A_n'(0) = 0 \text{ for } n > 0.$$

The function f_n, $n > 0$, is a polynomial in $A_0, \ldots, A_{n-1}'$, $\cos t$, $\sin t$, $A_n(t)$ is a polynomial in t, $A_n(\xi, \eta; \sin t, \cos t; t)$ one in the five variables, and $A_o(t) = \xi \cos t + \eta \sin t$.

7. The solution (6.3) just obtained is a general solution of (6.2) with the independent parameters ξ, η. Suppose now that we seek a solution of period 2π, or a so-called *harmonic* solution. We must have then

$$x(2\pi) = x(0),\ y(2\pi) = y(0)\,. \tag{7.1}$$

Upon substituting the solution (6.3) in (7.1) and setting for simplicity

$$A_n(\xi, \eta, 2\pi) = B_n(\xi, \eta),\ A_n'(\xi, \eta, 2\pi) = C_n(\xi, \eta),\ n > 0\,,$$

one obtains a system

$$\begin{cases} H(\xi, \eta, \mu) = B_1(\xi, \eta) + \mu B_2(\xi, \eta) + \ldots = 0\,, \\ K(\xi, \eta, \mu) = C_1(\xi, \eta) + \mu C_2(\xi, \eta) + \ldots = 0\,. \end{cases} \tag{7.2}$$

This is a so-called "algebroid" system and there is no difficulty in discussing its solution completely (see VIII, § 8). However, this would scarcely be profitable here and we shall confine our attention to the following special case:—The algebraic curves $B_1(\xi, \eta) = C_1(\xi, \eta) = 0$ of the ξ, η plane intersect in points such that if ξ_0, η_0 is one of them

$$\left|\frac{\partial(H, K)}{\partial(\xi_0, \eta_0)}\right|_{\mu=0} = \left|\frac{\partial(B_1, C_1)}{\partial(\xi_o, \eta_o)}\right| \neq 0\,. \tag{7.3}$$

This makes it possible to apply the implicit function theorem to (7.2) and obtain a solution

$$\xi^* - \xi_o = g(\mu),\ \eta^* - \eta_o = h(\mu) \tag{7.4}$$

where g and h are ascending power series in μ and $g(0) = h(0) = 0$. Correspondingly the general solution (6.3) of (6.2) will be periodic.

(7.5) *Subharmonics*. Suppose that f has actually the period $2\pi/n$, $n > 1$. This will be the case for instance if f is a polynomial in cos nt and sin nt. Then the same procedure that we have used will continue to yield a solution with the period 2π, i.e., n times the period of the forcing term. This solution is then in an evident sense a subharmonic.

8. *Stability of the periodic solution.* Let us consider the stability of the periodic solution corresponding to (7.4). We cannot use the method of (VIII, 11) since the matrix P there considered has pure imaginary characteristic roots. What we shall do really is to apply a variant of Poincaré's method of sections (VIII, 5).

If one starts at a point $P(\xi, \eta)$ at time $t = 0$ near $F(\xi^*, \eta^*)$ and reaches at time $t = 2\pi$ a point $P'(\xi', \eta')$ then the periodic solution will be orbitally stable if the ratio of distances $P'F/PF$ is < 1 and unstable if it is > 1.

Now the coordinates of P' are $x(\xi, \eta, \mu, 2\pi)$, $y(\xi, \eta, \mu, 2\pi)$ and we have

$$(8.1)\quad \begin{cases} x(\xi, \eta, \mu, 2\pi) - \xi^* = \xi - \xi^* + \mu B_1(\xi, \eta) + \dots, \\ y(\xi, \eta, \mu, 2\pi) - \eta^* = \eta - \eta^* + \mu C_1(\xi, \eta) + \dots. \end{cases}$$

Since ξ^*, η^* correspond to a periodic solution

$$x(\xi^*, \eta^*, \mu, 2\pi) = \xi^*, \; y(\xi^*, \eta^*, \mu, 2\pi) = \eta^*.$$

Hence

$$\mu B_1(\xi^*, \eta^*) + \mu^2 B_2(\xi^*, \eta^*) + \dots = 0,$$
$$\mu C_1(\xi^*, \eta^*) + \mu^2 C_2(\xi^*, \eta^*) + \dots = 0,$$

and therefore (8.1) may be written

$$(8.2)\quad \begin{cases} x(\xi,\eta,\mu,2\pi) - \xi^* = \xi - \xi^* + \mu(B_1(\xi,\eta) - B_1(\xi^*,\eta^*)) + \dots, \\ y(\xi,\eta,\mu,2\pi) - \eta^* = \eta - \eta^* + \mu(C_1(\xi,\eta) - C_1(\xi^*,\eta^*)) + \dots, \end{cases}$$

Let $B_{1\xi}, \dots,$ denote the partials of $B_1, \dots,$ as to $\xi, \dots$ and set $B_{1\xi_o} = b$, $B_{1\eta_o} = b'$, $C_{1\xi_o} = c$, $C_{1\eta_o} = c'$ $\xi - \xi^* = \Delta\xi$, $\eta - \eta^* = \Delta\eta$. Finally to indicate the fact that terms in $\mu^2, (\Delta\xi)^2, \dots,$ are dropped we shall replace $=$ by $\doteq$. We then have

$$B_1(\xi, \eta) - B_1(\xi^*, \eta^*) \doteq (\xi - \xi^*) B_{1\xi^*} + (\eta - \eta^*) B_{1\eta^*}.$$

Recalling now that $\xi^* - \xi_o = O(\mu)$, $\eta^* - \eta_o = O(\mu)$, we have:

$$B_1(\xi, \eta) - B_1(\xi^*, \eta^*) = (\xi - \xi^*)\{B_{1\xi_o} + (\xi^* - \xi_o) B_{1\xi_o{}^2}$$
$$+ (\eta^* - \eta_o) B_{1\xi_o\eta_o} + O(\mu^2)\} + (\eta - \eta^*)\{B_{1\eta_o}$$
$$+ (\xi^* - \xi_o) B_{1\xi_o\eta_o} + (\eta^* - \eta_o) B_{1\eta_o{}^2} + O(\mu^2)\},$$

and similarly with C_1 in place of B_1. Hence (8.2) yields

$$(8.3)\qquad \begin{cases} x(\xi, \eta, \mu, 2\pi) - \xi^* \doteq (1 + \mu b)\,\Delta\xi + \mu b'\,\Delta\eta\,, \\ y(\xi, \eta, \mu, 2\pi) - \eta^* \doteq \mu c \Delta\xi + (1 + \mu c')\,\Delta\eta\,. \end{cases}$$

If the roots of

$$(8.4)\qquad \begin{vmatrix} 1 + \mu\, b - r, & \mu\, b' \\ \mu\, c & , 1 + \mu\, c' - r \end{vmatrix} \doteq 0$$

are both < 1 in absolute value the periodic solution under consideration is orbitally asymptotically stable; if unity separates the absolute values the periodic solution is orbitally conditionally stable; if the absolute values are both greater than one it is unstable. For this comparison one may clearly replace, for μ small, (8.4) by

$$(8.5)\qquad \begin{vmatrix} 1 + \mu\, b - r, & \mu\, b' \\ \mu\, c & , 1 + \mu\, c' - r \end{vmatrix} = 0.$$

If we set $\dfrac{r - 1}{\mu} = s$ then s satisfies

$$(8.6)\qquad \begin{vmatrix} b - s, & b' \\ c & , c' - s \end{vmatrix} = 0,$$

which is the characteristic equation of the Jacobian matrix

$$(8.7)\qquad \frac{\partial (B_1, C_1)}{\partial (\xi_0, \eta_0)}.$$

Let s_1, s_2 be the roots of (8.6) and $r_j = 1 + \mu\, s_j$, $j = 1, 2$ those of (8.5). Suppose first s_1 and s_2 real and of the same sign. Since $\mu^2\, s_1\, s_2 = (r_1 - 1)\,(r_2 - 1)$, r_1 and r_2 are on the same side of 1. Since $(r_1 - 1) + (r_2 - 1) = s_1 + s_2$, if s_1 and s_2 are both negative [positive] r_1 and r_2 are both < 1 [> 1] and the periodic solution is orbitally asymptotically stable [is unstable].

Suppose now s_1 and s_2 complex: $s_1 = \alpha + i\beta$, $s_2 = \alpha - i\beta$. Then $|r_1|^2 = |r_2|^2 = (1 + \mu\, \alpha)^2 + \mu^2\, \beta^2 = 1 + 2\mu\, \alpha + O(\mu^2)$. Hence if $a < 0\,[> 0]$ we have $|r_1|, |r_2| < 1\,[> 1]$ and so the periodic solution is again orbitally asymptotically stable [is unstable].

Finally if s_1, s_2 are real and of opposite sign then r_1, r_2 are separated by unity and we have orbital conditional asymptotic stability.

To sum up then we may state:

(8.8) *If the real parts of the characteristic roots s_1, s_2 of the Jacobian matrix* (8.7) *are both negative [positive] the periodic solution generated by the circle $x = \xi_0 \cos t$, $y = \eta_0 \sin t$ is orbitally asymptotically stable [is unstable]; if s_1, s_2 are of opposite sign the periodic solution is conditionally orbitally asymptotically stable.*

Referring to (IX, 6.1) we may also state directly:

(8.9) *Let*

$$J = \left|\frac{\partial (B_1, C_1)}{\partial (\xi_0, \eta_0)}\right|, \; S = B_{1\xi_0} + C_{1\eta_0}\,.$$

The periodic solution is orbitally asymptotically

stable when $J > 0,\ S < 0$;

conditionally stable when $J < 0$;

it is unstable when $J > 0,\ S > 0$.

(8.10) *Generalization.* Suppose that $p \geq 1$ is the first integer such that $B_p(\xi, \eta)$ and $C_p(\xi, \eta)$ are not both identically zero. If ξ_0, η_0 is a point of intersection of the curves

$$B_p(\xi, \eta) = 0, \; C_p(\xi, \eta) = 0$$

such that

$$J_p(\xi_0, \eta_0) = \left|\frac{\partial (B_p, C_p)}{\partial (\xi_0, \eta_0)}\right| \neq 0$$

then everything that we have said goes through as before with J replaced by J_p.

9. *The stroboscopic method according to Minorsky.* In some recent articles Minorsky [2], [3], has developed a rapid process of approximation for dealing with periodic solutions. The general idea may be explained thus:—If a phenomenon has a slowly

varying small period differing little from T, then by a flash illumination at times T, $2T$, ..., one sees it varying slowly and one may examine it more leisurely. This is the stroboscopy of the physicist. Minorsky as it were examines the motion solely at times 2π, 4π, ... In other words he starts from the system (8.1), writes

$$\Delta\xi_1 = x - \xi,\ \Delta\eta_1 = y - \eta\,,$$
$$\Delta\xi = \xi - \xi_0,\ \Delta\eta = \eta - \eta_0\,,$$

and gives (8.1) the aspect:

$$\Delta\xi_1 = \mu(b\Delta\xi + b'\Delta\eta + \ldots) + \mu^2(\quad) + \cdots,$$
$$\Delta\eta_1 = \mu\,(c\Delta\xi + c'\Delta\eta + \ldots) + \mu^2(\quad) + \cdots.$$

Since μ and $\Delta\xi, \ldots$, are all small the variable μ is assimilated to a time differential $d\tau$. Set $\Delta\xi = u$, $\Delta\eta = v$. Neglecting higher degree terms the system takes the well known form

$$\text{(9.1)} \qquad \begin{cases} du/d\tau = bu + b'v\,, \\ dv/d\tau = cu + c'v\,. \end{cases}$$

If the characteristic roots of the matrix

$$\begin{pmatrix} b, & b' \\ c, & c' \end{pmatrix}$$

have negative real parts the vector $du/d\tau$, $dv/d\tau$: variation of u, v after one period, points back to ξ_0, η_0 and we have orbital stability. If the real parts are both positive instability takes place, while if the roots are of opposite signs conditional orbital stability takes place. Thus the orbital stability properties are reduced to those of the differential system (9.1). This is essentially the result developed in (8).

10. *Applications.* A number of practical applications and their consequences are discussed in Minorsky [1], pp. 323–352. We borrow the simplest: a van der Pol system with forcing term,

$$\text{(10.1)} \quad d^2x/dt^2 + \mu\,(x^2 - 1)\,dx/dt + x = \mu\,a\sin t,\ a \neq 0\,.$$

We set then

$$(10.2) \qquad \begin{cases} x = A_o\ (\xi, \eta, t) + \mu\, A_1\ (\xi, \eta, t) + \ldots, \\ y = A_o'\ (\xi, \eta, t) + \mu\, A_1'\ (\xi, \eta, t) + \ldots. \end{cases}$$

It is already known that

$$A_o = \xi \cos t + \eta \sin t,\ A_o' = -\xi \sin t + \eta \cos t\,.$$

We must then determine A_1 by:

$$(10.3) \qquad A_1'' + A_1 = -\frac{1}{3}\frac{d}{dt}(A_o^3 - 3A_o) + a \sin t\,,$$

with the initial conditions

$$(10.4) \qquad A_1(0) = A_1'(0) = 0\,.$$

By means of the classical relations

$$4 \cos^3 t = \cos 3t + 3 \cos t,\ 4 \sin^3 t = -\sin 3t + 3 \sin t$$

we find

$$(10.5) \quad A_o^3 - 3A_o = 3M \cos t + 3N \sin t + P \cos 3t + Q \sin 3t$$

where

$$(10.6) \quad M = \frac{\xi}{4}(\varrho^2 - 4),\ N = \frac{\eta}{4}(\varrho^2 - 4),\ \varrho^2 = \xi^2 + \eta^2\,,$$

and the expressions of P and Q are not required. As a consequence (10.3) becomes

$$(10.7) \quad A_1'' + A_1 = (a + M) \sin t - N \cos t + P \sin 3t - Q \cos 3t = f(t)\,.$$

We have now from (III, 10.5), and since $A_1(0) = A_1'(0) = 0$, the values

$$A_1(t) = \int_0^t f(u) \sin (t-u)\, du,\ A_1'(t) = \int_0^t f(u) \cos (t-u)\, du\,.$$

Hence the periodicity conditions $A_1(2\pi) = A_1'(2\pi) = 0$ yield here

$$(10.8)\qquad \begin{cases} B_1(\xi, \eta) = \int_0^{2\pi} f(t) \sin t \, dt = -\pi(M + \alpha) = 0\,, \\ C_1(\xi, \eta) = \int_0^{2\pi} f(t) \cos t \, dt = -\pi N = 0\,, \end{cases}$$

that is to say

$$(10.9)\qquad \xi(\varrho^2 - 4) + 4\alpha = 0, \quad \eta(\varrho^2 - 4) = 0\,.$$

The only solution is given by

$$(10.10)\qquad \eta = 0, \quad \xi(\xi^2 - 4) + 4\alpha = 0\,.$$

The Jacobian matrix

$$(10.11)\qquad \frac{\partial(B_1, C_1)}{\partial(\xi, \eta)} = \begin{pmatrix} 3\xi^2 - \eta^2 + 4, & -2\xi\eta \\ 2\xi\eta & , -\xi^2 - 3\eta^2 + 4 \end{pmatrix}.$$

Its characteristic roots are $4 - 3\xi^2$, $4 - \xi^2$, where ξ is a root of (10.10), and they are both negative if $\xi^2 > 4$. Thus the periodic motion corresponding to (ξ, η) defined by (10.10) is orbitally stable when and only when $\xi^2 > 4$. We have to determine there-

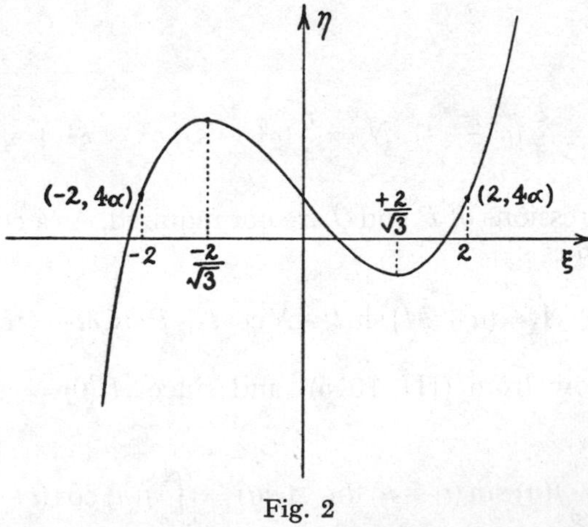

Fig. 2

fore the position relative to the segment $-2 \leq \xi \leq +2$, of the roots of $f(\xi) = \xi(\xi^2 - 4) + 4a$. This is best done by means of the graph of $f(\xi)$. Its extrema are at $\xi = \pm 2/\sqrt{3}$ and $f(-2) = f(+2) = 4a$. The graph has thus the appearance of fig. 2. Since upon varying a the graph is merely translated vertically, it is clear that there is one and only one root outside the segment, hence one and only one orbitally stable oscillation.

11. As a second application of the perturbation method let us consider an example studied by Reuter [1]. The equation under discussion is

$$(11.1)\quad d^2x/dt^2 + a\mu^2 dx/dt + (1 + b\mu^2)x(1 + \mu x) = 3c\mu\cos 2t,$$

where $a, b, c > 0$, arises in the theory of loudspeakers. We propose to show the existence of a stable subharmonic of period 2π, i.e., double the period π of the forced oscillation (subharmonic of order two).

We seek then a solution of the customary form

$$x = A_0 + \mu A_1 + \mu^2 A_2 + \dots,$$

and period 2π. The situation as regards A_0 is the same as before and so

$$A_0 = \xi\cos t + \eta\sin t.$$

We then have

$$A_1'' + A_1 = -A_0{}^2 + 3c\mu\cos 2t,\ A_1(0) = A_1'(0) = 0,$$

which yields

$$A_1 = -\frac{\beta\varrho^2}{2} + \left(\frac{\beta\varrho^2}{2} + \frac{R}{3}\right)\cos t + \frac{2}{3}(S - R)\sin t$$

$$-\frac{R}{3}\cos 2t - \frac{S}{3}\sin 2t,$$

$$\varrho^2 = \xi^2 + \eta^2,\ R = \frac{\eta^2 - \xi^2}{2} + 3c,\ S = -\xi\eta.$$

One verifies readily that

$$B_1(\xi, \eta) = C_1(\xi, \eta) \equiv 0\,,$$

and so in view of (8.10) one must calculate $B_2(\xi, \eta)$ and $C_2(\xi, \eta)$, i.e., consider A_2. It satisfies the equation

$$A_2'' + A_2 = -aA_0' - 2A_0A_1 - bA_0\,.$$

The formulas (III, 10.5) yield then after some simplification

$$\frac{A_2(2\pi)}{\pi} = \frac{B_2(\xi, \eta)}{\pi} = -a\,\xi + (b+c)\,\eta - \frac{5}{6}\eta\,\varrho^2\,,$$

$$\frac{A_2'(2\pi)}{\pi} = \frac{C_2(\xi, \eta)}{\pi} = -a\,\eta - (b-c)\,\xi + \frac{5}{6}\xi\,\varrho^2\,.$$

Let us set

$$\frac{6\,a}{5} = a^*,\ \frac{6\,b}{5} = b^*,\ \frac{6\,c}{5} = c^*, \frac{c}{a} = d.$$

Thus ξ, η are to satisfy

$$(11.2) \qquad \begin{cases} \eta\varrho^2 - (b^* + c^*)\,\eta + a^*\,\xi = 0\,, \\ \xi\varrho^2 - (b^* - c^*)\,\xi - a^*\,\eta = 0\,, \end{cases}$$

and the appropriate stability Jacobian is up to a positive numerical factor

$$(11.3) \qquad \begin{pmatrix} -2\,\eta\,\xi - a^*\,, & -(\xi^2 + 3\,\eta^2) + b^* + c^* \\ 3\,\xi^2 + \eta^2 - (b^* - c^*)\,, & 2\,\eta\xi - a^* \end{pmatrix}.$$

Leaving aside the obvious solution $\xi = \eta = 0$, we deduce from (11.2):

$$\varrho^2 = 2\,d\,\xi\eta,\ (\xi^2/\eta^2) - 2\,d\,(\xi/\eta) + 1 = 0\,,$$

and hence

$$\xi/\eta = d + \delta,\ \delta = \pm\sqrt{d^2 - 1}\,.$$

More or less laborious calculations yield now for the determinant of (11.3) the value

$$\Delta = -(b^* - a^*\delta)(b^* + 3a^*\delta) - 4c^*\delta(d+\delta)\varepsilon$$
$$+ (a^{*2} + b^{*2} - c^{*2}),$$
$$\varepsilon = \frac{(b^* - a^*\delta)}{1 + (d+\delta)^2}.$$

Let us take now δ with the — sign. The second term in the expression of Δ will then be positive. In order that the first term be positive we must have

$$|\delta| > |b^*/3a^*|, \text{ or } c^2/a^2 - 1 > b^2/9a^2, c^2 - a^2 > b^2/9.$$

In order that the third term be positive we require $b^2 > c^2 - a^2$. Thus $\Delta > 0$ if

(11.4) $$b^2/9 < c^2 - a^2 < b^2.$$

The sum of the diagonal terms in (11.3) is $-2a^*$ which is < 0 since $a > 0$.

If (11.4) holds there are two corresponding real solutions of (11.2) such as (ξ, η) and $(-\xi, -\eta)$, and they both correspond to orbitally asymptotically stable subharmonic periodic solutions of the given system (11.1) for μ small.

§ 3. Approximations for Quasi-Harmonic Systems

12. A noteworthy method of approximation for quasi-harmonic systems was introduced by van der Pol in connection with his equation. An interesting deviation was introduced later by Krylov and Bogoliubov and intensely exploited by them (see [1]). We will first describe the method on a general equation and then discuss an application.

Consider then the equation

(12.1) $$d^2x/dt^2 + \omega^2 x = f(x, dx/dt),$$

where f is in some sense essentially small. For example f could be a polynomial in x and dx/dt, whose coefficients are small. One does not require that f be of the form μf_1, with μ small: the smallness of f need not be controlled by a single multiplier.

Following van der Pol we endeavor to solve (12.1) or rather its equivalent system

$$dx/dt = y,\ dy/dt = -\omega^2 x + f(x, y) \tag{12.2}$$

by a substitution

$$\begin{aligned} &\text{(a)}\ x = \xi \cos \omega t + \eta \sin \omega t\,, \\ &\text{(b)}\ y = -\omega\, \xi \sin \omega t + \omega\, \eta \cos \omega t\,. \end{aligned} \tag{12.3}$$

If we had $\omega = 1$, ξ and η would merely be the coordinates of the point (x, y) relative to axes rotating around the origin with unit velocity. Since the motion is not to deviate too much from the harmonic it justifies the expectation that ξ and η vary but slowly with the time. Be it as it may, as a consequence of (12.3):

$$\xi' \cos \omega t + \eta' \sin \omega t = 0\,. \tag{12.4}$$

Differentiating again (12.3b) we have in view of (12.1):

$$-\omega\, \xi' \sin \omega t + \omega\, \eta' \cos \omega t = f(x, x')\,. \tag{12.5}$$

Hence by combining with (12.3):

$$\begin{cases} \dfrac{d\xi}{dt} = \dfrac{-1}{\omega} f(\xi \cos \omega t + \eta \sin \omega t, -\omega\, \xi \sin \omega t + \omega \eta \cos \omega t) \sin \omega t\,, \\[2ex] \dfrac{d\eta}{dt} = \dfrac{1}{\omega} f(\xi \cos \omega t + \eta \sin \omega t, -\omega\, \xi \sin \omega t + \omega\, \eta \cos \omega t) \cos \omega t\,. \end{cases} \tag{12.6}$$

Since f is small we find that indeed ξ and η vary but slowly with the time.

Apparently all that has been accomplished is to replace the autonomous system (12.2) by the non-autonomous (12.6) thus increasing the complication. Now comes van der Pol's main idea:—Since the right hand sides of (12.6) are small one may as well replace them by their time averages. This replaces (12.6) by an autonomous system

$$d\xi/dt = H(\xi, \eta),\ d\eta/dt = K(\xi, \eta) \tag{12.7}$$

to which one may hope to apply the methods of Poincaré. This is the essence of the story.

Instead of the tentative solution (12.3) Krylov and Bogoliubov try the solution

$$(12.8) \qquad x = a\cos(\omega t + \varphi)\,,\ y = -a\omega\sin(\omega t + \varphi)\,,$$

thus placing at the center of discourse the amplitude a and the phase φ (both all important in the applications) which are to vary slowly. Except for that the rest of their treatment is the same. One must say also that they consider higher approximations, as well as a noteworthy process of linearization which they have pushed very far indeed.

13. We shall now discuss an application which is not strictly of the preceding type but makes use of the van der Pol basic approximate solution (12.3). The example is a van der Pol equation

$$(13.1) \qquad d^2x/dt^2 + \mu(x^2 - 1)\,dx/dt + x = E\sin\omega t$$

with a forced oscillation that is not necessarily small. This equation was discussed by van der Pol [1] and much more amply by Andronov and Witt [1]. Considerable information, both mathematical and physical is given by Minorsky [1], p. 344, and by Stoker [1], p. 149. The importance of (13.1) as regards so-called *frequency entrainement* is fully developed in these various references.

We look then this time for a solution (12.3) with the ω of (13.1), i.e., of the forcing term and we now have (12.4) or

$$(13.2) \qquad \xi'\cos\omega t + \eta'\sin\omega t = 0$$

together with

$$-\xi'\sin\omega t + \eta'\cos\omega t = (\omega^2 - 1/\omega)(\xi\cos\omega t + \eta\sin\omega t)$$

$$-\frac{\mu}{3\omega}\frac{d}{dt}\{(\xi\cos\omega t + \eta\sin\omega t)^3 - 3(\xi\cos\omega t + \eta\sin\omega t)\}$$

$$+\frac{E}{\omega}\sin\omega t\,.$$

Referring to (10.5) this last relation becomes

(13.3) $$-\xi' \sin \omega t + \eta' \cos \omega t = \left(\frac{\omega^2 - 1}{\omega} \xi - \mu N\right) \cos \omega t$$

$$\left(+ \frac{\omega^2 - 1}{\omega} \eta + \mu M + \frac{E}{\omega}\right) \sin \omega t + \mu (P \sin 3\omega t - Q \cos 3\omega t) .$$

We now solve (13.2) and (13.3) for ξ' and η' and, with van der Pol, average at once over a period $2\pi/\omega$. The result is

(13.4) $$\begin{cases} 2\dfrac{d\xi}{dt} = -\dfrac{\omega^2-1}{\omega}\eta + \mu\dfrac{\xi(4-\varrho^2)}{4} - \dfrac{E}{\omega}, \\ 2\dfrac{d\eta}{dt} = \dfrac{\omega^2-1}{\omega}\xi + \mu\dfrac{\eta(4-\varrho^2)}{4}; \; \varrho^2 = \xi^2 + y^2 . \end{cases}$$

If we make the change of variables $\tau = \dfrac{\mu t}{8}$, and set $4(\omega^2 - 1)/\mu\omega = \sigma$, $4E/\mu\omega = F$, the system (13.4) assumes the simple form

(13.5) $$\begin{cases} d\xi/d\tau = -\sigma\eta + \xi(4 - \varrho^2) - F , \\ d\eta/d\tau = \sigma\xi + \eta(4 - \varrho^2) , \end{cases}$$

which is of the Poincaré type, and was given essentially by van der Pol [1]. The analysis that follows is based on Andronov and Witt [1].

The critical points of the system (13.5) satisfy

(13.6) $$\begin{cases} -\sigma\eta + \xi(4 - \varrho^2) = F , \\ \sigma\xi + \eta(4 - \varrho^2) = 0 . \end{cases}$$

Solving for ξ, η in terms of $r = \varrho^2$ we find

(13.7) $$\begin{cases} \xi = \dfrac{(4 - r) F}{(4 - r)^2 + \sigma^2}, \\ \eta = \dfrac{-\sigma F}{(4 - r)^2 + \sigma^2}. \end{cases}$$

On the other hand by squaring and adding the relations of (13.6) there comes

$$r(\sigma^2 + (4 - r)^2) = F^2 . \tag{13.8}$$

Thus to obtain the critical points one must solve (13.8) for its positive roots in r and substitute in (13.7).

The stability of a given critical point depends upon the Jacobian matrix of the right hand sides of (13.5) if its determinant is not zero. The matrix is found to be

$$\begin{pmatrix} 4 - r - 2\,\xi^2, & -\sigma - 2\,\eta\xi \\ \sigma - 2\,\eta\xi & , \ 4 - r - 2\,\eta^2 \end{pmatrix}.$$

Its determinant is

$$D(r, \sigma) = (4 - r)(4 - 3r) + \sigma^2 ,$$

and the sum of the diagonal terms is

$$S(r) = 4(2 - r) .$$

Thus we have approximately:

$D > 0$, $S(r) < 0$: orbitally stable periodic motion;

$D > 0$, $S(r) > 0$: unstable periodic motion;

$D < 0$: conditionally orbitally stable periodic motion.

Consider F as fixed and σ as variable. In the upper half of the

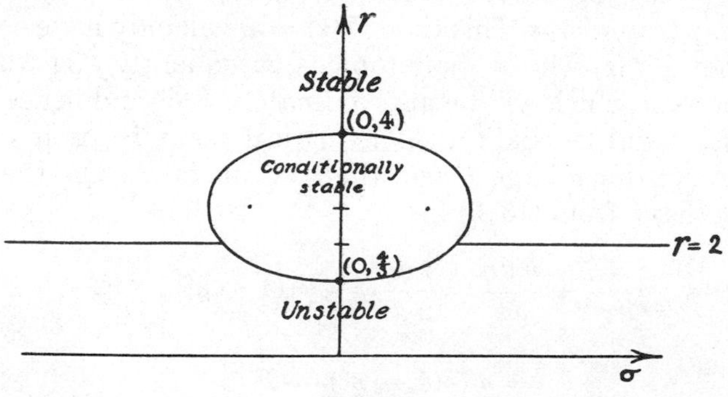

Fig. 3

plane σ, r the curve $D = 0$ represents an ellipse cutting the r axis at the points (0,4) and (0,4/3). The ellipse is also symmetrical with respect to the r axis. The length of the two semi-axes is $4\sqrt{3}/3$ and 4/3. The stability disposition is indicated in fig. 3.

14. An interesting observation may now be made regarding the paths of the system (13.4).

I. *Suppose σ very small and F very large.* That is to say ω^2 is very near 1, which is the basic self-excited harmonic (for $\mu = 0$). Let us plot the graph

$$f(r) = r(\sigma^2 + (4 - r)^2) - F^2 .$$

Its extrema occur for

$$f'(r) = 3r^2 - 4r + \sigma^2 + 1 = 0 ,$$

which gives approximately $r = 4/3, 4$. We also find $f(4) =$ almost $(-F^2)$ and $f(4/3) =$ almost $(256/27 - F^2)$. Hence $f(r) < 0$ for $r < 4/3$ and for $r > 4$, f is monotone and has exactly one root. Under the circumstances then f has just one root. As this root $\rightarrow +\infty$ with F^2 it will be > 4 and so it will fall in the stable region. Thus there is only one critical point P and it is stable. At all events solutions not too far from P will tend to P. If ξ_0, η_0 are the coordinates of P, the solutions of our differential equation will show a marked tendency to tend toward

$$x = \xi_0 \cos \omega t + \eta_0 \sin \omega t ,$$

that is to say the forced oscillation will tend to predominate.

II. *Suppose σ large.* This time $f'(r) = 0$ will only have complex roots, hence $f(r)$ will be monotonic. Consequently $f(r)$ will have only one root which will be approximately F/σ^2 and hence small. Thus there will be again a single critical point P. As it will fall below $r = 2$ for σ large P will then be unstable. On the other hand we have from (13.5)

$$\frac{\xi\, d\xi}{d\tau} + \frac{\eta\, d\eta}{d\tau} = \frac{1}{2}\frac{d\varrho^2}{d\tau} = \varrho^2(4 - \varrho^2) - F\xi$$

$$= \varrho\left\{\varrho(4 - \varrho^2) - F\frac{\xi}{\varrho}\right\}.$$

Since $\xi/\varrho < 1$, the bracket is negative for ϱ large, and hence for ϱ large, $d\varrho^2/d\tau < 0$. Thus along a large circle C, in particular surrounding P, the paths all point inward. If D is a small circle around the point P then on the boundary of the annular region Ω between C and D, the paths all point inward. Since Ω contains no critical point it must contain a limit-cycle and at least one Γ surrounding P, which will be stable (Poincaré, see X, 9.3). Thus ξ, η will show a marked tendency toward periodicity with a new frequency ω^*, which is very small compared with ω, since ξ and η are expected to vary but slowly during the period $2\pi/\omega$.

One must emphasize again that mathematically these results are very tentative. They do appear however to show good agreement with experiment.

§ 4. Equations of Mathieu and of Hill

15. One understands by the equation of Mathieu the following:

$$d^2x/dt^2 + (\lambda + \mu \cos t)\, x = 0\,, \tag{15.1}$$

where λ, μ are constants, and by the equation of Hill:

$$d^2x/dt^2 + p\,(t)\, x = 0\,, \tag{15.2}$$

where $p(t)$ is periodic. These equations have numerous practical applications. They arise also as variation equations of periodic solutions of non-linear systems and may thus be exploited in questions of stability.

We shall only touch briefly upon Mathieu's equation as an additional illustration of the perturbation method. Our treatment is inspired mainly by Stoker [1]. See also McLachlan [1] for various practical applications.

According to our general theory for the equation or rather for its equivalent

$$dx/dt = y,\ dy/dt = -(\lambda + \mu \cos t)\, x, \tag{15.3}$$

if $X(t)$ is a non-singular matrix solution of the associated matrix equation then

$$X\,(t + 2\pi) = X\,(t)\, C$$

where C is a constant matrix (III, 11). Since the matrix

$$\begin{pmatrix} 0 & , 1 \\ -(\lambda + \mu \cos t) & , 0 \end{pmatrix}$$

has for trace zero we have from (III, 11.4): $|C| = 1$. Hence if

$$C = \begin{pmatrix} \alpha, & \beta \\ \gamma, & \delta \end{pmatrix},$$

we have $\alpha\delta - \beta\gamma = 1$ and so the characteristic exponents r_1, r_2 satisfy an equation

$$r^2 - a\,r + 1 = 0\,, \tag{15.4}$$

where $a = \alpha + \delta$. If r_1, r_2 are the roots of (15.4)—the characteristic exponents—then

$$r_1 r_2 = 1\,. \tag{15.5}$$

Hence:

(15.6) *If r_1 and r_2 are real and distinct they have the same sign and one of $|r_1|$, $|r_2| < 1$ while the other > 1. If they are equal they are both $+1$ or both -1. If they are complex then $r_1 = e^{ia}$, $r_2 = e^{-ia}$, $a \neq k\pi$.*

We recall also that for linear homogeneous equations boundedness of all solutions is equivalent to stability (IV, 3.3). Hence the following various possibilities:

I. *r_1 and r_2 are real distinct and positive.* If $s_j = 1/2\pi \log r_j$ then by (III, 12.6) there is a fundamental system of solutions

$$x_1 = \varphi_1(t)\, e^{s_1 t},\ y_1 = \psi_1(t)\, e^{s_1 t}\,,$$

$$x_2 = \varphi_2(t)\, e^{s_2 t},\ y_2 = \psi_2(t)\, e^{s_2 t}\,.$$

where $\varphi_1, \ldots, \psi_2$ are continuous with period 2π. Since say $0 < r_1 < 1 < r_2$, we have $s_2 > 0$ and so (x_2, y_2) is an unbounded solution. Hence the origin is unstable.

II. *r_1 and r_2 are real distinct and negative.* With s_j as above $|r_j|$, the result obtained is the same as under I.

III. $r_1 = r_2 = \pm 1$. The fundamental solutions are now

$$x_1 = \pm \varphi_1(t), \quad y_1 = \pm \psi_1(t),$$

$$x_2 = \pm t\, \varphi_2(t), \quad y_2 = \pm \psi_2(t).$$

One of them is periodic with period 2π for $+1$, 4π for -1, and the other unbounded, hence we have again instability.

IV. $r_1 = e^{ia}$, $r_2 = e^{-ia}$, $a \neq k\pi$. We still have two fundamental solutions but they are conjugate complex:

$$x_1 = \varphi_1(t)\, e^{iat}, \quad y_1 = \psi_1(t)\, e^{iat},$$

$$x_2 = \bar{\varphi}_1(t)\, e^{-iat}, \quad y_2 = \bar{\psi}_1(t)\, e^{-iat}.$$

By linear combination they yield the two real fundamental solutions

$$x_1 = \varphi_1(t) \cos at, \; y_1 = \varphi_1'(t) \cos at - a\varphi_1(t) \sin at,$$

$$x_2 = \varphi_2(t) \sin at, \; y_2 = \varphi_2'(t) \sin at + a\varphi_2(t) \cos at,$$

where φ_1, φ_2 are real and of course with period 2π. This time the solutions are both bounded and so we have stability.

To sum up then we have proved:

(15.7) *A n.a.s.c. for the stability of the Mathieu equation is that the characteristic exponents be imaginary.*

16. The transition from stability to instability is through equality of the characteristic exponents or equivalently through the existence of a periodic solution with period 2π or 4π. This property will now be exploited systematically. We shall assume μ small and look for a solution of (15.1) which has the period 2π or 4π. Upon adding the relation

$$d\lambda/dt = 0$$

to (15.3) one may apply the method of (6) (or of VIII, 10) to the new system. The solution will admit an expansion

$$x = x_0 + \mu\, x_1 + \mu^2 x_2 + \dots, \tag{16.1}$$

$$\lambda = \lambda_o + \mu\, \lambda_1 + \mu^2 \lambda_2 + \dots, \tag{16.2}$$

where the λ_j are constants. Upon substituting in (15.1) one obtains the recurrent relations

$(16.3)_o \quad x_o''(t) + \lambda_o x_o(t) = 0,$

$(16.3)_1 \quad x_1''(t) + \lambda_o x_1(t) = -(\lambda_1 + \cos t)\, x_o,$

$(16.3)_2 \quad x_2''(t) + \lambda_o x_2(t) = -\lambda_2 x_o - (\lambda_1 + \cos t)\, x_1,$

. . . .

The initial conditions are those of $x_o(t)$ so that $x_p(0) = x_p'(0) = 0, p > 0$.

Take first the periodic solutions with period 2π. From $(16.3)_o$ we find that $\lambda_o = n^2$.

Suppose first $n = 0$. Here $\lambda_o = 0$, $x_o(t) = 1$. In order that x_1 be periodic we must have $\lambda_1 = 0$. Then $x_1 = \cos t + c$. The condition $x_1(0) = x_1'(0) = 0$ yields $c = -1$, hence

$$x_1(t) = -1 + \cos t.$$

It follows that

$$x_2'' = -\lambda_2 - \cos t(\cos t - 1)$$
$$= -\lambda_2 - \frac{1 + \cos 2t}{2} + \cos t.$$

Here periodicity requires that $\lambda_2 = -1/2$. Then

$$x_2 = \frac{\cos 2t}{8} - \cos t + d.$$

Again $x_2(0) = x_2'(0) = 0$ demands $d = 7/8$ and so

$$x_2(t) = \frac{7}{8} - \cos t + \frac{\cos 2t}{8}.$$

Thus:

$$(16.4) \qquad n = 0: \qquad \lambda = -\frac{\mu^2}{2} + O(\mu^3),$$

$$x(t) = 1 + \mu(-1 + \cos t) + \mu^2\left(\frac{7}{8} - \cos t + \frac{\cos 2t}{8}\right) + O(\mu^3).$$

Take now $n > 0$. Owing to linearity we may as well try separately $\cos nt$ and $\sin nt$ for x_o.

Take first $n = 1$ and $x_0 = \cos t$. This time $(16.3)_1$ yields

$$x_1''(t) + x_1(t) = -(\lambda_1 + \cos t)\cos t\,,$$

$$x_1 = -\frac{\lambda_1 t}{2}\sin t - \frac{(3 - \cos 2t)}{6} + A\cos t + B\sin t\,.$$

If $x_1(t)$ is to be periodic we must have $\lambda_1 = 0$. Then $x_1(0) = x_1'(0) = 0$ yields $A = \frac{1}{3}$, $B = 0$. Thus

$$x_1(t) = -\frac{1}{2} + \frac{\cos t}{3} + \frac{\cos 2t}{6}\,.$$

Then $(16.3)_2$ yields

$$x_2''(t) + x_2(t) = \lambda_2 x_0(t) + x_1(t)\cos t$$

$$= \left(\lambda_2 - \frac{1}{2}\right)\cos t + \frac{\cos t}{6}(2\cos t + \cos 2t).$$

Periodicity requires now that $\lambda_2 = \frac{1}{2}$. Thus

$$n = 1: \qquad \lambda = 1 + \frac{1}{2}\mu^2 + O(\mu^3)$$

$$x = \cos t - \mu\left(\frac{1}{2} - \frac{\cos t}{3} - \frac{\cos 2t}{6}\right) + O(\mu^2)\,.$$

Similarly for any $n > 0$

(16.5) $$\lambda = n^2 + \lambda_{n+1}\mu^{n+1} + O(\mu^{n+2})\,.$$

The same result would follow for $\sin nt$, and likewise for $\lambda_0 = n^2/4$, $n > 1$, corresponding to period 4π. For $\lambda_0 = 1/4$ we find by calculations such as above that there are the two solutions

(16.6) $$\lambda = 1/4 \pm 1/2\,\mu + O(\mu^2)\,.$$

It is convenient to represent the couples λ, μ as points of an Euclidean plane. If the corresponding solutions of (15.3) are all bounded one refers to the point as *stable*, otherwise as *unstable*.

Consider in particular the points on the λ axis. If $\lambda > 0$ the fundamental solutions are $\sin \sqrt{\lambda}\, t$ and $\cos \sqrt{\lambda}\, t$ and hence the points are stable. For $\lambda = 0$ there is the solution t, and for $\lambda < 0$ the solution $e^{\sqrt{-\lambda}\, t}$ both unbounded. Hence $\lambda \leq 0$ consists solely (on the λ axis) of unstable points.

In the whole plane the transition curves are those corresponding to periodic solutions and they separate the stable from the unstable regions. From (16.4), (16.5), (16.6) and the details given regarding the λ, μ relations one infers the configuration indicated

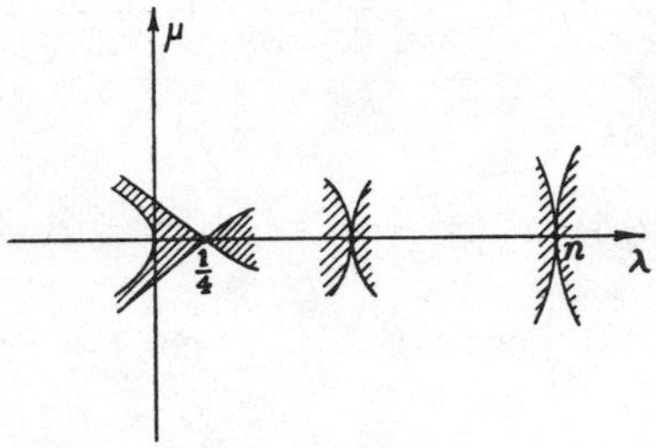

Fig. 4

in fig. 4, where the shaded portions are the stable regions. More complete details regarding the figure will be found in Stoker [1], p. 205.

§ 5. The Limiting Position of Limit-Cycles

17. Let us return to Liénard's equation

$$d^2x/dt^2 + \mu f(x)\, dx/dt + g(x) = 0 \tag{17.1}$$

of (XI, 2) under the same assumptions I, . . ., IV, as made there, with this complement: $g(x)$ is a monotonic function of x. We propose to consider the limiting position of the limit-cycle of the equivalent pair of first order equations for μ large. This problem has already been solved for van der Pol's equation and hence for (17.1) by Stoker-Flanders [1], and also in a somewhat different manner by Lasalle [1]. The treatment that follows rests more or less upon the work of these authors.

Let us set $\mu = 1/\varepsilon$, so that ε is positive and small. This replaces (17.1) by

$$\varepsilon\, d^2x/dt^2 + f(x)\, dx/dt + \varepsilon g(x) = 0\,. \tag{17.2}$$

In the notations of (XI, 2) we choose as equivalent system

$$\begin{cases} \varepsilon\, dx/dt = y - F(x)\,, \\ dy/dt = -\varepsilon g(x)\,. \end{cases} \tag{17.3}$$

It is known from Liénard's theorem (XI, 2.2) that (17.3) has a unique limit-cycle $\Gamma(\varepsilon)$. We propose to prove:

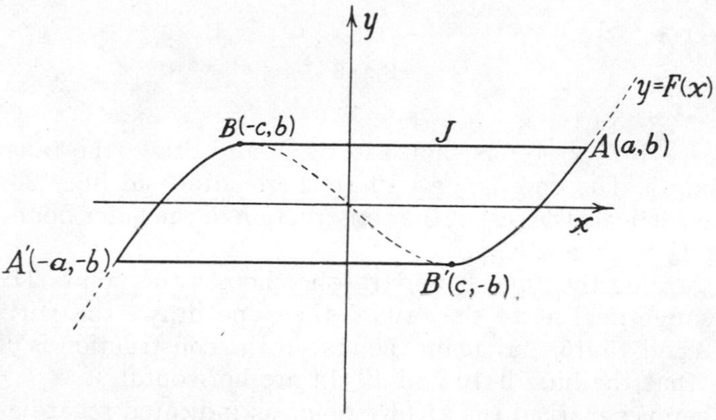

Fig. 5

(17.4) *As* $\varepsilon \to 0$ *the limit-cycle tends towards the piecewise analytic Jordan curve* J *drawn in heavy line in fig. 5.*

The method of proof could not be simpler: we construct a closed annular region Ω containing J, arbitrarily near J as $\varepsilon \to 0$ and show that for ε small enough the paths of (17.3) all cross the boundary of Ω inward. Poincaré's theorem asserts then that Ω contains a limit-cycle and since $\Gamma(\varepsilon)$ is the only limit-cycle of the system, it is the one inside Ω. Thus everything is reduced to the construction of Ω. Naturally it will be carried out symmetrically relative to the origin.

The construction is shown in fig. 6. One takes the arcs of

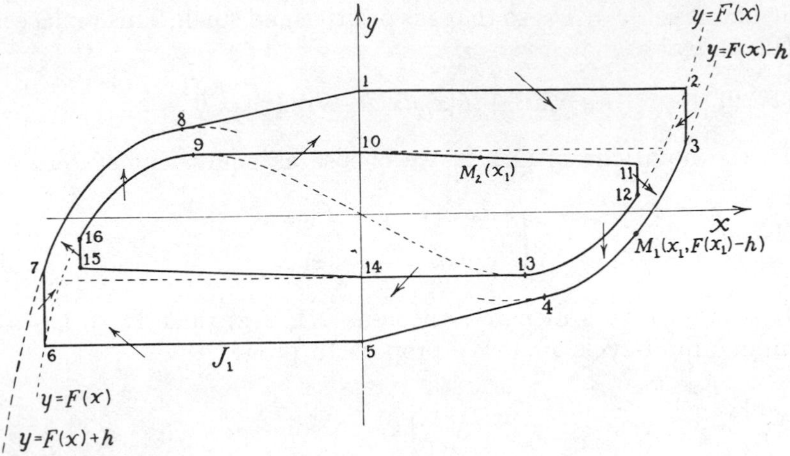

Fig. 6

curves $y = F(x) \pm h$ as shown in fig. 6 and draws the tangents 81 and 45. The lines 56 and 12 are horizontal and lines 23 and 67 are vertical. The rest of the construction of the outer boundary J_1 of Ω is then obvious.

Regarding the inner boundary one chooses the points 11 and 15 symmetrical as to the origin, then one draws the verticals 11–12 and 15–16, and again the rest of the construction is clear. Note that the lines 9–10 and 13–14 are horizontal.

Wherever marked the arrows point as indicated regardless of ε. This is found at once by reference to the system (17.3). We must show that for ε small enough the pointing along the other arcs is likewise inward.

Arcs 34 and 78. Let us take 34. The slope of the arc at a point M_1 is $f(x_1) > f(4)$. The slope of the path at M_1 is

$$\frac{dy}{dx} = \frac{-\varepsilon^2 g(x_1)}{-h} < \frac{\varepsilon^2 g(3)}{h}$$

which $\rightarrow 0$ with ε. Hence for ε small enough it is $< f(4) < f(x_1)$ on the arc. Since $dy/dt < 0$ along the arc, for ε small the pointing is inward. Similarly for 78.

Arcs 45 and 81. Along 45: $|y - F(x)| > h$, hence the slope of the path $< \varepsilon^2 g(4)/h$ which $\to 0$ with ε. Since again $dy/dt < 0$ on this arc, for ε small enough the pointing will be inward. Similarly for 81.

Thus, for ε small enough, along the outer boundary the pointing is inward relative to Ω.

Arcs 10–11 and 14–15. Let k be the vertical distance 12–11. For k small enough along 10–11: $|y - F(x)| > k$. Hence the slope of the path at a point $M_2(x_2)$ of 10–11 is $< \varepsilon^2 g(x_2)/k < \varepsilon^2 g(11)/k$ which $\to 0$ with ε. Since $dx/dt > 0$ on this arc, for ε small enough the pointing will be inward. Similarly for 14–15.

Thus, for ε small, the pointing along the inner boundary is likewise inward relative to Ω.

We conclude that our annular region contains the limit-cycle $\Gamma(\varepsilon)$. The construction depends upon four parameters: h, k, and the slopes of 45 and 10–11. A glance at the figure shows that by taking all four sufficiently small the region Ω will be arbitrarily near to the Jordan curve J of fig. 5. Therefore $\Gamma(\varepsilon) \to J$ as $\varepsilon \to 0$ and this proves (17.4).

18. Let us make estimates of the limiting period and amplitude. Referring to fig. 5 the limiting amplitude is simply a. To obtain it we take the only positive root c of the even function $f(x)$ and solve the equation $F(x) = F(-c) = b$ for its only real root a other than $-c$.

Regarding the period we see that AB and $A'B'$ are described with velocity $\to \infty$ and hence in arbitrarily small time. Hence the period T is approximately twice the time of description of the arc AB'. Now along this arc approximately

$$dy/dt = f(x)\, dx/dt = -\varepsilon g(x) .$$

Hence for the period

$$T = \frac{2}{\varepsilon} \int_c^a \frac{f(x)dx}{g(x)} .$$

Let us apply these results to van der Pol's equation. We find at once $c = 1$, $b = 2/3$. Hence a is the root $\neq -1$ of

$$x^3 - 3x = 2$$

and this root is readily verified to be 2. Thus the limiting amplitude has the same value 2, as the initial amplitude for $\mu = 0$ ($\varepsilon = \infty$), which is the radius 2 of the generating circle.

Regarding the period our approximation is

$$T = \frac{2}{\varepsilon}\int_1^2 \frac{x^2 - 1}{x}\,dx = \frac{3 - 2\log 2}{\varepsilon}.$$

A treatment leading to more accurate estimates is given at the end of Lasalle's paper. See also Cartwright [1], Stoker [1], p. 140, J. Haag [1]. For purposes of comparison with Lasalle it should be noted that his time $\tau = \varepsilon t$ so that his result assumes the form $\varepsilon^2 \times \text{period} = 3 - 2\log 2$.

APPENDIX I

Complement on Matrices

Basic references: Bellman [4]; Gantmacher [1]; Wedderburn [1].

§ 1. Reduction to Normal Form

1. We will first prove Theorem (I, 8.4). Let then T be a complex transformation of the complex vector n-space $\mathfrak{B}$, e a base for $\mathfrak{B}$, $A = (a_{jk})$ the matrix of the transformation operating on e:

$$Te_j = \Sigma\, a_{jk}\, e_k\,. \tag{1.1}$$

It is clear that the correspondence $T \leftrightarrow A$ is one-one. Furthermore let T^*, A^* correspond in the same way. Then this holds also for T^*T and AA^*. Moreover if a, a^* are any complex scalars then $aA + a^*A^*$ correspond to a unique transformation written $aT + a^*T^*$. Finally, if $A = 0$ we write $T = 0$. Thus the algebras W_T of the transformations of $\mathfrak{B}$ into itself, and W_A of the complex square matrices of order n are isomorphic. It is also readily verified that a change of base in $\mathfrak{B}$ does not affect W_T. Notice that if I denotes the identity transformation and if

$$f(x) = a_0 + a_1 x + \ldots + a_p x^p \tag{1.2}$$

is a complex polynomial then the transformation

$$a_0 I + a_1 T + \ldots + a_p T^p$$

is uniquely defined, and it is written $f(T)$.

Suppose that there is a direct sum decomposition

(1.3) $$\mathfrak{V} = \mathfrak{V}_1{}^{d_1} \oplus \mathfrak{V}_2{}^{d_2} \oplus \dots \oplus \mathfrak{V}_s{}^{d_s}$$

such that $T\mathfrak{V}_h \subset \mathfrak{V}_h$. If $\{e_1{}^h, \dots, e_{d_h}{}^h\}$ is a base for $\mathfrak{V}_h$, then (the d's are the dimensions):

(1.4) $$Te_j{}^h = \Sigma\, a_{jk}{}^h\, e_k{}^h\,.$$

Thus $A_h = (a_{jk}{}^h)$ determines a transformation T_h of $\mathfrak{V}_h$ into itself which is written $T|\mathfrak{V}_h$. Notice that $\{e_k{}^h\}$ (all h, k) is a base for $\mathfrak{V}$. Hence if A is the initial matrix of T we have $A \sim \operatorname{diag}(A_1, \dots, A_s)$. Conversely if such a similitude exists we can write (1.3), define the $e_k{}^h$, write (1.4) and in short reproduce the above situation.

Under the above circumstances we shall say more or less appropriately: T is *reducible*, and the $\mathfrak{V}_h$, A_h are the associated subspaces and blocks. If T_h is irreducible we shall also refer to the subspace $\mathfrak{V}_h$ as irreducible.

2. Let $\lambda_1, \dots, \lambda_n$ be the characteristic roots of A. We will first assume that no λ_j is zero. The case where a λ_j is zero will be readily disposed of afterwards.

(2.1) *One may choose the base e, and hence replace A by an equivalent matrix such that the new matrix assumes the triangular form:*

(2.1a) $$\begin{pmatrix} \lambda_1, 0 & . & . & . & . & . & 0 \\ a_{21}, \lambda_2, 0, & & . & . & . & . \\ . & . & . & . & . & . & . \\ . & . & . & . & . & . & \lambda_n \end{pmatrix}.$$

If $n = 1$, the asserted property is obvious. We proceed then by induction on n. By a well-known argument and since A is nonsingular, there exists a vector $e_1{}^*$ such that $Te_1{}^* = e_1{}^{*\prime} = \lambda_1 e_1{}^*$. We take it as first element of the base and call it e_1. Then for any $h > 1$:

$$Te_h = a_h e_1 + \sum_{j>1} b_{hj} e_j\,.$$

Thus

$$A \sim \begin{pmatrix} \lambda_1, 0, & . & . & . & ., 0 \\ a_h, b_{h2}, & & . & ., & b_{hn} \end{pmatrix}.$$

Hence $B = (b_{hj})$ has the characteristic roots $\lambda_2, \ldots, \lambda_n$. Now B is a matrix of a transformation on the $\mathfrak{B}^{n-1}$ spanned by $e_2, \ldots, e_n$. By the hypothesis of the induction it may be reduced to the form (2.1a) and this implies (2.1).

Observe that the characteristic roots are merely the values of λ which make $T - \lambda I$ singular. Let in particular λ_1 have the multiplicity k, i.e., it is repeated k times among the characteristic roots.

(2.2) *There is a direct sum decomposition* $\mathfrak{B}^n = \mathfrak{B}_1{}^k \oplus \mathfrak{B}_2{}^{n-k}$ *where*: (a) $T\mathfrak{B}_j = \mathfrak{B}_j$; (b) $T|\mathfrak{B}_1$ *has the sole characteristic root* λ_1 (*and therefore to multiplicity* k); (c) $T|\mathfrak{B}_2$ *has the remaining characteristic roots each taken with its multiplicity.*

In the reduction of A to the form (2.1a) one may suppose that the first k roots λ_j are λ_1 repeated k times. Thus the space $\mathfrak{B}_1{}^k$ spanned by $e_1, \ldots, e_k$ is already such that $T\mathfrak{B}_1 = \mathfrak{B}_1$ and $T_1 = T|\mathfrak{B}_1$ has for characteristic roots λ_1 repeated k times.

Let now $Te_{k+1} = \lambda_{k+1} e_{k+1} + \eta$, $\eta \in \mathfrak{B}_1$ and take $e_{k+1}{}^* = e_{k+1} + \zeta$, $\zeta \in \mathfrak{B}_1$. Thus $e_{k+1}{}^*$ is not in $\mathfrak{B}_1$. Now

$$\begin{aligned} Te_{k+1}{}^* &= \lambda_{k+1} e_{k+1} + \eta + T_1 \zeta \\ &= \lambda_{k+1} e_{k+1}{}^* - \lambda_{k+1} \zeta + \eta + T_1 \zeta . \end{aligned}$$

Since λ_{k+1} is not a characteristic root of T_1, $(T_1 - \lambda_{k+1} I)$ is non-singular. One may then choose $\zeta = -(T_1 - \lambda_{k+1} I)^{-1} \eta$ and as a consequence $Te_{k+1}{}^* = \lambda_{k+1} e_{k+1}{}^*$. We take $e_{k+1}{}^*$ as new e_{k+1}.

Suppose that we have already obtained $e_{k+1}, \ldots, e_{j-1}$ none in $\mathfrak{B}_1$ and such that if $\mathfrak{W}_{j-1}$ is the vector space which they span, then $T\mathfrak{W}_{j-1} = \mathfrak{W}_{j-1}$. Then $Te_j = \eta + \zeta + \lambda_j e_j$, $\eta \in \mathfrak{B}_1$, $\zeta \in \mathfrak{W}_{j-1}$. Take $e_j{}^* = e_j + \theta$, $\theta \in \mathfrak{B}_1$. Thus $e_j{}^*$ is not in $\mathfrak{B}_1 \oplus \mathfrak{W}_{j-1}$. And now

$$\begin{aligned} Te_j{}^* &= \lambda_j e_j + \eta + \zeta + T_1 \theta \\ &= \lambda_j e_j{}^* + (T_1 - \lambda_j I)\, \theta + \eta + \zeta , \end{aligned}$$

and as before we can choose θ such that the two middle terms disappear. Then $Te_j{}^* = \lambda_j e_j{}^* + \zeta \in \mathfrak{W}_{j-1} \oplus$ the space spanned by $e_j{}^*$. Hence if we take $e_j{}^*$ as new e_j and $\mathfrak{W}_j$ has its obvious meaning, then $T\mathfrak{W}_j = \mathfrak{W}_j$. Pursuing this we will arrive at $\mathfrak{W}_{n-k} = \mathfrak{B}_2$ and $\mathfrak{B}_1$, $\mathfrak{B}_2$ behave in accordance with (2.2).

Let $\lambda_1, \ldots, \lambda_r$ denote this time the distinct characteristic roots and let k_h be the multiplicity of λ_h. By applying (2.2) to $\mathfrak{B}_2$, etc, or else by an evident induction we obtain:

(2.3) *There is a direct sum decomposition* $\mathfrak{B}^n = \mathfrak{B}_1^{k_1} \oplus \ldots \oplus \mathfrak{B}_r^{k_r}$ *where* $T_h = T|\mathfrak{B}_h$ *has the sole characteristic root* λ_h *(and therefore to multiplicity* k_h*). Moreover since* $\lambda_h \neq 0$ *we have* $T_h\mathfrak{B}_h = \mathfrak{B}_h$.

3. As we have seen (2) if A_h is the block of T_h then the initial matrix $A \sim \operatorname{diag}(A_1, \ldots, A_r)$. Hence to prove (I, 8.4) it is sufficient to prove it for the triple T_h, $\mathfrak{B}_h$, A_h, i.e., for a matrix with a single characteristic root $\lambda \neq 0$. Let then the initial matrix A already have that property. It is not ruled out that T may still be reducible. Let it have associated subspaces $\mathfrak{B}_1, \ldots, \mathfrak{B}_s$ with blocks $A_1, \ldots, A_s$ where this time $T|\mathfrak{B}_h$ is irreducible. Once more (I, 8.4) will follow if we prove it for $T|\mathfrak{B}_h$ i.e., for an irreducible T.

Suppose then that T is irreducible with sole characteristic root $\lambda \neq 0$. Let us write $e_h' = Te_h$. We may select a base such that $e_1' = \lambda e_1$. Then for $h > 1$:

$$e_h' = a_{h1}\, e_1 + \ldots,$$

and the a_{h1} cannot all be zero since T would then be reducible. Let us range $e_2, e_3, \ldots,$ in such order that $a_{21} \neq 0$. Upon replacing e_2 by e_2/a_{21} the situation will be the same save that $a_{21} = 1$. Replacing then e_k, $k > 2$, by $e_k - a_{k1}\, e_2$ the situation will still be the same save that every $a_{k1} = 0$ for $k > 2$. Thus the matrix A assumes now the form

$$\begin{pmatrix} \lambda, & 0, & \ldots & \ldots, & 0 \\ 1, & a_{22}, & \ldots & \ldots & \ldots \\ 0, & a_{32}, & \ldots & \ldots & \ldots \\ \ldots & \ldots & \ldots & \ldots & \ldots \\ 0, & a_{n2}, & \ldots & \ldots, & a_{nn} \end{pmatrix}. \tag{3.1}$$

One may consider the matrix

$$A^* = (a_{jk});\ j, k = 2, \ldots, n$$

as matrix of a transformation T^* of the space $\mathfrak{B}^*$ spanned by $e_2, \ldots, e_n$. Since its characteristic roots are also characteristic roots of (3.1) it has λ as only such root. One may now choose a base $e_2, \ldots, e_n$ for $\mathfrak{B}^*$ such that $a_{22} = \lambda$, $a_{2h} = 0$ for $h > 2$. Repeating the same reasoning we may choose the base so that $a_{32} = 1$, $a_{3h} = 0$ for $h > 3$, etc. In the last analysis a base will be obtained such that (3.1) assumes the normal form

$$(3.2) \qquad C(\lambda) = \begin{pmatrix} \lambda, 0, \cdot & \cdot & \cdot & \cdot & \cdot & \cdot & \cdot & \cdot \\ 1, \lambda, \cdot & \cdot & \cdot & \cdot & \cdot & \cdot & \cdot \\ \cdot & \cdot & \cdot & \cdot & \cdot & \cdot & \cdot & \cdot \\ \cdot & \cdot & \cdot & \cdot & \cdot & 0, 1, \lambda \end{pmatrix}.$$

To complete the proof of theorem (I, 8.4) when all the characteristic roots are non-zero there remains to show that the direct sum decomposition (1.3) is unique. Suppose that

$$\mathfrak{B} = \mathfrak{B}_1 \oplus \ldots \oplus \mathfrak{B}_p = \mathfrak{B}_1' \oplus \ldots \oplus \mathfrak{B}_q'.$$

Then

$$\mathfrak{B}_1 = \mathfrak{B}_1 \cap \mathfrak{B}_1' \oplus \ldots \oplus \mathfrak{B}_1 \cap \mathfrak{B}_q'.$$

Since $\mathfrak{B}_1$ is irreducible one of the terms at the right is $\mathfrak{B}_1$ and the others are zero. If say $\mathfrak{B}_1 = \mathfrak{B}_1 \cap \mathfrak{B}_1'$, $\mathfrak{B}_1 \cap \mathfrak{B}_2' = \ldots = 0$, then $\mathfrak{B}_1 = \mathfrak{B}_1'$. This together with an evident induction proves that the decomposition is unique and that (I, 8.4) holds.

Suppose now that zero is a characteristic root of A. Take $T_1 = T - \lambda I$, where λ is not a characteristic root of A. The matrix of T_1 is $A_1 = A - \lambda E$ whose characteristic roots $\lambda_j - \lambda$ are all $\neq 0$. Hence there is a choice of base e such that $C_1 = PA_1P^{-1}$ is in normal form. The corresponding matrix C for T is $C = C_1 + \lambda E = P(A_1 + \lambda E) P^{-1} = PAP^{-1} \sim A$ and it is also in normal form. Thus (I, 8.4) holds also in the present case, and its proof is completed.

(3.3) *Remark.* Let α be any number $\neq 0$. Then one may modify the reduction process so as to replace the block $C(\lambda)$ by a block

$$C(\lambda, \alpha) = \begin{pmatrix} \lambda, 0, \cdot & \cdot & \cdot & \cdot & \cdot & \cdot, 0 \\ \alpha, \lambda, \cdot & \cdot & \cdot & \cdot & \cdot & \cdot \\ \cdot & \cdot & \cdot & \cdot & \cdot & \cdot & \cdot \\ \cdot & \cdot & \cdot & \cdot & \cdot & \cdot, \alpha, \lambda \end{pmatrix}.$$

All that is necessary for instance at the beginning is to replace

e_2 by $\alpha e_2/a_{21}$. The remaining modifications required are obvious. Evidently also α may vary from block to block. Thus when A is real for blocks corresponding to pairs of conjugate roots $\lambda, \bar{\lambda}$ one may choose conjugate pairs α, $\bar{\alpha}$. It is simpler, however, to take just one real $\alpha \neq 0$, the same for all the blocks, and this will suffice for the applications in the text.

§ 2. Normal Form for Real Matrices

4. We proceed now to the proof of the complement (I, 9.1) for real transformations of real spaces. Thus there is a real base e relative to which the transformation T has a real matrix A and

$$Te_j = \Sigma\, a_{jk}\, e_k\, .$$

It follows that real vectors have real transforms. However a complex vector

$$f = \Sigma x_j e_j,\ x_j \text{ complex}\, ,$$

has a conjugate

$$\bar{f} = \Sigma\, \bar{x}_j e_j\, .$$

and now

$$Tf = \Sigma\, x_j a_{jk} e_k;\ T\bar{f} = \Sigma\, \bar{x}_j a_{jk} e_k\, .$$

It follows that if instead of e we have a complex base e^* then

$$Te_j{}^* = \Sigma\, b_{jk}\, e_k{}^*,\ T\bar{e}_j{}^* = \Sigma\, \bar{b}_{jk}\, \bar{e}_k{}^*\, .$$

Thus the matrix of transformation of $\bar{e}^*$ is the conjugate of that of e^*.

5. For convenience let the distinct characteristic roots λ_j be ranged in such an order that $\lambda_{2k+2} = \bar{\lambda}_{2k+1}$, $k \leq p$, and that $\lambda_{2p+1}, \ldots, \lambda_q$ are real. Thus the first $2p$ are the p pairs of complex conjugate roots. Take also the decomposition

$$\mathfrak{V} = \mathfrak{V}_1{}^{d_1} \oplus \ldots \oplus \mathfrak{V}_q{}^{d_q}$$

where $\mathfrak{V}_h$ is associated with λ_h. Observe that the $\mathfrak{V}_{2p+j}$ are all real. For in the argument leading to (2.1a) if one makes "real transformation" part of the induction process, when λ_1 is real then the associated space is also real. Let A_h be the matrix of $T|\mathfrak{B}_h$ and $\{e_{h1}, \ldots, e_{hd_h}\}$ a base for $\mathfrak{B}_h$. When λ_h is real the e_{hj} may then likewise be chosen real. We may assume also that $A = \operatorname{diag}(A_1, \ldots, A_q)$ where A is the matrix of T.

Notice now that the matrix $A_h - \lambda E_{d_h}$ (E_{d_h} has its obvious meaning), in normal form and hence in any form, is nilpotent when and only when $\lambda = \lambda_h$ and non-singular otherwise. Hence

(5.1) $$(T - \lambda_h I)^\nu \mathfrak{B}_h = 0, \text{ for some } \nu;$$

(5.2) $$(T - \lambda_h I)^\nu \mathfrak{B}_k = \mathfrak{B}_k,\ k \neq h, \text{ for every } \nu.$$

Let us set

$$\mathfrak{B}' = \mathfrak{B}_1 \oplus \mathfrak{B}_3 \oplus \ldots \oplus \mathfrak{B}_{2p-1},\ \mathfrak{B}'' = \mathfrak{B}_2 \oplus \mathfrak{B}_4 \oplus \ldots \oplus \mathfrak{B}_{2p},$$
$$\mathfrak{B}''' = \mathfrak{B}_{2p+1} \oplus \ldots \oplus \mathfrak{B}_q.$$

Observe that $\{e_{2k-1,j}\}$, $k \leq p$, is a base for $\mathfrak{B}'$. Suppose now that there is a non-trivial relation

(5.3) $$\sum_{k \leq p} \alpha_{2k-1,j}\, e_{2k-1,j} + \sum_{k \leq p} \beta_{2k-1,j}\, \bar{e}_{2k-1,j} + \sum \gamma_{2p-h,j}\, e_{2p+h,j} = 0.$$

Let also

$$f(x) = (x - \lambda_1)(x - \lambda_3) \ldots (x - \lambda_{2p-1}),\ \varphi(x) = f(x)\bar{f}(x).$$

Owing to (5.1), (5.2) upon applying $\varphi^\nu(T)$ for some ν to (5.3) the first two sums are annulled while if the third is non-trivial it is transformed into a non-trivial relation between the $e_{2p+h,j}$. Since they are linearly independent every $\gamma_{2p+h,j} = 0$. Applying now $f^\nu(x)$ for some ν the first sum is annulled and if the second sum is non-trivial it becomes a non-trivial relation

$$\sum \delta_{2k-1,j}\, \bar{e}_{2k-1,j} = 0.$$

This implies a non-trivial relation

$$\sum \bar{\delta}_{2k-1,j}\, e_{2k-1,j} = 0,$$

which is ruled out since the $e_{2k-1,j}$ form a base for $\mathfrak{B}'$. Hence every $\beta_{2k-1,j} = 0$. Thus (5.3) reduces to its first sum and so by the remark just made every $\alpha_{2k-1,j} = 0$. Thus (5.3) is trivial.

Since the total number of elements e_{hk} in the three sums in (5.3) is the dimension of $\mathfrak{V}$, these elements make up a base for $\mathfrak{V}$.

We have then a direct sum decomposition

$$\mathfrak{V} = \mathfrak{V}' \oplus \overline{\mathfrak{V}}' \oplus \mathfrak{V}'''.$$

Consider now a decomposition of $\mathfrak{V}'$ into irreducible subspaces

$$\mathfrak{V}' = \mathfrak{V}_1' \oplus \ldots \oplus \mathfrak{V}_s'.$$

One may suppose the $e_{2k-1,j}$ so chosen that they consist of a set of bases $\{e_{hj}'\}$ for the $\mathfrak{V}_h'$. The $\bar{e}_{hj}'$ (h fixed) span the irreducible subspace $\overline{\mathfrak{V}}_h'$ of $\overline{\mathfrak{V}}'$ and we have

$$\overline{\mathfrak{V}}' = \overline{\mathfrak{V}}_1' \oplus \ldots \oplus \overline{\mathfrak{V}}_s'.$$

Thus the conjugate pairs $\mathfrak{V}_h'$, $\overline{\mathfrak{V}}_h'$ together with the real irreducible subspaces of a decomposition of $\mathfrak{V}'''$ make up a decomposition of $\mathfrak{V}$ whose subspaces behave in accordance with (I, 9.1).

Since the bases $\{e_{hj}'\}$, $\{\bar{e}_{hj}'\}$ of $\mathfrak{V}_h$, $\overline{\mathfrak{V}}_h'$ consist of conjugate elements, referring to (4) the associated blocks in the final transformation matrix C are likewise conjugate. Since those associated with the irreducible subspaces of $\mathfrak{V}'''$ are manifestly real, the proof of (I, 9.1) is completed.

§ 3. Normal Form of the Inverse of a Matrix

6. As an application of the preceding considerations let us prove the following property which is required in the text:

(6.1) *If a non-singular matrix* $A \sim \operatorname{diag}(C_1(\lambda_1), \ldots, C_r(\lambda_r))$ *where* $C_h(\lambda)$ *is of the type of* (3.2), *then* $A^{-1} \sim \operatorname{diag}(C_1(\lambda_1^{-1}), \ldots, C_r(\lambda_r^{-1}))$.

Let V_h, T_h correspond as before to C_h, so that both are irreducible. Thus in association with T, where $T_h = T|\mathfrak{V}_h$, we have

$$\mathfrak{V} = \mathfrak{V}_1 \oplus \ldots \oplus \mathfrak{V}_r. \tag{6.2}$$

It follows that $T\mathfrak{V}_h = \mathfrak{V}_h$. Hence also $T^{-1}\mathfrak{V}_h = \mathfrak{V}_h$. Thus the direct sum decomposition (4.2) corresponds also to T^{-1} and $T^{-1}|\mathfrak{B}_h = T_h^{-1}$. From this follows that it is sufficient to prove (6.1) for $r = 1$, or that:

(6.3) *If* $A \sim C(\lambda)$, $\lambda \neq 0$, *then* $A^{-1} \sim C(\lambda^{-1})$.

Thus we only need to show that if $\lambda \neq 0$ is the unique characteristic root of A then λ^{-1} is the unique characteristic root of A^{-1}.

Let R denote the matrix $C(0)$ of order n. We recall that R^j is R but with its diagonal of units moved j steps down. Hence $R^n = 0$. Now $C(\lambda) = \lambda E + R$. Hence if

$$D(\lambda) = \frac{E}{\lambda} - \frac{R}{\lambda^2} + \dots + (-1)^{n-1}\frac{R^{n-1}}{\lambda^n}$$

then $C(\lambda)\, D(\lambda) = E$, and hence $D(\lambda) = C^{-1}(\lambda)$. But $D(\lambda)$ is merely a matrix with $1/\lambda$ along the main diagonal and zeros above. Thus $D(\lambda)$ has for sole characteristic root $1/\lambda$. We already know that it is irreducible and so $D(\lambda) = C^{-1}(\lambda) \sim C(\lambda^{-1}) \sim A^{-1}$. This proves (6.3) and hence also (6.1).

§ 4. Determinations of log A

7. Let A be a non-singular matrix and let B be a matrix defined by $A = e^B$. If B is reduced to normal form: $B = \text{diag}\,(C(\lambda_1), \dots, C(\lambda_r))$, then A is reduced to $\text{diag}\,(e^{C(\lambda_1)}, \dots, e^{C(\lambda_r)})$. Thus we may suppose that $B = C(\lambda)$, $A = e^{C(\lambda)}$. Let n be the common order of the matrices A, B and let $C(\lambda) = \lambda E + D$ so that $D^n = 0$. Thus if $\lambda = \log \mu$ then $A = e^{\lambda E} \cdot e^D = e^\lambda \cdot e^D = \mu e^D = \mu \left(E + \frac{D}{1!} + \dots + \frac{D^{n-1}}{(n-1)!}\right)$. Now the matrix $F = \mu^{-1} A - E$ is known and is nilpotent: $F^n = 0$. The relation

$$F = D/1! + \dots + D^{n-1}/(n-1)!$$

may be inverted as if F, D were analytical variables and it yields

$$D = F - \frac{F^2}{2} + \dots + (-1)^n \frac{F^{n-1}}{n-1}.$$

Since $\lambda = \log \mu$ we have

$$\log A = (\log \mu + 2k\pi i)\, E + F - \frac{F^2}{2} + \dots + (-1)^n \frac{F^{n-1}}{n-1}.$$

Returning to the general case if B is as above with $\lambda_h = \log \mu_h$, and $C(\lambda_h)$ is of order n_h, and if $F_h = \dfrac{e^{C(\lambda_h)}}{\mu_h} - E_h$ then finally

$$\log A \sim \operatorname{diag}\Big(\ldots, (\log \mu_h + 2\, k_h \pi i)\, E_h + F_h - \frac{F_h^2}{2} + \ldots + (-1)^{n_h} \frac{F_h^{n_h - 1}}{n_h - 1}, \ldots\Big).$$

The main observation to be made is that each λ_h corresponding to every block of A is only known up to a multiple $2k_n\pi i$, where k_h may vary from block to block, even though the corresponding μ_h remains the same.

§ 5 A Certain Matrix Equation.

8. In the text (VI, 7) the solution of a certain partial differential equation has been reduced to the solution in B of a matrix system

$$(8.1) \qquad \begin{aligned} B' &= B \\ A'B + BA &= \varepsilon C \end{aligned}$$

under the following assumptions: $\varepsilon = \pm 1$, εA is a known stable matrix of order n, and $C > 0$ (notations of VI, 7). It is asserted that *under these conditions* (8.1) *has a unique solution* B *and* $B > 0$. This is what we propose to prove.

Two preliminary properties are required.

Property I. A transformation of coordinates does not affect the form of (8.1). Let P be the (non-singular) matrix of the transformation. Using the vector and quadratic form notations of (VI, 7), a quadratic form $x'Dx$ will become, under $x \to Px$, a quadratic form $x'P'DPx$, so that the effect of the transformation on the quadratic form is expressed by $D \to P'DP$.

On the other hand the effect on the basic differential equation

$$\frac{dx}{dt} = Ax$$

is representable by $A \to P^{-1}AP$. Hence let $A^* = P^{-1}AP$, $B^* = P'BP$, $C^* = P'CP$. Upon multiplying both sides of (8.1) by P' to the left and by P to the right we obtain

$$P'A'BP + P'BAP = \varepsilon P'CP = \varepsilon C^*$$

or

$$P'A'(P^{-1})'P'BP + P'BPP^{-1}AP = \varepsilon C^*$$

which is

$$A^{*\prime}B^* + B^*A^* = \varepsilon C^*.$$

That is, the form of (8.1) has not been affected by the transformation of coordinates.

Property II. The characteristic roots of $-A$ *are the negatives of those of* A. In fact if $|A - \lambda E| = f(\lambda)$, then $|-A - \lambda E| = -|A-(-\lambda E)| = -f(-\lambda)$ from which our assertion follows.

9. Returning now to our problem we must first distinguish two cases:

I. *The normal form of the matrix* A *is diagonal.* Thus if λ_i are the characteristic roots then $A \sim \text{diag}(\lambda_i, \ldots, \lambda_n)$. Choose coordinates such that A is normal. Upon writing (8.1) as a system of linear equations between the terms of the matrices we obtain

$$(9.1)\qquad \begin{cases} b_{ij} = b_{ji} \\ (\lambda_i + \lambda_j)b_{ij} = \varepsilon c_{ij}. \end{cases}$$

Since the λ_i have real parts of the sign of ε, the $\lambda_h + \lambda_j$ are all $\neq 0$ and so (8.2), hence also (9.1), has a solution and it is unique in the case under consideration.

II. *The normal form of the matrix* A *is not diagonal.* Referring to (3.3) we may assume that A is in a normal form with the λ_i in the main diagonal and underneath it a subdiagonal consisting (in fixed places) of a number $a \neq 0$ and of zeros. Upon writing again the analogue of (8.2) for the b_{ij} there results a linear system with a determinant which is a polynomial $\Delta(a)$ in a. Now $\Delta(0)$ corresponds to the system with $A = \text{diag}(\lambda_1, \ldots, \lambda_n)$. By the preceding result $\Delta(0) \neq 0$. Hence $\Delta(a) \not\equiv 0$. If we take for a some value not a root of the polynomial we will have $\Delta(a) \neq 0$. Hence the system (8.1) will again have a unique solution.

Observe that if a_0 is a root of $\Delta(a)$, then $\Delta(a_0) = 0$, and so, seemingly, we have a system with a non-unique solution. This contradiction proves that $\Delta(a)$ has no roots, i.e. $\Delta(a) = \Delta(0)$ is just a constant $\neq 0$.

10. There remains to show that the unique matrix solution B of (8.1) is > 0. Suppose first that A *is stable*. Then the unique matrix solution of (8.1) is given by

$$B = \int_0^{+\infty} e^{A't} \, C e^{At} dt. \tag{10.1}$$

One must first prove that B exists. Since A is stable so is A'. Now e^{At} being a solution of

$$dx/dt = Ax$$

its terms and those of the integrand are finite sums of expressions of the form $H(t) = e^{\alpha t} (\cos \beta t + i \sin \beta t) \, g(t)$, where g is a polynomial, and $\alpha + i\beta$ is a characteristic root of A. Hence

$$\int_0^{+\infty} H(t) \, dt$$

exists and so does (10.1).

We must now prove that B satisfies (5.1). We have (by integration by parts)

$$\begin{aligned} A'B &= \int_0^{+\infty} A'e^{A't}Ce^{At}dt \\ &= [e^{A't}Ce^{At}]_0^{+\infty} - \int_0^{+\infty} e^{A't}Ce^{At}A\,dt = -C - BA, \end{aligned} \tag{10.2}$$

which is (8.1) in the present instance.

Let now $-A$ be stable. Then

$$(-A') \, B + B \, (-A) = -C$$

and this is (8.1) for the present case. Thus the property asserted in (8) is now completely proved.

§ 6. **Another Matrix Problem.**

11. We propose to prove two properties of real matrices utilized in (VI, 7) and to be described presently. Let $\lambda_i, \ldots, \lambda_n$ be the characteristic roots of a real non-singular matrix A. Suppose that the real parts of $\lambda_1, \ldots, \lambda_p$ are positive while those of $\lambda_{p+1}, \ldots, \lambda_n$ are negative. On the strength of the reduction to a normal form we have $A \sim \text{diag}\,(A_1, A_2)$ where

A has for characteristic roots, $\lambda_1, \ldots, \lambda_p$ and A_2, the characteristic roots $\lambda_{p+1}, \ldots, \lambda_n$. We wish to prove:

(a) *One may choose A_1 and A_2 real.*

(b) *The transformation matrix P from A to diag (A_1, A_2) may also be chosen real.*

If λ_h, $h \leqq p$, is complex, then $\bar{\lambda}_h$ is likewise among the first p characteristic roots. Assuming first A_1 in normal form and writing λ for λ_h, if the block $C(\lambda)$ is in A_1 so is $C(\bar{\lambda})$. Let $\mathfrak{V}$, $\overline{\mathfrak{V}}$ be the corresponding conjugate vector subspaces such as in (1), and let e_h, $\bar{e}_h$ be two conjugate elements of the associated bases. Then $e'_h = e_h + \bar{e}_h$ and $e''_h = i(e_h - \bar{e}_h)$ are real and we may replace in the joint bases for $\mathfrak{V}$, $\overline{\mathfrak{V}}$ the element pair e_h, $\bar{e}_h$ by the pair e'_h, e''_h. Proceeding thus, one will arrive at a real base for the space $\mathfrak{V} \oplus \overline{\mathfrak{V}}$ and hence this space is real. Upon treating all the blocks for A_1 in the same manner, we will have, say

$$A_1 \sim \text{diag}(B_1, \ldots, B_q)$$

where the B_h are all real. A similar treatment applies of course to A_2. Hence (a) follows.

Regarding (b) it is sufficient to prove

(c) *If $A \sim B$ and both matrices are real there is a real transformation matrix P such that $B = PAP^{-1}$.*

Consider $P = (p_{jk})$ as unknown. The relation $BP = PA$ is equivalent to a real linear homogeneous system S of relations between the p_{jk}. Now all the solutions of S are linear combinations of a finite real system of solutions $P^1, \ldots, P^r$. Take a general solution $P(c) = \Sigma\, c_h P^h$. Its determinant $D(c)$ is a form of degree n in the c_h and this form $\not\equiv 0$ since there is a $P(c) \neq 0$ for some values of the c_h. If we apply a real linear transformation U to $c_1, \ldots, c_h$ the result is the same as changing the set of independent matrices P^h. One may choose in particular a transformation U such that in the new $D(c)$ the coefficient of the term of highest degree in c_1 be $\neq 0$. For in the contrary case $D(c) = 0$ for all real c_h and therefore $D(c) \equiv 0$, which is not the case. Suppose then that the highest coefficient in c_1 alone in $D(c)$ be $\neq 0$. As a consequence $D(1, 0, \ldots, 0) \neq 0$ and so there exists a real non-singular matrix P. This proves (c) and hence also (b).

APPENDIX II

Some Topological Complements

In the present appendix we shall give more complete topological details regarding a number of questions considered in the text, more particularly in Chs. IX and XI. We shall especially develop the properties of the index.

§ 1. The Index in the Plane

1. For a proper development of the theory of the index one requires the freedom offered by a deformation of a Jordan curve. This is best provided through the concept of *circuit*.

A circuit Γ in a space $\mathfrak{R}$ is a pair (f, J) consisting of an oriented Jordan curve J and of a mapping $f: J \to \mathfrak{R}$. One agrees that if J' is a second Jordan curve and φ a topological orientation preserving mapping $J' \to J$ then the circuit $(f\varphi, J')$ is still Γ. Notice that J is the circuit $(1, J)$. We also denote by Γ for short the set fJ.

Let three arcs λ, μ, ν form a θ curve (curve like a θ) with common end points a, b and let them be oriented from a to b. Thus $J_1 = \lambda - \nu$, $J_2 = -\mu + \nu$, $J = \lambda - \mu$ are all three oriented Jordan curves. Let f map the figure into $\mathfrak{R}$, producing circuits Γ_1, Γ_2, Γ. Then one defines $\Gamma = \Gamma_1 + \Gamma_2$, and similarly $\Gamma = \Gamma_1 + \ldots + \Gamma_s$.

Let Γ_0, Γ_1 be two circuits in $\mathfrak{R}$. Consider the cylinder $J \times l$, where l is the segment $0 \leq u \leq 1$. If there exists a mapping Φ: $J \times l \to \mathfrak{R}$ such that $\Gamma_0 = (\Phi, 0 \times J)$, $\Gamma_1 = (\Phi, 1 \times J)$, then Γ_0 and Γ_1 are said to be *homotopic* in $\mathfrak{R}$ and Φ is a *homotopy*. The pair $(\Phi, J \times l)$ is the *homotopy cylinder* and if $M \epsilon J$ then $\Phi(l \times M)$ is the *path* of M in the homotopy. If $\Gamma_0 = (1, J)$ then the homotopy is known as a *deformation*. If Γ_1 is a point one says: Γ_0 is homotopic to or deformable into a point in $\mathfrak{R}$. Thus:

(1.1) *A Jordan curve in a 2-cell E^2 is deformable into a point of E^2.*

We have an *ε-homotopy* or *ε-deformation* whenever every path is of diameter $< \varepsilon$.

Evidently:

(1.2) *Take two circuits in $\mathfrak{R}$: $\Gamma_0 = (f_0, J)$, $\Gamma_1 = (f_1, J)$. Suppose that whatever $M \epsilon J$ the points f_0M, f_1M may be joined by a segment in $\mathfrak{R}$ varying continuously and which reduces to the points themselves when they coincide. Then Γ_0 and Γ_1 are homotopic in $\mathfrak{R}$.*

2. The preceding considerations may be applied to the orientation of an Euclidean plane Π and of its Jordan curves. Generally Π is oriented by assigning a concordant sense to all its circles. Let J be a Jordan curve in Π and Ω its interior. Take a circumference C contained in Ω and draw the Jordan curve H sufficiently clearly indicated by the construction of Fig. 1. One orients now J so that in the resulting orientation for C this circumference is described negatively. This process yields a unique positive orientation for J. Conversely given a preassigned orientation of J specified as positive the process yields a positive orientation for C and hence for Π. The orienting curve J is called an *indicatrix* of the plane.

The process is also applicable: (*a*) To a 2-cell E^2. One maps E^2 topologically on Π oriented and assigns to the Jordan curves of

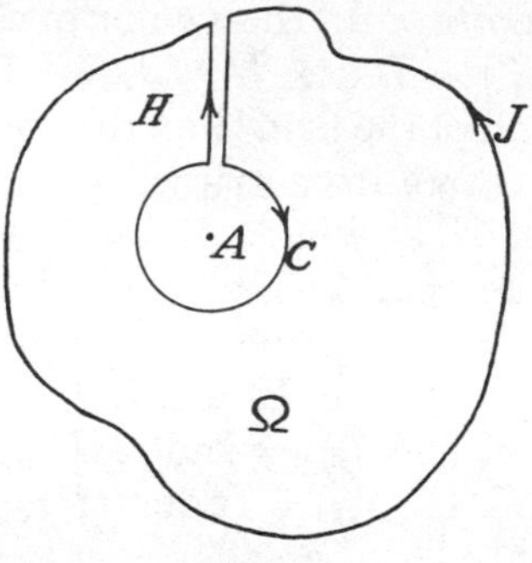

Fig. 1

E^2 the orientation specified by that of their images in E^2. (*b*) To a differentiable orientable manifold M^2. One orients directly the Jordan curves in one cell of the basic covering of M^2 and then step by step those of the other cells, operating from cell to overlapping cell.

3. Consider now a field $\mathfrak{F}$ defined at all points of a circuit $\Gamma = (f, J)$ in the Euclidean plane Π and free from critical points on Γ. As $M \epsilon J$ describes J once the angle of the vector $V(M)$ at the point fM with a fixed direction varies by $m \cdot 2\pi$. The number m is the *index of Γ relative to* $\mathfrak{F}$, written Index $(\Gamma, \mathfrak{F})$ or merely Index Γ. In particular one may define Index $(J, \mathfrak{F})$.

(3.1) *Given two fields $\mathfrak{F}$, $\mathfrak{F}'$ defined and free from critical points on Γ, if their vectors $V(M)$, $V'(M)$ are never in opposition then* Index $(\Gamma, \mathfrak{F})$ = Index $(\Gamma, \mathfrak{F}')$.

Let α be the greatest angle, $|\alpha| < \pi$, of both vectors. Since $|\alpha| < \pi$, their angular variations differ by less than 2π. Since it is $m \cdot 2\pi$, $m = 0$, hence (3.1) follows. On the other hand evidently:

(3.2) *If $V(M)$, $V'(M)$ are always in opposition then* Index $(\Gamma, \mathfrak{F})$ = Index $(\Gamma, \mathfrak{F}')$.

The central property of the index of circuits is:

(3.3) *Let $\mathfrak{F}$ be a field defined in a plane region Ω and free from critical points in Ω. If two circuits Γ_0, Γ_1 are homotopic in Ω then they have the same index as to $\mathfrak{F}$.*

Let Φ, u, l, J have the same meaning as in (1), where now $\Phi(l \times J) \subset \Omega$. Let the segment $0 \leq u \leq 1$ be decomposed into n equal parts by the subdivision points $1/n, 2/n, \ldots$. Let Γ^h denote the circuit $(\Phi, h/n \times J)$. If $M \epsilon J$ let $\mathfrak{F}'$ be the field defined on Γ^h by transferring to $h/n \times M$ the vector of $\mathfrak{F}$ at $(h + 1/n) \times M$. Clearly Index $(\Gamma^h, \mathfrak{F}')$ = Index $(\Gamma^{h+1}, \mathfrak{F})$. However, owing to the continuity of Φ and of the field $\mathfrak{F}$ on the compact set $\Phi(l \times J)$, and since the latter is free from critical points, for n sufficiently large, the vectors of $\mathfrak{F}$ at $h/n \times M$ and $(h + 1/n) \times M$ for all $M \epsilon J$ and all h, may be made to make an arbitrarily small largest angle. Hence $\mathfrak{F}$ and $\mathfrak{F}'$ will never be in opposition on any Γ^h. Applying now (3.1) we have Index $(\Gamma^h, \mathfrak{F}')$ = Index $(\Gamma^h, \mathfrak{F})$. Hence for any h: Index $(\Gamma^h, \mathfrak{F})$ = Index $(\Gamma^{h+1}, \mathfrak{F})$ and therefore Index $(\Gamma_0, \mathfrak{F})$ = Index $(\Gamma^0, \mathfrak{F})$ = Index $(\Gamma^n, \mathfrak{F})$ = Index $(\Gamma_1, \mathfrak{F})$, which proves (3.3).

Applications. (3.4) *If the field* $\mathfrak{F}$ *is defined over a closed two-cell* Ω *and is free from critical points in* Ω, *then the index of every circuit* Γ *in* Ω *is zero.*

This follows at once from the fact that Γ is homotopic to a point in Ω.

(3.5) THE BROUWER FIXED POINT THEOREM. *Every mapping of a closed two-cell into itself has at least one fixed point.*

Take as the closed two-cell a closed circular region $\mathfrak{R}$. Let C be the boundary circumference of $\mathfrak{R}$ and O the center of C. If φ is the mapping, M a point of $\mathfrak{R}$ and $M' = \varphi M$ then the directed segment $V(M) = MM'$ defines a vector distribution $\mathfrak{F}$ on $\mathfrak{R}$ whose critical points are the fixed points of φ. Suppose that φ has no fixed point on C. Then $V(M) \neq 0$ on C. Let it be replaced on C by $V'(M) = MO$, thus resulting in a new field $\mathfrak{F}'$ on C. Since $V(M)$ and $V'(M)$ are never in opposition on C, Index $(C, \mathfrak{F}) =$ Index $(C, \mathfrak{F}') = 1$. Hence $\mathfrak{F}$ must have a critical point (3.4) and φ must have a fixed point.

As a last property of the index of circuits we have evidently:

(3.6) *Let* $\Gamma = \Gamma_1 + \Gamma_2$ *where the circuits are in* Π. *Let the field* $\mathfrak{F}$ *be defined and be free from critical points on* Γ_1 *and* Γ_2. *Then it behaves likewise relative to* Γ *and Index* $\Gamma =$ *Index* $\Gamma_1 +$ *Index* Γ_2.

4. Let a field $\mathfrak{F}$ be defined in a plane region Ω and let A be a point of Ω. Let $\mathfrak{R}$ be a closed 2-cell in Ω containing A and suppose that in $\mathfrak{R}$ the field $\mathfrak{F}$ has no other critical point than possibly A. Draw in $\mathfrak{R}$ a positive convex curve C surrounding A. If C' is another such curve then C and C' are homotopic in $\mathfrak{R} - A$. Hence Index $(C, \mathfrak{F}) =$ Index $(C', \mathfrak{F})$. Thus Index $(C, \mathfrak{F}')$ is the same for all convex curves such as C surrounding A in $\mathfrak{R}$. Its value, clearly independent of $\mathfrak{R}$, is by definition the *index of the point* A *relative to the field* $\mathfrak{F}$, written Index $(A, \mathfrak{F})$ or merely Index A.

Let J be a positively oriented Jordan curve in $\mathfrak{R}$. Its interior $\mathfrak{R}_1$ is then in $\mathfrak{R}$. Let $A \in \mathfrak{R}_1$ and let C be a circumference of center A contained in $\mathfrak{R}_1$. Let H be the composite Jordan curve in heavy lines in Fig. 1. The interior of H is in $\mathfrak{R} - A$. Hence Index $(H, \mathfrak{F}) = 0$. The construction can be so carried out that $|\text{Index } H - \text{Index } C + \text{Index } J|$ is arbitrarily small. Since the expression is an integer it is zero. Hence Index $J =$ Index $C =$

Index A. Thus the latter may be obtained from any positively oriented Jordan curve surrounding A and no other critical point, as previously from a small convex curve surrounding the point.

Notice that if one reverses the orientation of the plane Π one must also reverse that of J, and so its index, hence that of A, is unchanged. Thus:

(4.1) *The index of a point is independent of the orientation of the plane.*

(4.2) If Σ^2 is a differentiable cell one may treat it in the same manner without reference to Π and again its orientation (by means of its Jordan curves) does not affect the indices of the points.

Referring to (2.4) we also have at once:

(4.3) *The index of a non-critical point is zero.*

The central property of the indices of points is the following, which recalls the classical residue theorem of complex variables:

(4.4) THEOREM. *Let Ω be the closed interior of a Jordan curve in the plane Π oriented concordantly with Π and $\mathfrak{F}$ a field defined over Ω, free from critical points on J and having at most a finite number $A_1, \ldots, A_n$ in Ω. Then*

$$\text{Index } J = \Sigma \text{ Index } A_i .$$

Make the construction indicated in Fig. 2. As above Index $H = 0$. One may carry out the scheme so that

$$|\text{Index } J - \text{Index } H - \Sigma \text{ Index } A_i| < 1/2 .$$

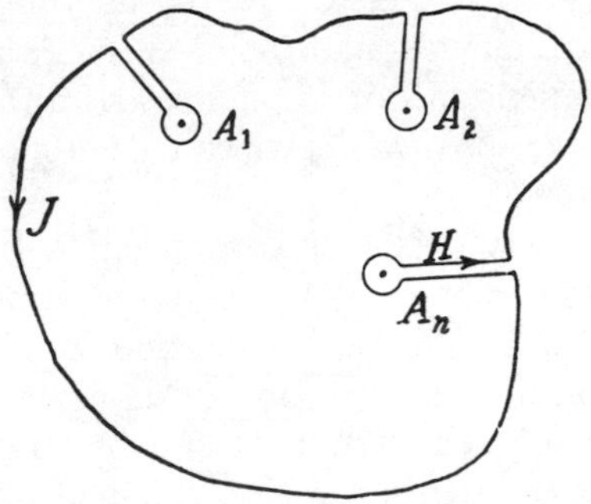

Fig. 2

Since this expression must be an integer, it is zero and this yields (4.4).

(4.5) *Remark.* Although our treatment of the index is tinged with considerable topological flavor, it does not follow that the concept as defined is topological. Indeed this cannot be the case, since vectors, a non-topological notion, have played an essential role in the definitions.

The following mode of defining say the index of a plane Jordan curve J is more topological and may also be readily extended to higher dimensions: Take a circumference C in the plane of J and from the center O of C draw a ray M' parallel to and sensed like the vector $V(M)$ at $M \epsilon J$. Then $M \to M'$ defines a mapping φ: $J \to C$ whose degree, in the sense of Brouwer (see Lefschetz [1], p. 124) is Index $(J, \mathfrak{F})$. The identification with the previous definition is immediate. Since the degree can be defined for any dimension, this new formulation carries over to higher dimensions. In particular by means of a "higher dimensional index" one may prove the Brouwer fixed point theorem for any closed cell.

5. The following proposition has been proved by Jaime Lifshitz [1]:

(5.1) *Let J be a planar positively oriented Jordan curve. Let T be a topological transformation of J into itself which is free from fixed points. If M is a point of J then $M' = TM \neq M$ and so the vector distribution $\mathfrak{F}$ given by $V(M) = MM'$ is defined and free from critical points on J. The assertion is that the index of this distribution is unity.*

It is clear that if J is an oriented closed path for a differential system such as (VIII, 1.1) then as limiting case (5.1) yields the Poincaré index theorem (VIII, 11.1), and indeed this is the main merit of (5.1). In outline the proof runs as follows. A polygon K_1 with sufficiently small sides is first inscribed in J. It may have multiple crossings but these are readily suppressed reducing K_1 to a simple closed polygon K. The polygon is so constructed that there is a point P interior to both J and K. An ε-homotopy $J \to K$ in $\Pi - P$ is established and a topological transformation T_1: $K \to K$ is defined with vector distribution $\mathfrak{F}_1$ analogous to $\mathfrak{F}$. One shows that Index $(J, \mathfrak{F})$ = Index $(K, \mathfrak{F}_1)$. The second index is, however, readily computed and found to be unity, which proves (5.1).

§ 2. The Index of a Surface

6. We have had occasion to utilize the index of a sphere or a projective plane in (IX). Both are given by a general theorem on the index of a surface due to Poincaré ([2], p. 125) and which we shall prove in the present section. His proof based on his definition of the index (see IX, 7.4) was only given for orientable surfaces but the extension to any surface offers no difficulty.

Let us state that the term "surface" in the present context will stand for "compact differentiable two-dimensional manifold."

Let Φ be a surface. There is no difficulty in defining a field $\mathfrak{F}$ over Φ (See I, 15.4). Let us suppose that $\mathfrak{F}$ has at most a finite number of critical points $A_1, \ldots, A_s$. Let $\{U_h\}$ be the finite open covering by differentiable 2-cells which serves to define Φ and let U_j contain A_h. If x, y are the parameters serving to define U_j then $\mathfrak{F}$ determines a field $\mathfrak{F}(x, y)$ over the region U_h of the Euclidean plane of the coordinates x, y. This field has A_h as an isolated critical point and it has then an associated index:

$$\text{Index}\,(A_h, \mathfrak{F}(x,y)).$$

The argument of (4) shows that if $A_h \in U_k$ and one defines the index through the medium of U_k, its value is the same as before. This value is now taken as Index $(A_h, \mathfrak{F})$ and we define

$$\text{Index}\,(\Phi, \mathfrak{F}) = \Sigma\, \text{Index}\,(A_h, \mathfrak{F})\,.$$

(6.1) POINCARÉ'S INDEX THEOREM. *The index of a surface relative to any field $\mathfrak{F}$ with at most a finite number of critical points, is independent of the field and equal to the Euler-Poincaré characteristic $\chi(\Phi)$ of the surface.*

7. We will first treat the case where Φ is a *sphere*. The tetrahedron is a suitable triangulation in which there are

$$\alpha_0 = 4 \text{ vertices}, \alpha_1 = 6 \text{ edges}, \alpha_2 = 4 \text{ triangles}\,.$$

Hence $\chi(\Phi) = \alpha_0 - \alpha_1 + \alpha_2 = 2$. We thus have to prove

(7.1) *The index of a vector field $\mathfrak{F}$ with a finite number of critical points, on the sphere, is two.*

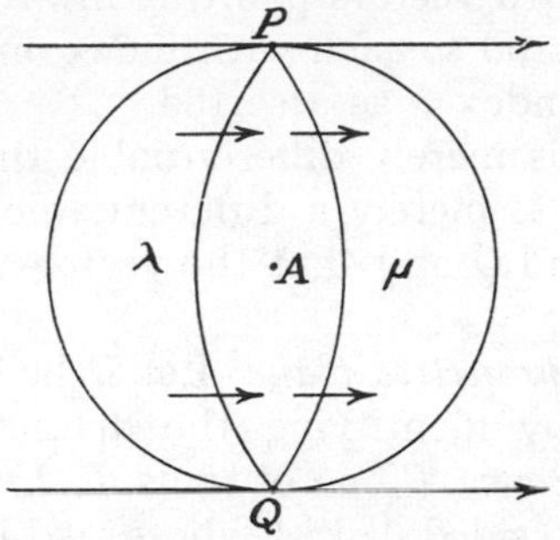

Fig. 3

Let us suppose first that the sphere Φ is Euclidean. We follow now Gomory. Take a non-critical point A on Φ and draw a circle C of center A so small that within C the angular variation is very small. In a certain circular region of center A containing C we may modify the vectors of $\mathfrak{F}$ and make them parallel. As a consequence they will be tangent to the circle at two points P, Q. We now join P to Q by two arcs without contact λ, μ in C disposed as in the figure. Let $J = \bar{\lambda} \cup \mu$. Then Index $\Phi =$ Index J relative to the outside of J, i.e., to the component of $S - J$ which does not contain the point A. This outside component Δ together with its boundary has the appearance of Fig. 4. Join P to Q by an arc ν of great circle outside C and hence in Δ and complete the distribution as indicated in Fig. 4. There

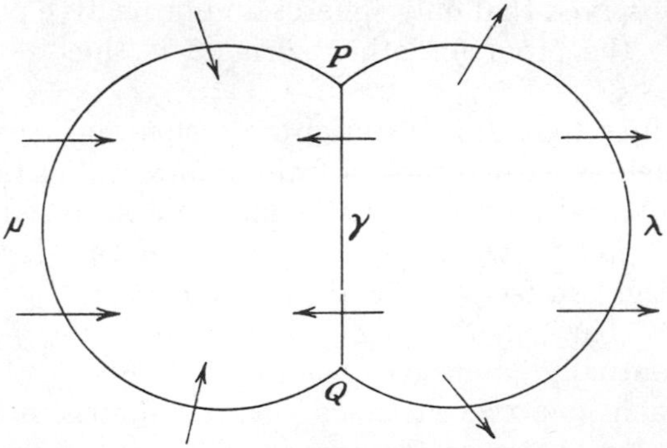

Fig. 4

result two 2-cells with vectors pointing inward on one and outward on the other, and so each with index one. Hence, the sum is two and this is Index Φ as asserted.

If the sphere Φ is merely differentiable the reasoning is the same save that PQ is merely a differentiable arc orthogonal to the vectors at P and Q and that the vectors along ν are normal to ν.

Let now Φ be a *projective plane*. Let Ω be the closed circular region such that by identifying the diametral points of the boundary circumference Γ one obtains Φ. Let Ω be considered as one half of a two sided disk Δ whose two faces Ω and Ω_1 are matched along Γ. If $\mathfrak{F}$ is a field on Φ one obtains by natural extension across Γ a field $\mathfrak{F}^*$ on Δ which has twice the number of critical points of $\mathfrak{F}$, each going into two points of the same nature on Δ. Therefore as Δ is a sphere Index $\Phi = 1/2$ Index $\Delta = 1$. On the other hand

$$\chi(\Phi) = 1/2\,\chi(\Delta - \Gamma) + \chi(\Gamma) = 1\,.$$

Hence (6.1) holds also for the projective plane.

8. To complete the proof of the Poincaré theorem we must consider surfaces other than the sphere and the projective plane. In this connection we will liberalize the term "surface" by omitting the condition of differentiability in the definition. This is merely done to "speed" up the treatment that follows and that is based on a most ingenious argument due to Gomory. It may be observed that only spheres and projective planes occur in the text: the proof for other surfaces is thus a topological luxury.

Suppose first that Φ is an orientable surface of genus $p > 0$ carrying a field $\mathfrak{F}$ with at most a finite number of critical points. Take as canonical model of Φ a planar double faced disk with p holes. (See Lefschetz [1] p. 83). Let J be the Jordan curve separating the two faces of the disk and $J_1, \ldots, J_p$ the borders of the holes. Thus, in the plane, J surrounds the J_k, and the latter are mutually exterior to one another (Fig. 5). Let us now make for each hole two incisions such as shown in the figure, taking care that between the two incisions for each hole there

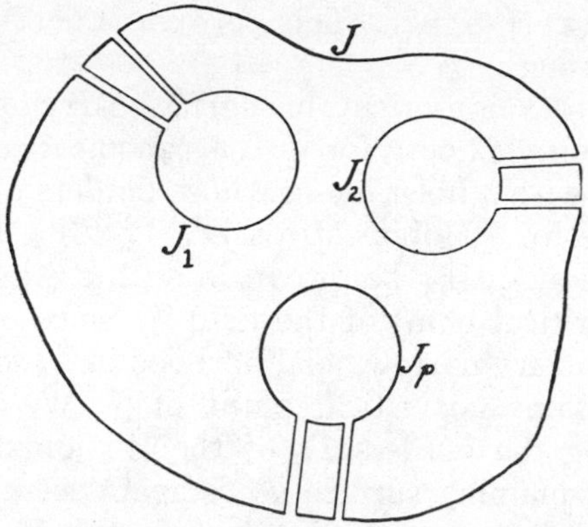

Fig. 5

are no critical points of the field $\mathfrak{F}$. Let the borders of all the incisions all be contracted to points. Let Φ_1 be the surface outside of all incisions. It is readily shown that Φ_1 is still a surface, and is in effect a sphere since it is a double faced disk bounded by a single Jordan curve. Let $\mathfrak{F}_1$ be the field on Φ_1. On the part common to Φ_1 and Φ, $\mathfrak{F}_1$ coincides with $\mathfrak{F}$. The two outside incision borders say for the k-th hole yield two critical points B_k', B_k'' of $\mathfrak{F}_1$. The two inside incision borders and the part between make up a sphere S_k with field $\mathfrak{F}_{1k}$ which has two and only two critical points C_k', C_k'' corresponding to the two incisions. Evidently

$$\text{Index}\,(B_k', \mathfrak{F}_1) = \text{Index}\,(C_k'', \mathfrak{F}_{1k}),$$
$$\text{Index}\,(B_k'', \mathfrak{F}_1) = \text{Index}\,(C_k', \mathfrak{F}_{1k}).$$

Hence

$$\text{Index}\,(B_k', \mathfrak{F}_1) + \text{Index}\,(B_k'', \mathfrak{F}_1) = 2.$$

Thus

$$2 = \text{Index}\,(\Phi_1, \mathfrak{F}_1) = \text{Index}\,(\Phi, \mathfrak{F})$$
$$+ \Sigma\,(\text{Index}\,(B_k', \mathfrak{F}_1) + \text{Index}\,(B_k'', \mathfrak{F}_1)),$$

and therefore

$$\text{Index}\,(\Phi, \mathfrak{F}) = 2 - 2p = \chi(\Phi).$$

(See Lefschetz [1], p. 82). This proves the theorem for an orientable surface.

Let now Φ be a non-orientable surface carrying the same type of field as before. Choose for Φ the canonical model consisting of a sphere S with q holes along whose borders one has matched the boundaries of q Möbius strips $W_1, \ldots, W_q$. Let the median line L_k of the strip W_k be drawn in such a way that it passes through no critical point of the field $\mathfrak{F}$. Since one may replace W_k by an arbitrary narrow neighborhood of L_k, one may assume that W_k contains no critical point of $\mathfrak{F}$. We will now make incisions along the borders Γ_k of the W_k and reduce these to points. The remaining surface Φ_1 is again seen to be a sphere. The new field $\mathfrak{F}_1$ on Φ_1, will coincide with $\mathfrak{F}$ on the portion common to Φ_1 and to Φ, and in place of Γ_k it will acquire a critical point B_k. The Möbius strip W_k will have become a closed surface Ψ_k and a ready calculation shows that $\chi(\Psi_k) = 1$. Hence Ψ_k is a projective plane. The new field $\mathfrak{F}_{1k}$ on Ψ_k has a single critical point C corresponding to B_k and with equal index. Thus

$$1 = \text{Index}\,(C_k, \mathfrak{F}_{1k}) = \text{Index}\,(B_k, \mathfrak{F}_1)\,.$$

Hence

$$1 = \text{Index}\,(\Phi_1, \mathfrak{F}_1) = \text{Index}\,(\Phi, \mathfrak{F}) + \Sigma\,\text{Index}\,(B_k, \mathfrak{F}_1)\,.$$

Therefore

$$\text{Index}\,(\Phi, F) = 1 - q = \chi(\Phi)\,,$$

and this completes the proof of Poincaré's theorem.

§ 3. A Property of Planar Jordan Curves

9. In the present section we propose to prove the following property utilized in the text (XI, 15.1):

(9.1) THEOREM. *Let $J_1, \ldots, J_n$ be Jordan curves in the Euclidean plane Π and let U_i be the interior of J_i, If U_i, U_{i+1} intersect for $i = 1, 2, \ldots, n-1$, then the infinite component* V of $\Pi - (J_1 \cup \ldots \cup J_n)$ *has for boundary a Jordan curve J contained in $J_1 \cup \ldots \cup J_n$.*

The basic part of the proof will follow the one contributed by Floyd to Cartwright [1] (p. 175) but some preliminaries are required.

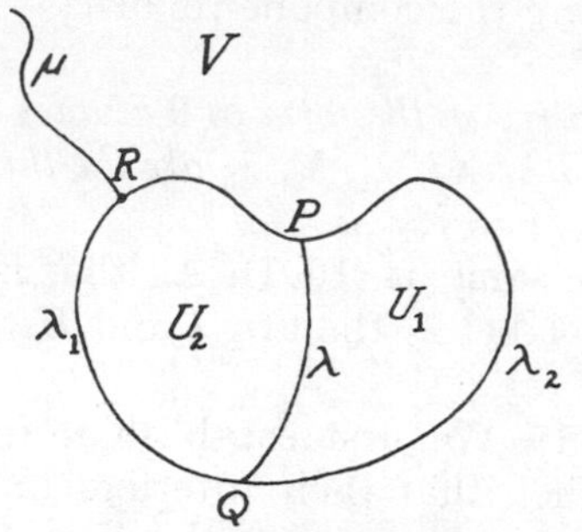

Fig. 6

10. (10.1) *Let J be a Jordan curve in the plane Π, U the interior of J, λ an arc joining in U two distinct points P, Q of J. Then λ divides U into two distinct regions with the common boundary λ in U.*

Let λ_1, λ_2 be the two arcs of J with the common end-points P, Q and let $J_1 = \lambda_2 \cup \bar{\lambda}$, $J_2 = \lambda_1 \cup \bar{\lambda}$. Denote also by U_i, V_i the interior and exterior of J_i and by V the exterior of J itself. Since V is unaffected by λ, a point R of λ_1 can be joined to infinity by an arc μ, which does not meet $J \cup \lambda$. Similarly if R is any point of V. Hence $V \subset V_1$. Moreover since R is not in J_1, $\lambda_1 \subset V_1$ also. Since there are points of U_2 in any neighborhood of R, U_2 has points in V_1. Since it is connected U_2 which is in $\Pi - J_1$, cannot meet U_1, for otherwise the two components U_1, V_1 of $\Pi - J_1$ would meet the same connected set in $\Pi - J_1$. Thus U_1 and U_2 are disjoint.

Since $V \subset V_1$ necessarily $U_1 \subset U - \lambda$ and similarly $U_2 \subset U - \lambda$.

Suppose now that, in addition to U_1 and U_2, $U - \lambda$ contains some open set W disjoint from both. Let $M \in W$ and let L be a line through M which does not contain the points P, Q. Let the extension of L to one side of M meet $\overline{W}$ in a last point N. This point cannot be in λ_1 since then intervals of L ending in N would belong to U_2 or V and not to W. Similarly N is not in λ_2. Since N is neither P nor Q and is not in V, $N \in U$. Since N is in one of $\mathfrak{B}\, U_1$, $\mathfrak{B}\, U_2$, necessarily $N \varepsilon \lambda$. Thus a full neighborhood of N in $\Pi - J_1$ consists of U_1 and V_1. But the only points of V_1 near N are those of U_2. Hence $U_1 \cup U_2$ is a full neighborhood of N in $U - \lambda$ and so N is not in W. This contradiction proves the non-existence of W. Thus $U - \lambda = U_1 \cup U_2$.

Hence U_1 and U_2 are the components of $U - \lambda$ and this proves (10.1).

(10.2) *A plane curve in the form of* θ *divides the plane into three regions* U_1, U_2, U_3 *where if* λ_1, λ_2, λ_3 *are the three arcs of the curve then* $\mathfrak{B}U_i = \lambda_j \cup \bar{\lambda}_k$, $i \neq j \neq k$.

This is really the same as (10.1). All that is necessary for the identification is to label λ_3 the arc λ and U_3 the region V.

11. *Proof of* (9.1). We first establish it for $n = 2$. Thus J_1 and J_2 in Π are such that their interiors U_1 and U_2 intersect. Now if $J_2 \subset \bar{U}_1$ or $J_1 \subset \bar{U}_2$ we are through. Otherwise then J_2 enters and leaves U_1. Let x be a point of $\mathfrak{B}V$ contained in J_2.

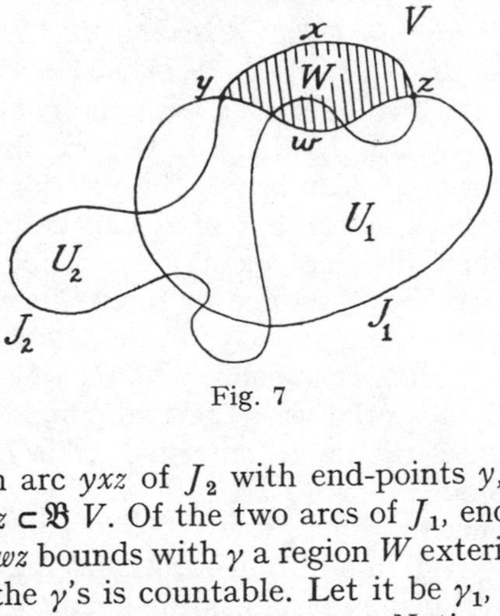

Fig. 7

Thus x is in an arc yxz of J_2 with end-points y, z in J_1, where the arc $\gamma = yxz \subset \mathfrak{B}\,V$. Of the two arcs of J_1, ending at y and z, only one $\delta = ywz$ bounds with γ a region W exterior to U_1 (10.2). The set of all the γ's is countable. Let it be $\gamma_1, \gamma_2, \ldots$, and let W_i, δ_i be the W, δ corresponding to γ_i. Notice that γ_i and γ_j, $i \neq j$, are disjoint, for otherwise the corresponding regions W_i would overlap and hence coincide. Thus $\gamma_i \leftrightarrow \delta_i$ is one-one. Define now a topological mapping τ_i: $\gamma_i \to \delta_i$ such that the common end-points remain fixed. Let τ be a mapping $\mathfrak{B}\,V \to J_1$ which is the identity on the points of $\mathfrak{B}\,V$ in J_1 and coincides with τ_i on γ_i, $i = 1, 2, \ldots$. It is clear that τ is topological. Hence $\mathfrak{B}\,V$ is a Jordan curve. As it is manifestly contained in $J_1 \cup J_2$, (9.1) is proved for $n = 2$. The proof for $n > 2$ follows now by an obvious induction.

PROBLEMS

1. *Peano's existence theorem*. Consider the vector equation $dx/dt = X(x; t)$ and let X be merely continuous in a region $D \subset \mathfrak{V}_x \times \mathfrak{T}$. Then through every point $P(x_0; t_0)$ of D there passes at least one trajectory of the system. Show by an example that there may even pass an infinity of trajectories through P.

(Method: subdivide the segment $t_0 \leqq t \leqq t_1$ into q equal parts; approximate linearly by replacing the differential equation by a difference equation in each interval, then pass to the limit as $q \to +\infty$. See notably Kamke [1] and Bellman [2]).

2. The equation being the same as in the preceding problem and real or complex, if X is holomorphic in D, prove by exhibiting the power series $x(t)$ that there is a unique solution holomorphic in t at t_0 and such that $x(t_0) = x_0$, ($(x_0, t_0) \in D$).

3. *Theorem of Cauchy-Kovalevsky*. Consider the real or complex system of partial differential equations in the p-vector u and q-vector x:

$$\partial u/\partial t = A(u) \cdot \partial u/\partial x ,$$

where A is a $p \times q$ matrix holomorphic at u_0. Let $\varphi(x)$ be a given p-vector holomorphic at x_0 and such that $\varphi(x_0) = u_0$. Then the system has a unique solution $u(x; t)$ holomorphic at $(x_0; t_0)$ and such that $u(x; t_0) = \varphi(x)$.

4. Derive the analogue of theorem (VI, 21.2) for $X(x; t)$ periodic with period ω in t and show that the associated function $V(x; t)$ may be chosen with period ω in t (Massera [2]).

5. Obtain the analogue of theorem (VI, 21.2) for a merely continuous function $X(x; t)$. Show also that one may then choose $V(x; t)$ of any class C^k (Kurzweil [1]).

6. Prove the converse of Četaev's instability theorem (Krassovskii [1]).

7. Prove the expansion theorem (V, 9.1) when the characteristic numbers are replaced by the negatives of the Liapunov numbers and the system is regular in the sense of Liapunov (see his paper [1] p. 32).

8. Consider an analytical autonomous system

$$dx/dt = Px + q(x), q = [x]_2$$

where $P \sim \operatorname{diag}(\lambda_1, \ldots, \lambda_n)$, the λ_i having only negative real parts. There is a regular transformation $x = f(y)$ which reduces the system to the form

$$dy_i/dt = \lambda_i y_i + g_i(y_1, \ldots, y_{i-1})$$

where g_i is a polynomial containing solely terms in $y_1^{m_1} \ldots y_{i-1}^{m_{i-1}}$ such that there exists a relation

$$\lambda_i = m_1 \lambda_1 + \ldots + m_{i-1} \lambda_{i-1}.$$

(Dulac [1]).

9. Consider the two systems

$$(\alpha)\ dx/dt = Px; \quad (\beta)\ dx/dt = \{P + Q(t)\} x$$

where P is a constant matrix and $\int^{+\infty} ||Q(t)||\, dt$ converges. If all the solutions of (α) are bounded as $t \to +\infty$ so are those of (β). If $||Q(t)||$ is bounded for $t \geqq t_0$ and the solutions of (α) all $\to 0$ as $t \to +\infty$ (i.e., the characteristic roots of P all have negative real parts) so do the solutions of (β). (Bellman [3], Ch. 4).

10. Consider the system

$$dx/dt = Px + q(x; t),\ q(0; t) = 0 \text{ for } t \geqq \tau,$$

where P is a constant matrix and q is continuous in $||x|| \leqq A$, $t \geqq \tau$, and satisfies a Lipschitz condition in that set. Suppose also that: (a) all solutions of the first approximation are bounded; (b) $||q(x; t)|| \leqq a\, g(t)\, ||x||$ for $||x|| \leqq A$ and $\int^{+\infty} g(t)\, dt$ is convergent. Under these conditions the origin is stable (Bellman [3], Ch. 4).

11. Prove that the origin is unstable for the system in two variables

$$dx/dt = -ax$$
$$dy/dt = \{\sin \log t + \cos \log t - 2a\} y + x^2$$

but that it is stable for the first approximation. (Perron [3]).

12. Investigate the stability of an autonomous system with two purely conjugate complex characteristic roots, the rest having negative real parts. (Liapunov [1], Malkin [1]).

13. Discuss the complete phase-portrait of a system $dx/dt = P(x, y)$, $dy/dt = Q(x, y)$ where P and Q are relatively prime real polynomials of degree two. (Büchel [1]).

14. Discuss the periodic solutions (harmonic and subharmonic) of

$$d^2x/dt^2 + \mu(x^2 - 1)\, dx/dt + x = \nu x + \varrho\, dx/dt + \sigma \cos \omega t$$

where μ, ν, ϱ, σ are all small parameters. (Obi [1]).

15. Discuss the complete behavior in the projective plane of the system with constant coefficients

$$dx/dt = ax + by, \; dy/dt = cx + dy, \; ad - bc \neq 0 .$$

16. *Duffing's equation.* Investigate the possible values of ω for which

$$d^2x/dt^2 + x + \mu x^3 = \mu \cos \omega t$$

has solutions of period $2\pi/\omega$. (See Stoker [1]).

17. Prove the theorem on structural stability stated in (X, 29.1).

18. Consider the system of the third order

$$\begin{aligned} dx/dt &= f(x) - (1 + \alpha)\, x - z \\ dy/dt &= -\beta\,(f(x) - x - z) \\ dz/dt &= -\gamma\,(y + z), \\ \alpha, \beta, \gamma &> 0 \end{aligned}$$

which arises in a vacuum tube circuit problem. Prove that with suitable values of the constants α, β, γ the system is self-oscillatory.
(L. L. Rauch [1]. See also K. O. Friedrichs [1]).

19. Derive a more accurate estimate for the period of oscillations of van der Pol's equation for μ large than that given in (XII, 18).
(Mary Cartwright [1], Lasalle [1], Haag [1]).

BIBLIOGRAPHY

Aizerman, M. A.

[1] *Theory of automatic regulation*, (in Russian), Gosizdat Fiz. **Mat.** Lit., Moscow, **1958**.

[2] On a problem concerning the stability of dynamical systems in the large, *Uspehi Mat. Nauk*, IV, **1949**, (Russian).

Andronov, A. A., and Chaikin, S. E.

[1] Theory of Oscillations. English language edition. Princeton University Press, **1949**.

Andronov, A. A., and Witt, A.

[1] Zur Theorie des Mitnehmens von van der Pol. *Archiv für Elektrotechnik* **24**: **99–110** (**1930**).

Andronov, A. A., and Pontrjagin, L. S.

[1] Systèmes grossiers. *Doklady Akad. Nauk* **14**: **247–251** (**1937**).

Antosiewicz, H. A.

[1] Forced periodic solutions of systems of differential equations. *Annals of Math.* (2) **57**: **314–317** (**1953**).

[2] A survey of Liapunov's second method, **141–166** in *Contributions to the Theory of Nonlinear Oscillations*, Princeton University Press, **1958** (Annals of Mathematics Studies, no. **41**).

Barocio, Samuel

[1] On certain critical points of a differential system in the plane, in *Contributions to non-linear oscillations*, vol. **3**, Princeton University Press, **1955** (Annals of Mathematics Studies, no. **36**).

[2] Singularidades de sistemas analiticos en el plano, Mexico City, thesis, *Boletín de la Soc. Mat. Mexicana*, **1–25** (**1959**).

Bellman, Richard

[1] On the boundedness of solutions of nonlinear differential and difference equations. *Trans. Amer. Math. Soc.* **62**: **357–386** (**1947**).

[2] *Lectures on Differential Equations*. Princeton, 1947 (typewritten).
[3] *Stability Theory of Differential Equations*. New York, McGraw-Hill, 1953.
[4] *Introduction to Matrix Analysis*, McGraw-Hill, New York, 1959.

Bendixson, Ivar
[1] Sur les courbes définies par des équations différentielles. *Acta Mathematica* 24: 1–88 (1901).

Birkhoff, G. D.
[1] *Collected Mathematical Papers*, 3 volumes, American Mathematical Society, 1950.

Bochner, S., and Martin, W. T.
[1] *Several Complex Variables*. Princeton University Press, 1948 (Princeton Mathematical Series, vol. 10).

Bogoliubov and Krylov. See Krylov and Bogoliubov.

Bogoliubov, N. N. and Mitropolskii, Iu. A.
[1] *Asymptotic Methods in the Theory of Nonlinear Oscillations* (in Russian). Moscow, Gos. Izd. Tekh.—Teor. Lit., 1955.

Büchel, Wilhelm
[1] Zur Topologie der durch eine gewöhnliche Differentialgleichung erster Ordnung und ersten Grades definierten Kurvenschar. *Mittheil. der Math. Gesellsch. in Hamburg* 4: 33–68 (1904).

Bulgakov, B. V.
[1] *Oscillations* (in Russian). Moscow, Gos. Izd. Tekh.—Teor. Lit., 1954.

Cartwright, M. L.
[1] Forced oscillations in nonlinear systems, 149–241 in *Contributions to the Theory of Nonlinear Oscillations*, Princeton University Press, 1950 (Annals of Mathematics Studies, no. 20).
[2] Van der Pol's equation for relaxation oscillations, 3–18 in *Contributions to the Theory of Nonlinear Oscillations*, vol. 2, Princeton University Press, 1952 (Annals of Mathematics Studies, no. 29).

Cartwright, M. L., and Littlewood, J. E.
[1] On non-linear differential equations of the second order. II. *Annals of Math.* (2) 48: 472–494 (1947).

Četaev, N. G.
[1] *Stability of Motion* (in Russian). Moscow, Gos. Izd. Tekh.—Teor. Lit., 1955.

Chaikin, S. E. See Andronov and Chaikin.

Coddington, E. A. and Levinson, N.
[1] *Theory of ordinary differential equations*, McGraw-Hill, New York, 1955.

Conti, R. See Sansone and Conti.

De Baggis, Henry
[1] Dynamical systems with stable structure. 37–59 in *Contributions to the Theory of Nonlinear Oscillations*, vol. 2, Princeton University Press, 1952 (Annals of Mathematics Studies, no. 29).

Diliberto, S. P.
[1] On systems of ordinary differential equations, 1–38 in *Contributions to the Theory of Nonlinear Oscillations*, Princeton University Press, 1950 (Annals of Mathematics Studies, no. 20).

Duff, G. F.
[1] Limit-cycles and rotated vector fields. *Annals of Math.* (2) 57: 15–31 (1953).

Dulac, H.
[1] Solutions d'un système d'équations différentielles dans le voisinage de valeurs singulières. *Bulletin Soc. Math. de France* 40: 324–392 (1912).

Dykhman, E. I.
[1] On a reduction principle (in Russian). *Izvestiia Akad. Nauk Kazakh. SSR*. Ser. math. mekh. 1950, issue 4; 73–84.

Friedrichs, K. O.
[1] On nonlinear vibrations of the third order. *Studies in nonlinear vibration theory*. Institute for mathematics and mechanics, New York University, 65–103 (1946).
[2] Fundamentals of Poincaré's theory. *Proceedings of the symposium on nonlinear circuit analysis*. 60–67, New York (1953).

Frommer, Max
[1] Die Integralkurven einer gewöhnlichen Differenzialgleichung erster Ordnung in der Umgebung rationaler Unbestimmtheitsstellen, *Math. Ann.* 99, 222–272 (1928).

Gantmacher, F. R.
[1] *The Theory of Matrices*, English translation, 2 volumes, Chelsea, New York, 1959.

Gomory, R. E.
[1] Critical points at infinity and forced oscillations, *Contributions to the theory of nonlinear oscillations, III*, Princeton University Press, **1956** (Annals of Mathematics Studies, no. **46**).

Goursat, Édouard
[1] *Cours d'Analyse Mathématique*, 5th ed., vol. **2**. Paris, Gauthier-Villars, **1927**.

Graffi, Dario
[1] Forced oscillations for several nonlinear circuits. *Annals of Math.* (2) **54**: **262–271** (**1951**).

Gronwall, T. H.
[1] Note on the derivatives with respect to a parameter of the solutions of a system of differential equations. *Annals of Math.* (2) **20**: **292–296** (**1919**).

Haag, Jules
[1] *Les Mouvements Vibratoires*. Paris, Presses Universitaires de France, **1952**.

Hayashi, C.
[1] *Forced Oscillations in Non-linear Systems*. Osaka, Nippon printing and publ. Co. **1953**.

Hurewicz, W.
[1] *Ordinary Differential Equations in the Real Domain with Emphasis on Geometric Methods*. Providence, Brown University, **1943** (mimeographed).

Kamke, Erich
[1] *Differentialgleichungen: Lösungsmethoden und Lösungen*. Leipzig, Akademische Verlagsgesellschaft, **1943**.

Kerekjarto, B. von
[1] *Vorlesungen über Topologie*. Berlin, Springer, **1923**.

Krassovskii, N. N.
[1] *On some problems in the theory of stability of motion*, (in Russian) Gosizdat Fiz. Mat. Lit., Moscow, **1959**.

Krylov, N. and Bogoliubov, N. N.
[1] *An introduction to nonlinear mechanics*, Princeton University Press, **1947** (Annals of Mathematics Studies, vol. **2**).

Kurzweil, Jaroslav

[1] On the inversion of the second theorem of Liapunov on the stability of motion, (in Russian and English) *Cechoslovak Mat. Journal*, 6, 217–259, 455–484.

Langenhop, C. E.

[1] Note on Levinson's existence theorem for forced periodic solutions of a second order differential equation. *Journal of Math. and Phys.* 30: 36–39 (1951).

Lasalle, J.

[1] Relaxation oscillations. *Quarterly of appl. math.* 7: 1–19 (1949).

LaSalle, J. P., and Lefschetz, S.

[1] *Stability by Liapunov's direct method and applications*, Academic Press, (1962).

Lefschetz, S.

[1] Existence of periodic solutions for certain differential equations. *Proc. Nat. Acad. Sci.* 29: 29–32 (1943) (Also reproduced at the end of [2], p. 204–209).

[2] *Lectures on Differential Equations*. Princeton University Press (Annals of Mathematics Studies, no. 14).

[3] *Introduction to Topology*. Princeton University Press, 1949 (Princeton Mathematical Series, vol. 11).

[4] Notes on differential equations, p. 61–73 in *Contributions to the Theory of Nonlinear Oscillations*, vol. 2, Princeton University Press, 1952 (Annals of Mathematics Studies, no. 29).

[5] Complete families of periodic solutions of differential equations. Commentarii Helvet. 28: 341–345 (1954).

[6] On a theorem of Bendixson. *Boletín de la Sociedad Matemática Mexicana*. (2) 1: 13–27 (1956).

Lefschetz, S., and Lasalle, J. See Lasalle and Lefschetz.

Leontovic, E., and Mayer, A.

[1] Sur les trajectoires qui déterminent la structure qualitative de la division de la sphère en trajectoires. *Doklady Akad. Nauk* 14: 251–254 (1937).

Levinson, N.

[1] On the existence of periodic solutions for second order differential equations with a forcing term. *Journal of Math. and Phys.* 22: 41–48 (1943).

[2] Transformation theory of nonlinear differential equations of the second order. *Annals of Math.* (2) 45: 723–737 (1944).
[3] On stability of non-linear systems of differential equations. *Colloquium Mathematicum* (Wroclaw) 2: 40–45 (1949).

Levinson, N., and Coddington, E. A. see Coddington and Levinson.

Levinson, N., and Smith, O. K.
[1] A general equation for relaxation oscillations. *Duke Math. Jour.* 9: 382–403 (1942).

Liapunov, A. M.
[1] *Problème Général de la Stabilité du Mouvement.* Princeton University Press, 1947 (Annals of Mathematics Studies, no. 17). (Reproduction of the french translation in 1907 of a russian mémoire dated 1892).

Liénard, A.
[1] Étude des oscillations entretenues. *Revue Générale de l'Électricité* 23: 901–912, 946–954 (1928).

Lifshitz, Jaime
[1] Un teorema sobre transformaciones de curvas cerradas sobre si mismas. *Boletin de la Sociedad Mat. Mex.* 3: 21–25 (1946).

Littlewood, J. E. See Cartwright and Littlewood.

Malkin, I. G.
[1] Certain questions on the theory of stability of motion in the sense of Liapunov (in Russian). *Sbornik Nauchnykh Trudov Kazanskovo Aviatsionnovo Instituta,* no. 7, 1937, 103 p. (Amer. Math. Soc. Translation no. 20). On stability in the first approximation (in Russian). Ibid. No. 3 (1935).
[2] On the stability of motion in the sense of Liapunov (in Russian). *Mat. Sbornik* n.s. 3: 47–100 (1938). (Amer. Math. Soc. Translation no. 41).
[3] Some basic theorems of the theory of stability of motion in critical cases (in Russian). *Prikl. Mat. i Mekh.* 6: 411–448 (1942) (Amer. Math. Soc. Translation no. 38).
[4] Oscillations of systems with one degree of freedom (in Russian). *Prikl. Mat. i Mekh.* 12: 561–596 (1948) (Amer. Math. Soc. Translation no. 22).
(5) Oscillations of systems with several degrees of freedom (in Russian). *Prikl. Mat. i Mekh.* 12: 673–690 (1948) (Amer. Math. Soc. Translation no. 21).
[6] *Methods of Liapunov and Poincaré in the Theory of Nonlinear Oscillations* (in Russian). Moscow-Leningrad, Gos.Izdat, 1949.

[7] On a method for solving the stability problem in the critical case of a pair of pure complex roots (in Russian). *Prikl. Mat. i Mekh.* 15: 473–484 (1951).
[8] *Some Problems in the Theory of Nonlinear Oscillations* (in Russian). Moscow, Gos. Izd. Tekh.—Teor. Lit., 1956.
[9] *Theory of Stability of Motion* (in Russian). Moscow, Gos. Izd. Tekh.—Teor. Lit., 1952.

Martin, W. T. See Bochner and Martin.

Massera, J. L.
[1] The number of subharmonic solutions of non-linear differential equations of the second order. *Annals of Math.* (2) 50: 118–126 (1949).
[2] On Liapounoff's conditions of stability. *Annals of Math.* 50: 705–721 (1949).
[3] Contributions to stability theory. *Annals of Math.* 64: 182–206 (1956).

Mayer, A. See Leontovič and Mayer.

McLachlan, N. W.
[1] *Theory and Application of Mathieu Functions*. Oxford, Clarendon Press, 1947.
[2] *Ordinary Non-linear Differential Equations in Engineering and Physical Sciences*. Oxford, Clarendon Press, 1950.

Minorsky, N.
[1] *Introduction to Non-linear Mechanics*. Ann Arbor, J. W. Edwards, 1947.
[2] Parametric excitation. *Jour. Appl. Phys.* 22: 49–54 (1951).
[3] Stationary solutions of certain nonlinear differential equations. *Jour. Franklin Inst.* 254: 21–42 (1952).
[4] On interaction of non-linear oscillations. *Jour. Franklin Inst.* 256: 147–165 (1953).

Mitropolskii, Iu. A.
[1] *Non-stationary Processes in Nonlinear Oscillating Systems* (in Russian). Kiev, Izdatelstvo Akad. Nauk. Ukrainskoi SSR., 1955.

Mitropolskii, Iu. A. See Bogoliubov and Mitropolskii.

Mizohata, Sigeru, and Yamaguti, Masaya
[1] On the existence of periodic solutions of the non-linear differential equation $\ddot{x} + a(x) \cdot \dot{x} + v(x) = p(t)$. *Memoirs College of Science, Univ. of Kyoto, Ser. A Mathematics* 27: 109–113 (1952).

Niemytski, V. V., and Stepanov, V. V.
[1] *Qualitative Theory of Differential Equations* (in Russian), 2d ed. Moscow, Gos. Izd. Tekh.–Teor. Lit., 1949.

Obi, Chike
[1] Subharmonic solutions of non-linear d.e. of the second order. *Jour. London Math. Soc.* 25: 217–226 (1950); Periodic solutions of non-linear d.e. of the second order. *Proc. Cambridge Phil. Soc.* 47: 741–751, 752–755 (1951); Periodic solutions of nonlinear d.e. of order 2n. *Jour. London Math. Soc.* 28: 163–171 (1953). Researches on the equation $\ddot{x} + (e_1 + e_2 x)\dot{x} + x + e_3 \nu x^2 = 0$. *Proc. Cambridge Phil. Soc.* 50: 26–32 (1954).

Osgood, W. F.
[1] *Lehrbuch der Funktionentheorie*, 2d ed. 2 vol. Leipzig, Teubner, 1912–32.

Peixoto, M. M.
[1] On structural stability, *Annals of Math.* 69, 199–222 (1959).

Perron, O.
[1] Die Ordnungszahlen linearer Differentialgleichungssysteme. *Math. Zeits.* 31: 748–766 (1930).
[2] Über eine Matrixtransformation. *Math. Zeits.* 32: 465–473 (1930).
[3] Die Stabilitätsfrage bei Differentialgleichungen. *Math. Zeits.* 32: 703–728 (1930).

Persidskii, K. P.
[1] On the stability of motion as determined by the first approximation (in Russian). *Mat. Sbornik* 40: 284–293 (1933).
[2] Some critical cases in countable systems (in Russian). *Izvestiia Akad. Nauk Kazakh. SSR.* Ser. mat. mekh. 1951, issue 5: 3 24.
[3] On a theorem of Liapunov (in Russian), *Doklady Akad. Nauk*, 14: (1937).

Picard, Émile
[1] *Traité d'Analyse*, 3d ed. 3 vol. Paris, Gauthier-Villars, 1922–28.

Pliss, V. A.
[1] On certain problems in the theory of the stability of motion in the large, *Izdat. Leningradskovo Univ.*, 1958, (Russian).

Poincaré, Henri
[1] Sur les propriétés des fonctions définies par les équations aux différences partielles. Thèse, 1879. *Œuvres*, Paris, Gauthier-Villars, t.l, IL-CXXXII.

[2] Mémoire sur les courbes définies par une équation différentielle. *Jour. Math. Pures et Appl.* (3) 7: 375–422 (1881); 8: 251–296 (1882); (4) 1: 167–244 (1885); 2: 151–217 (1886). *Œuvres*, t.l., p. 3–84; 90–161; 167–221.
[3] Sur le problème des trois corps et les équations de la dynamique. *Acta Mathematica* 13: 1–270 (1890). *Œuvres*, t. 7, p. 262–479.
[4] *Les Méthodes Nouvelles de la Mécanique Céleste*, 3 vol. Paris, Gauthiers-Villars, 1892–99.

van der Pol, B.
[1] On oscillation hysteresis in a triode generator with two degrees of freedom. *Phil. Mag.* (6) 43: 700–719 (1922).
[2] On "relaxation-oscillations." *Phil. Mag.* (7) 2: 978–992 (1926).

Pontrjagin, L. S. See Andronov and Pontrjagin.

Rauch, L. L.
[1] Oscillations of a third order nonlinear autonomous system. *Contributions to the theory of nonlinear oscillations.* Ann. of Math. St. 20: 35–88, (1950).

Reuter, G. E. H.
[1] Subharmonics in a non-linear system with unsymmetrical restoring force. *Quarterly Jour. Mech. and Appl. Math.* 2: 198–207 (1949).
[2] A boundedness theorem for non-linear differential equations of the second order. I. *Proc. Cambridge Phil. Soc.* 47: 49–54 (1951).
[3] Boundedness theorems for nonlinear differential equations of the second order. II. *Journal London Math. Soc.* 27: 48–58 (1952).

Sansone, G.
[1] *Equazioni Differenziali nel Campo Reale*, 2d ed. vol. 1–2. Bologna, Zanichelli, 1948–49.

Sansone, G. and Conti, R.
[1] *Equazioni differenziali non-lineari.* Rome, Monografie matematiche III, Consiglio nazionale delle richerche, Edizione Cremonese, 1956.

Shimizu, Tatsujoro
[1] On differential equations for non-linear oscillations, I. *Mathematica Japonica* 2: 86–96 (1951).

Smith, O. K. See Levinson and Smith

Stepanov, V. V. See Niemytski and Stepanov.

Stoker, J. J.
[1] *Nonlinear Vibrations in Mechanical and Electrical Systems*. New York, Interscience, 1950.

Turritin, H. L.
[1] Asymptotic expansions of solutions of systems of ordinary linear differential equations containing a parameter, 81–115 in *Contributions to the Theory of Nonlinear Oscillations*, vol. 2, Princeton University Press, 1952. (Annals of Mathematics Studies, No. 29).

Urabe, Kojuro
[1] On the existence of periodic solutions for certain non-linear differential equations. *Mathematica Japonica* 2: 23–26 (1949).

Van der Waerden, B. L.
[1] *Moderne Algebra*, 2nd ed. vol. 1. Berlin, Springer.

Wedderburn, J. H. M.
[1] *Lectures on Matrices*, American Mathematical Society Coll. Publ. 17, 1934.

Witt, A. See Andronov and Witt.

Yamaguti, Masaya. See Mizohata and Yamaguti.

Zubov, V I.
[1] *Mathematical methods of investigation of automatic controls*, (in Russian) Gosizdat sudostroitel'noy promyshlenosti, Leningrad, 1959.

LIST OF PRINCIPAL SYMBOLS

ϵ is a member of; $\cup$ union; $\cap$ intersection; $\subset$ is contained in; $\supset$ contains; $\rightarrow$ implies; E^p p-cell; $\bar{A}$ closure of A; $\mathfrak{B}A$ boundary of A; $A \times B$ product of the spaces A, B; $\mathfrak{B}_x{}^n$ n dimensional vector space referred to coordinates x_h; $\oplus$ direct sum; δ_{jk} Kronecker deltas ($= 1$ for $j = k$, $= 0$ for $j \neq k$); $||x||$ norm of vector x; (a_{jk}) matrix with term a_{jk} in j — *th* row and k — *th* column; $x_1, x_2, \ldots,$ coordinates of vector x.

If $A = (a_{jk})$ then A' is its transpose (a_{kj}) and $\bar{A}$ its complex conjugate; if $a_{jk} = a_{jk}(t)$ then $dA/dt = (da_{jk}/dt)$ and $\int_{t_0}^{t} A\, dt = (\int_{t_0}^{t} a_{jk}(t)\, dt)$. If $y(x)$ is a vector function of a vector then its Jacobian matrix is $\partial y/\partial x = (\partial y_h/\partial x_k)$ and if $\mathfrak{B}_x$, $\mathfrak{B}_y$ have the same dimension then the Jacobian determinant is written $|\partial y/\partial x|$.

The Euler-Poincaré characteristic of a two dimensional manifold M^2 triangulated into α_2 triangles with α_1 sides and α_0 vertices is

$$\chi(M^2) = \alpha_0 - \alpha_1 + \alpha_2\,.$$

The block matrix of order r

$$\begin{pmatrix} \lambda, 0 & \cdot & \cdot & \cdot & \cdot \\ 1, \lambda & \cdot & \cdot & \cdot & \cdot \\ \cdot & \cdot & \cdot & \cdot & \cdot \\ \cdot & \cdot & \cdot & \cdot & 1, \lambda \end{pmatrix}$$

is written $C_r(\lambda)$.

One writes generally $f(x)$ for $f(x_1, \ldots x_n)$. In the system of convergent power series, $f(x)$ such that $f(0) \neq 0$, called a unit, is written $E(x)$. A series beginning with terms of degree p is denoted by $[x]_p$.

One denotes by (AB), $[AB)$, $(AB]$ $[AB]$, an arc with end-points A, B open at both ends, closed at A and open at B, closed at B and open at A, closed at both ends.

The half-path beginning at the point M of the path γ and formed by M and all the following [preceding] points is written $\gamma_M{}^+$ [$\gamma_M{}^-$], also $\gamma^+(M)[\gamma^-(M)]$. The limiting sets of $\gamma_M{}^+$, $\gamma_M{}^-$ are written $\Lambda^+(\gamma)$, $\Lambda^-(\gamma)$.

INDEX